D1640702

Frontiers in Biotransformation
Vol. 3

Frontiers in Biotransformation   Volume 3

Molecular Mechanisms of Adrenal Steroidogenesis and Aspects of Regulation and Application

Edited by
Klaus Ruckpaul and Horst Rein

Akademie-Verlag Berlin

Gesamt-ISBN 3-05-500456-6
Vol. 1-ISBN 3-05-500457-4
Vol. 2-ISBN 3-05-500458-2
Vol. 3-ISBN 3-05-500459-0

Erschienen im Akademie-Verlag Berlin, Leipziger Straße 3—4, O - 1086 Berlin
© Akademie-Verlag Berlin 1990
Printed in Germany
Gesamtherstellung: Maxim Gorki - Druck GmbH, O - 7400 Altenburg
Lektor: Christiane Grunow
Gesamtgestaltung: Martina Bubner
LSV 1315
Bestellnummer: 763 950 6 (9090/3)

Preface

K. Ruckpaul and H. Rein

Volumes 1 and 2 of the series "Frontiers in Biotransformation" were concerned with molecular mechanisms in the biotransformation of exogenous compounds. This may originate from the consequences of biotransformation of just these compounds leading to malignant processes such as carcinogenesis and mutagenesis which attract the interest of bioscientists of many disciplines as pharmacologists, toxicologists, physicians, biochemists and others. The proper physiological function of the cytochrome P-450 dependent biotransformation, however, is the conversion of endogenous compounds. Therefore, in discussing current problems in biotransformation one has also to include this aspect of cytochrome P-450 functions. The most extended P-450 dependent biotransformation of endogenous substrates concerns the steroids. Steroid metabolism (anabolic and catabolic processes likewise) proceeds in several tissues of the animal organism (testes, ovaries, liver and adrenals). The adrenals are of special importance in the biotransformation of steroids due to a number of reasons.

(i) Adrenals have a key function in steroidogenesis. Glucocorticosteroids and mineralocorticoids represent steroid hormones of vital importance.

(ii) In the adrenal steroidogenesis not only the endoplasmic but also the mitochondrial P-450 dependent systems are involved.

(iii) The biochemical composition of the mitochondrial P-450 dependent enzyme system differs from the endoplasmic system in having a further electron transferring protein between the electron donating flavoprotein and the electron accepting terminal oxygenase.

(iv) Differing from most liver microsomal P-450 isozymes which are characterized by a more or less broad substrate specificity, the adrenal mitochondrial steroid converting P-450 system is characterized by highly stereospecific reactions.

(v) The steroidogenesis from cholesterol to cortisol (cortison) is distinguished by a vectorial character shuttling between the mitochondrial matrix and the cytosol involving several mitochondrial and endoplasmic cytochromes P-450.

The molecular mechanisms of regulating the activities of the different adrenal steroidogenic systems are just under study and remarkable progress has been achieved within the last 5—10 years, including sequence analysis, hormonal regulation and mechanisms of insertion and processing.

But what may be the most remarkable aspect in this field is that due to the enormous practical role steroids play in therapy, biotransformation was used very early for conversion and production of useful compounds. After detection of the therapeutical properties of glucocorticoids in 1948/49 by HENCH and KENDALL only 4 years later the microbial catalysis was used in the pharmaceutical industry for transforming multistep chemical synthesis in only one elegant biochemical reaction step. This pioneering work was one of the milestones in introducing pragmatically based biotechnology in pharmaceutical production. The contributions of this volume try to cover that pretentious scope.

The editors would like to take this opportunity of thanking the authors of this volume for their excellent contributions. Although each review has been written individually the editors hope that when taken together they will provide a coherent picture of cytochromes P-450 involved in adrenal steroidogenesis, its regulation and application.

Contents

Chapter 1

Cytochrome P-450 Dependent Pathways of the Biosynthesis of Steroid Hormones

S. A. Usanov, V. L. Chashchin, and A. A. Akhrem

1. Introduction

Cholesterol utilization via several key hydroxylation steps to yield physiologically active metabolites (corticosteroids, mineralocorticoids, sex steroid hormones, bile acids, hydroxyvitamins D_3) has an important implication in modern science and biotechnology. Among different oxidative processes which take place in steroidogenic tissues (adrenal cortex, ovary, testis, corpus luteum, placenta) those responsible for steroid specific regio- and stereo-selective hydroxylations including cholesterol side chain cleavage and oxygenation of steroid nucleus are of special interest. Cholesterol conversion to pregnenolone via oxidative side chain cleavage is of special interest, since it was shown to be common to different steroidogenic tissues and pregnenolone is the common precursor for steroid hormones. It is thought that the cholesterol side chain cleavage reaction is the rate-limiting step in the steroid hormones biosynthesis and regulation of the level and activity of this form of cytochrome P-450 (cytochrome P-450scc) provides the general and universal mechanism for regulation of steroid hormones level.

The wide interest of scientists in cytochrome P-450-catalyzed reactions is connected with certain important fundamental and practical aspects:

1. elucidation of the molecular mechanisms of steroid hormones biosynthesis and understanding of the mechanisms by which optimal concentrations of physiologically active steroid hormones are maintained;

2. understanding of the molecular basis of differences in steroidogenic cytochrome P-450 gene expression to make a correlation between deficiencies of some steroidogenic enzymes and various disease states (congenital adrenal hyperplasia): use of specific oligonucleotides to predict some inherited deficiencies of steroidogenic enzymes;

3. until now cholesterol side chain cleavage reaction has no analogues in procariots;

4. use of highly specific stereo- and regio-selective hydroxylation reactions in biotechnology to prepare steroid hormones and their derivatives.

These are only some points which demonstrate the reasons why investigation of cytochrome P-450-dependent monooxygenases has become such an active area of research. Different questions of cytochrome P-450-dependent steroidogenesis have been recently reviewed: mechanism of steroidogenic electron transfer (LAMBETH et al., 1982; SHKUMATOV et al., 1985), role of ACTH in stimulation of stereoidogenesis (KIMURA, 1981), cellular organization of steroidogenesis (HALL, 1984), biosynthetic units, hormones (LIEBERMAN et al., 1984), regulation of stereoidogenesis (WATERMAN and SIMPSON, 1985; WATERMAN et al., 1986), role of cytochrome P-450 in the biosynthesis of physiologically active compounds (JEFCOATE, 1986). In this chapter we will concentrate on the structure and function of enzymes associated with corticosteroid hormones biosynthesis, the mechanism of specific protein-protein interactions, molecular organization, substrate specificity and perspectives of the application of steroid transforming enzymes in biotechnology.

2. Monooxygenase pathways in corticosteroid biosynthesis

Most reactions in steroid hormones biosynthesis are of the monooxygenase type catalyzed by various forms of cytochrome P-450 (Fig. 1). Cholesterol, as the starting material for steroid hormones biosynthesis, is transferred from blood and largely stored as cholesterol esters in the lipid droplets in adrenal cells. The first and rate-limiting step in the steroidogenesis is the conversion of cholesterol to pregnenolone. Pregnenolone must then reach the endoplasmic reticulum membranes where it undergoes conversion of 3β-ol structure to the Δ^4-3-ketone structure, catalyzed by 3β-hydroxysteroid dehydrogenase/Δ^5-Δ^4 isomerase, cytochrome P-450 independent enzyme, to form progesterone.

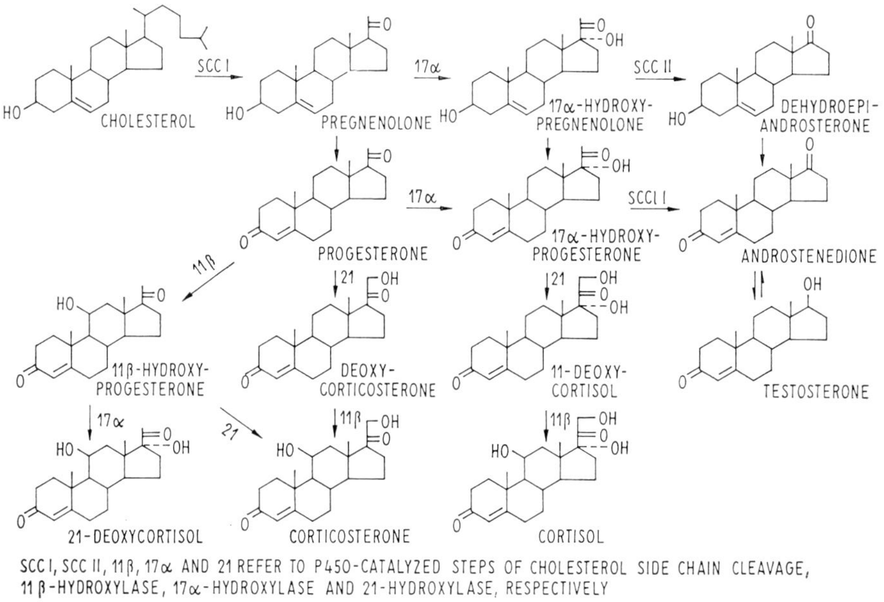

SCC I, SCC II, 11β, 17α AND 21 REFER TO P450-CATALYZED STEPS OF CHOLESTEROL SIDE CHAIN CLEAVAGE, 11 β-HYDROXYLASE, 17α-HYDROXYLASE AND 21-HYDROXYLASE, RESPECTIVELY

Fig. 1. The main cytochrome P-450 dependent pathways of steroid hormones biosynthesis. sccI and 11β-cholesterol side chain cleavage and 11β-hydroxylation reactions catalyzed by cytochrome P-450scc and cytochrome P-450$_{11\beta}$ from mitochondria; 17α and 21 — $C_{17\alpha}$ and C_{21} hydroxylation reactions catalyzed by cytochrome P-450$_{17\alpha}$ and cytochrome P-450$_{21}$ in microsomes; sccII — 17α-hydroxylase and 17,20-lyase reactions catalyzed by cytochrome P-450$_{17\alpha}$ in microsomes.

Pregnenolone may be hydroxylated in microsomes to form 17α-hydroxy-pregnenolone. Progesterone is usually hydroxylated at 21-position to form 11-deoxycorticosterone if 21-hydroxylase is more active than 17α-hydroxy-lase/17,20-lyase. Progesterone may also be hydroxylated at 17α-position if 17α-hydroxylase is too active. If 17α-hydroxylation takes place, it must precede 21-hydroxylation. Some of the 17α-hydroxyprogesterone is converted to androstenedione by 17α-hydroxylase/17,20-lyase activity.

Thereafter, 11-deoxycortisol or 11-deoxycorticosterone moves back to mitochondria which is necessary for 11β-hydroxylation, catalyzed by 11β-hydroxylating cytochrome P-450 (cytochrome P-450$_{11\beta}$) to form cortisol and corticosterone respectively and for 18-hydroxylation, catalyzed by the same cytochrome P-450$_{11}$, to give aldosterone. The mechanism by which some steroid metabolic intermediates should move back and forth between mitochondria, where some of the steroidogenic enzymes (cholesterol side chain cleavage, 11β-hydroxylation) were found to be localized to microsomes, where other enzymes (17α-hydroxylase, 17α-hydroxylase/17,20-lyase and 21-hydroxylase, 3β-hydroxysteroid dehydrogenase/Δ^5-Δ^4 isomerase) are situated, is not clear.

The classical scheme of corticosteroids biosynthesis (Fig. 1) indicates that some cytochrome P-450 isozymes may be involved in the biosynthesis of several hormones. Thus, it was shown that 17α-hydroxylase may be involved in corticosteroids as well as androgens biosynthesis. Furthermore, 21-hydroxylase is involved in reaction sequences leading to either corticosteroids or mineralcorticoids. Cytochrome P-450$_{11\beta}$ is involved in either corticosteroids or mineralcorticosteroids biosynthesis. Some enzymes of corticosteroids biosynthesis catalyze a complex sequence of reactions. Cholesterol side chain cleavage to form pregnenolone needs three sequential hydroxylation reactions. Conversion of pregnenolone to progesterone requires at least two reactions. An alternative pathway of 17α-hydroxyprogesterone metabolism to form androstenedione needs two monooxygenase reactions: 17α-hydroxylation and 17,20-cleavage of side chain. Different microsomal and mitochondrial substrate-specific cytochrome P-450 isozymes have the common electron transfer service proteins (ferredoxin reductase and ferredoxin for mitochondrial and flavoprotein for microsomal enzymes). The fact that steroidogenic enzymes have the common precursor (cholesterol) and hydroxylation reactions occur in both mitochondria and microsomes provides a unique possibility of controlling the total steroidogenesis via at least five different mechanisms:

1. control of electron transfer via efficiency of protein-protein interactions;

2. substrate control via availability of cholesterol;

3. hormonal control via ACTH;

4. allosteric control;

5. expression control.

Summarizing the reactions involved in corticosteroids biosynthesis one can conclude that at least five different activities are involved in this process:

1. hydroxylation at carbon atom, 11β- and 21-hydroxylations;
2. cleavage of C-C bond, cholesterol side chain cleavage and 17α-hydroxylation with subsequent cleavage of side chain of C_{21} steroids;
3. dehydrogenation, conversion of 3β-hydroxy to 3-keto group;
4. isomerization, conversion of pregnenolone to progesterone;
5. oxidation of hydroxy group, production of mineralcorticoids.

There are at least some steroidogenic organs and tissues: adrenal glands, testis, ovary, corpus luteum, placenta. The cytochrome P-450 isoenzymes distribution among these steroidogenic organs, activities of cytochrome P-450 isozymes as well as their expression and hormonal control are highly dependent on the type of steroid hormones which are synthesized by these steroidogenic tissues. Most information on the structure and function of steroidogenic enzymes has been obtained by studying enzymes of adrenal cortex cells. Therefore, the main subject of this chapter will be steroid hydroxylases of adrenal cortex.

The earlier studies (SABA et al., 1954) demonstrated that cholesterol is transformed by an unknown mechanism to give C_{21} steroid, pregnenolone. This reaction was shown to occur in mitochondria (CONSTANTOPOULOS and TCHEN, 1961), very similar to mixed function oxidation reactions, to require molecular oxygen and NADPH and to result in a reaction product which is removed from cholesterol and identified as isocaproic aldehyde (HALKERSTON et al., 1961; CONSTANTOPOULOS et al., 1966). Subsequent studies by SHIMIZU (SHIMIZU et al., 1961) indicated the incorporation of radioactive labels from (20S)-20-hydroxycholesterol to isocaproic acid, pregnenolone and progesterone. (20R,22R)-20,22-dihydroxycholesterol was shown to be a good substrate for cholesterol side chain cleavage reaction (SHIMIZU et al., 1962).

This data made it possible to postulate the so called sequential mechanism, according to which mono- and dihydroxyderivatives at 20- and 22-positions of the steroid nucleus are formed as the products of sequential hydroxylations, suggesting that an oxygen insertion step is also required for final desmolase reaction. Several other mechanisms of cholesterol side chain cleavage are currently well documented. The reactive intermediate mechanism with unspecified radicals or ionic compounds as intermediates was also proposed. Based on the results that some cholesterol derivatives, in which the C_{22} to C_{27} moeity of the side chain has been replaced (and consequently incapable of being oxygenated at the carbon atom corresponding to C_{22} of cholesterol) are hydroxylated at C_{20} and converted to pregnenolone it was concluded that the true intermediates are transient reactive species (ionic, free radicals or of a hybrid type) and the hydroxylated cholesterols are side products. The hydroperoxide mechanism implies 20S-perhydroxycholesterol and 20R,22R-dihydroxycholesterol as intermediates (VAN LIER and SMITH, 1970; VAN LIER and ROUSSEAU, 1976). The epoxy-diol mechanism implies 20,22-cholesterol, 20,22-epoxycholesterol and 20R,22R-dihydroxycholesterol as intermediates (KRAAIPOEL et al., 1975a; KRAAIPOEL et al., 1975b).

Evidence has been presented to show that the initial oxidative attack of cholesterol, catalyzed by cytochrome P-450scc, occurs at C_{22} with formation of 22-hydroxycholesterol, followed by an oxidative attack at C_{20} to form 20,22-dihydroxycholesterol (BURSTEIN and GUT, 1976). This sequence is based on kinetic studies on the relative rates of (20S)-20-hydroxycholesterol, (22R)-22-hydroxycholesterol and (20R,22R)-20,22-dihydroxycholesterol conversion to pregnenolone.

The geometrical isomers (E and Z) of 20,22-dehydrocholesterol and all diastereoisomers of their 20,22-epoxides and 20,22-glycols have been synthesized to prove any of the listed mechanisms (BURSTEIN and GUT, 1976; MORISAKI et al., 1976; TEICHER et al., 1978). Unlike cholesterol and its hydroxyderivatives, dehydrocholesterol and all its epoxides proved to be extremely poor substrates for the cholesterol side chain cleavage reaction. Thus, of the proposed models intended to explain the mechanism of cholesterol side chain cleavage, the "glycol" mechanism appears to find the most support.

The presence of cytochrome P-450 in mitochondria of adrenal glands was demonstrated by HARDING (HARDING et al., 1964). Carbon monoxide inhibition of deoxycorticosterone 11β-hydroxylation was also indicative of cytochrome P-450 involvement in this reaction (WILSON et al., 1965). Direct evidence for involvement of cytochrome P-450 in 11β-hydroxylation in bovine adrenocortical mitochondria was presented by photochemical action spectrum (ROSENTHAL and COOPER, 1967). Direct evidence for participation of cytochrome P-450 in cholesterol side chain cleavage has been obtained by a similar approach (SIMPSON and BOYD, 1966). The cytochrome P-450 dependent nature of 18-hydroxylation in adrenal mitochondria has been shown by GREENGARD et al. (1967).

In 1970—1973 the presence of two different isozymes of mitochondrial cytochrome P-450 with different catalytic and physico-chemical properties (cytochrome P-450scc and cytochrome P-450$_{11}$) was proved by a number of experimental approaches, including partial purification of these hemeproteins (JEFCOATE et al., 1970; SHIKITA and HALL, 1973; RAMSEYER and HARDING, 1973; SCHLEYER et al., 1972; SHIKITA and HALL, 1973). In 1975—1978 both cytochromes P-450scc and P-450$_{11\beta}$ were successfully purified to homogeneous form from bovine adrenocortical mitochondria (TAKEMORI et al., 1975; TAKEMORI et al., 1975). Subsequently, by using highly purified cytochrome P-450 isozymes (i) additional evidence for sequential hydroxylation mechanism of side chain cleavage reaction has been obtained; (ii) most critical physico-chemical parameters determined; (iii) the substrate specificity of reactions of

S. A. USANOV et al.

corticosteroid biosynthesis were described; (iv) the mechanism of steroidogenic electron transfer investigated. In 1980 great progress was made in studies
of protein-lipid interaction and cell free in vitro synthesis of the components
of steroid hydroxylases (HALL et al., 1979; SEYBERT et al., 1979; YAMAKURA
et al., 1981; NABI and OMURA, 1980; NABI et al., 1980).

In the middle of the 80s some progress had been reached in the determination of the structural characteristics of mitochondrial cytochrome P-450
isozymes and some progress made in molecular cloning and expression of
cytochrome P-450scc dependent enzymes.

In adrenal cells a number of alternative pathways of pregnenolone, formed
after side chain cleavage reaction, have been elucidated, including the metabolism to cortisol. Earlier studies on pregnenolone metabolism in adrenal cells
(HECHTER and PINCUS, 1954) presented evidence for its conversion to 17α-
hydroxypregnenolone, progesterone and 17α-hydroxyprogesterone. This finding allowed the conclusion that 17α-hydroxypregnenolone might also be an
intermediate in cortisol biosynthesis as well as an immediate precursor of 17α-
hydroxyprogesterone. Thus, both pregnenolone and progesterone may be used
as substrates for 17α-hydroxylation. The pathways leading to cortisol and
corticosterone in adrenal cortex show that 17α-hydroxy-20-keto steroids are
the intermediates in corticosteroids biosynthesis which are further either
21-hydroxylated or cleaved to form androgens.

3. Mitochondrial cytochrome P-450-dependent monooxygenases

The main pathways of steroid hormones biosynthesis are well understood at
present. As follows from Figure 1, in biosynthetic pathways of steroid hormones starting from cholesterol as substrate to form pregnenolone — pivotal
precursor of all steroid hormones — initial and final hydroxylations are realized in mitochondria. Both cytochrome P-450scc and cytochrome P-450$_{11\beta}$
are localized in the inner mitochondrial membrane of adrenal cortex and use
the same electron transport system, containing flavoprotein (adrenodoxin reductase) and ferredoxin (adrenodoxin). Thus, in the process of corticosteroids
biosynthesis the principles of compartmentation are realized that raise the
possibility of additional control and regulation.

3.1. Cholesterol side chain cleavage

Cholesterol side chain cleavage reaction was found to proceed in the inner
mitochondrial membrane of some steroidogenic tissues including adrenal cortex, ovary, testis, corpus luteum and placenta. Control of the rate of side chain
cleavage reaction provides the major mechanism for regulation of the level

of steroid hormones. Since conversion of cholesterol to pregnenolone, catalyzed by cytochrome P-450scc presents a complex series of reactions which may involve stable intermediates, a hypothesis has been proposed that cytochrome P-450scc represents a multienzyme complex in which intermediates are transferred from one isozyme to another without releasing oxysteroids into the reaction mixture (HALL and KORITZ, 1964). However, subsequent studies with highly purified cytochrome P-450scc showed that multistep catalysis is performed by the single cytochrome P-450scc through the complex interaction of heme-coordinated active oxygen species and cholesterol side chain (AKHREM et al., 1979; DUQUE et al., 1978). More recent studies indicate that cytochrome P-450scc utilizes thermodynamic stabilization of the enzyme-intermediate complex which prevents accumulation of intermediates during catalysis: the first hydroxylation reaction has a considerably higher activation energy than the subsequent oxidative step (LAMBETH et al., 1982). This is supported by the presence of (22R)-22-hydroxycholesterol and (20R,22R)-20,22-dihydroxycholesterol as native constituents of highly purified cytochrome P-450scc as enzyme-bound intermediates in the biosynthesis of pregnenolone (LARROQUE et al., 1981). The following sequential hydroxylation mechanism appear to be realized: (22R)-22-hydroxycholesterol → (20R,22R)-20,22-dihydroxycholesterol → pregnenolone.

The mitochondrial and highly purified cytochrome P-450scc as well have a wide substrate specificity with respect to the length of the side chain and position of the hydroxygroup (ARTHUR et al., 1976; MORISAKI et al., 1980). Nevertheless, from eight cholesterol analogues having modification in A/B rings, only 3-epicholesterol and cholestenone are cleaved to form epipregnenolone and progesterone, respectively (MORISAKI et al., 1982). The cholesterol analogue 25-doxyl-27-nor-cholesterol was found to be a substrate for cytochrome P-450scc which is transformed to pregnenolone (LANGE et al., 1988). Thus, modification of the cholesterol side chain at the 25 position by the rather large doxyl ring (about 0.5 nm diameter) does not change the stereospecificity of the oxidative C_{20}-C_{22} bond disruption reaction indicating that stereospecificity of the cytochrome P-450scc side chain cleavage activity is not dependent on the nature of the cholesterol side chain termination (C_{25} to C_{27}).

Based on spectral titration experiments using rat adrenocortical mitochondria JEFCOATE (1975) supposed that cytochrome P-450scc has two steroid binding sites. The first binds cholesterol and 25-hydroxycholesterol resulting predominantly in the high spin form, while the second site may be occupied without displacement of cholesterol from the first binding site by either (20S)-20-hydroxycholesterol or pregnenolone resulting predominantly in low spin form. Subsequently, it has been shown that cytochrome P-450scc in rat adrenal mitochondria exists in three spectroscopically detectable states: (i) a high spin cytochrome P-450scc cholesterol complex; (ii) a low spin form which is converted to the high spin state by both 25-hydroxycholesterol and

(20R,22R)-20,22-dihydroxycholesterol; (iii) a low spin state which is converted to a high spin state by (20R,22R)-20,22-dihydroxycholesterol not by 25-hydroxycholesterol (JEFCOATE and ORME-JOHNSON, 1975; JEFCOATE et al., 1976).

Two forms of cytochrome P-450scc have been found in rat adrenocortical mitochondria by careful difference spectroscopy, one of which is pH- and temperature sensitive (JEFCOATE, 1977). Both forms differ in the nature and number of steroid binding sites. One form has two binding sites termed as L_1 and L_{1-R}. The L_1 site binds cholesterol and 25-hydroxycholesterol resulting in the high spin form, while the second site (L_{1-R}) may bind without displacement of cholesterol, (22R)-hydroxycholesterol or pregnenolone, giving rise to the low spin form. High spin complexes of the first form undergo a change to the low spin state with decreased temperature or increased pH, but probably without substrate dissociation. The second form has a single site (L_2) and exhibits high affinity to (20R,22R)-20,22-dihydroxycholesterol forming high spin complexes, with low affinity to cholesterol and does not bind 25-hydroxycholesterol (JEFCOATE, 1977). To prove the presence of multiple forms of cytochrome P-450scc, ORME-JOHNSON (ORME-JOHNSON et al., 1979) studied the interaction of steroids with purified cytochrome P-450scc. Optical measurements, EPR-data and equilibrium dialysis indicate that the number of steroid binding sites for (20R)-20-hydroxycholestrol, (22R)-22-hydroxycholesterol and (20R,22R)-20,22-dihydroxycholesterol or pregnenolone equals 1 per cytochrome P-450scc molecule. Thus in contrast to membrane-bound cytochrome P-450scc, special forms of cytochrome P-450scc have not been detected. These differences may be a function of the peculiarities of the membrane environment.

Titration experiments of the substrate binding site can indicate the structural requirements of steroid derivatives for interaction with cytochrome P-450scc: (i) the β-configuration of the 3-hydroxy group is essential; (ii) the double bond between C_5 and C_6 is necessary for binding; (iii) cholesterol derivatives with hydroxy group at C_{25} position display a rapid binding to cytochrome P-450scc; (iv) 24-alkyl derivatives were inactive (KIMURA, 1981).

The equilibrium constant for interaction of low spin cytochrome P-450scc with cholesterol is $K_{eq} = 0.43 \cdot 10^6 \cdot M^{-1}$ (HUME et al., 1984). The first-order rate constant of cholesterol-induced transition from low to high spin state of cytochrome P-450scc is $k = 7.5 \cdot 10^{-4} \cdot s^{-1}$ which is similar for cholesterol association with cytochrome P-450scc in intact mitochondria. The activation energy for this process is $E = 88 \text{ kJ} \cdot \text{mol}^{-1}$ (HUME et al., 1984).

The steroid nucleus and hydrophobic side chain must have different affinity to and effect on the spin state of cytochrome P-450scc (KIDO et al., 1981). Binding of the nucleus portion of a steroid molecule was shown to induce a low spin state. Examples of such low spin effectors may be pregnenolone and progesterone. Oppositely, binding of the side chain portion of steroid to heme-

---

protein induces high spin state. In order to explain the abnormal temperature behavior of cytochrome P-450scc, the removal of the side chain of cholesterol from the heme rather than that of the whole cholesterol molecule was proposed when temperature increased, resulting in the low spin state (HASUMI et al., 1984).

A series of analogues of cholesterol with a modified side chain and primary amino group, were prepared and tested for their ability to interact with cytochrome P-450scc. The monohydroxylated steroids can be classified according to the position of the hydroxyl group and their efficiency in interacting with cytochrome P-450scc in the following way: 24- > 26- > 25-hydroxycholesterol (HUME et al., 1984). The configuration of the hydroxyl group at C_7 shows that 7β-hydroxycholesterol induces a greater maximal absorbance change than does 7α-hydroxycholesterol. Derivatives in which the side chain amino group was closer to steroid than the C_{22}-position, were found to be only very weak inhibitors and did not produce spectral changes when added to hemeprotein (SHEETS and VICKERY, 1982; SHEETS and VICKERY, 1983; SHEETS and VICKERY, 1983; VICKERY and KELLIS, 1983). Derivatives in which the amino group was attached at a greater distance from the steroid ring than the C_{22}-position, caused a progressive decrease in inhibitory tendency and a failure to produce spectral changes. The heme appear to be located sufficiently close to this position so that the side chain of cholesterol would allow a direct attack of an iron-bound oxidant to occur during hydroxylation and side chain cleavage. It was shown that the C_{22}-position of cholesterol molecule should lie 2.9 Å from the iron in the enzyme-substrate complex. Cytochrome P-450scc reveals a high degree of stereospecificity. The 22S-hydroxycholesterol binds to cytochrome P-450scc less efficiently than natural intermediate — 22R-hydroxycholesterol (HEYL et al., 1986). Spin echo studies with specifically deuterated cholesterol derivatives have revealed that the 22S deuterium of 22R-hydroxycholesterol is located at approximately 4 Å far from the heme iron (GROH et al., 1983). Furthermore, presence of carbon monoxide interferes with the binding of analogs with cytochrome P-450scc indicating that the 22R hydroxyl of steroid might be localized in a distance of $2-3$ Å from the heme iron of cytochrome P-450 (HEYL et al., 1986).

The nature of cytochrome P-450scc cholesterol interactions has also been studied using phospholipid-reconstituted cytochrome P-450scc (LAMBETH et al., 1980; YAMAKURA et al., 1981). The use of vesicle-reconstituted system showed that only cholesterol which was incorporated in the same membrane as cytochrome P-450scc was rapidly metabolized to pregnenolone, while cholesterol in different vesicle membranes was not accessible to cytochrome P-450scc (SEYBERT et al., 1979). Several lines of evidence indicate that the substrate binding site of cytochrome P-450scc is faced to the phospholipid membrane (SEYBERT et al., 1979). This is consistent with the finding that externally added cholesterol is not utilized by the intact mitochondria for steroidogenesis. Only cholesterol molecules of the inner mitochondrial membrane, but not completely, are accessible to cytochrome P-450scc (CHENG et al., 1985).

An important prerequisite for a better understanding of the catalytic mechanism was provided by the elucidation of the primary structure of bovine cytochrome P-450scc by protein sequencing (CHASHCHIN et al., 1986) as well as by sequencing its cDNA (MOROHASHI et al., 1984). Later the cDNA sequence of human adrenocortical (CHUNG, 1986) and placental (MOROHASHI et al., 1987) cytochrome P-450scc were elucidated. Figure 2 shows alignment of the primary structures of human and bovine cytochromes P-450scc. The amino acid sequence of human cytochrome P-450scc is 72% homologous to the bovine sequence while the coding sequences are 82% homologous. Both proteins are synthesized as a precursor molecule having an extension peptide at the N-terminal sequence. Since the mature form of bovine adrenocortical cytochrome P-450scc contains Ile as N-terminal sequence (AKHREM et al., 1980) this means that the extra peptide consists of 39 amino acids. Human cytochrome P-450scc contains 521 amino acids and appears to have an insertion of one extra amino acid (His) compared to bovine cytochrome P-450scc.

Alignment of the primary structure of bovine cytochrome P-450scc with sequences of some steroid-binding proteins indicates the homology in amino acid sequences which was interpreted as the cholesterol binding site of cytochrome P-450scc: I T N V M F G E R L G M (GOTOH et al., 1985).

Besides the substrate binding site, cytochrome P-450scc contains some functionally important units: (i) the catalytic center with protoporphyrine IX as prosthetic group; (ii) the site responsible for interaction of cytochrome P-450scc with adrenodoxin; (iii) hydrophobic regions responsible for interaction with phospholipid membrane. Limited proteolysis proved to be a useful technique to study the correlation between the structure and function of cytochrome P-450scc. This method promotes determination of compact globular structures in protein molecules which possess a definite function. Indication of the domain-like structure of cytochrome P-450scc was derived from limited trypsinolysis studies. In the presence of trypsin cytochrome P-450scc disappears with concomitant appearance of two polypeptide fragments, F_1 and F_2 with 29.8 kDa and 26.6 kDa, respectively (Fig. 3). These fragments are relatively stable in further proteolytic modification. Prolonged incubation of cytochrome P-450scc with trypsin was shown to result in a third stable polypeptide fragment referred to as F_3 with 16.8 kDa with concomitant loss of F_2 fragment, indicating that F_3 is the further proteolytically modified fragment F_2 (AKHREM et al., 1980; AKHREM et al., 1980; CHASHCHIN et al., 1984; CHASHCHIN et al., 1984). The observed pattern of proteolytic modification of cytochrome P-450scc indicates that proteolytic modification results in an equimolar ratio of F_1 and F_2 fragments, with the total molecular weight equal to that of native enzyme, suggesting that these fragments might represent func-

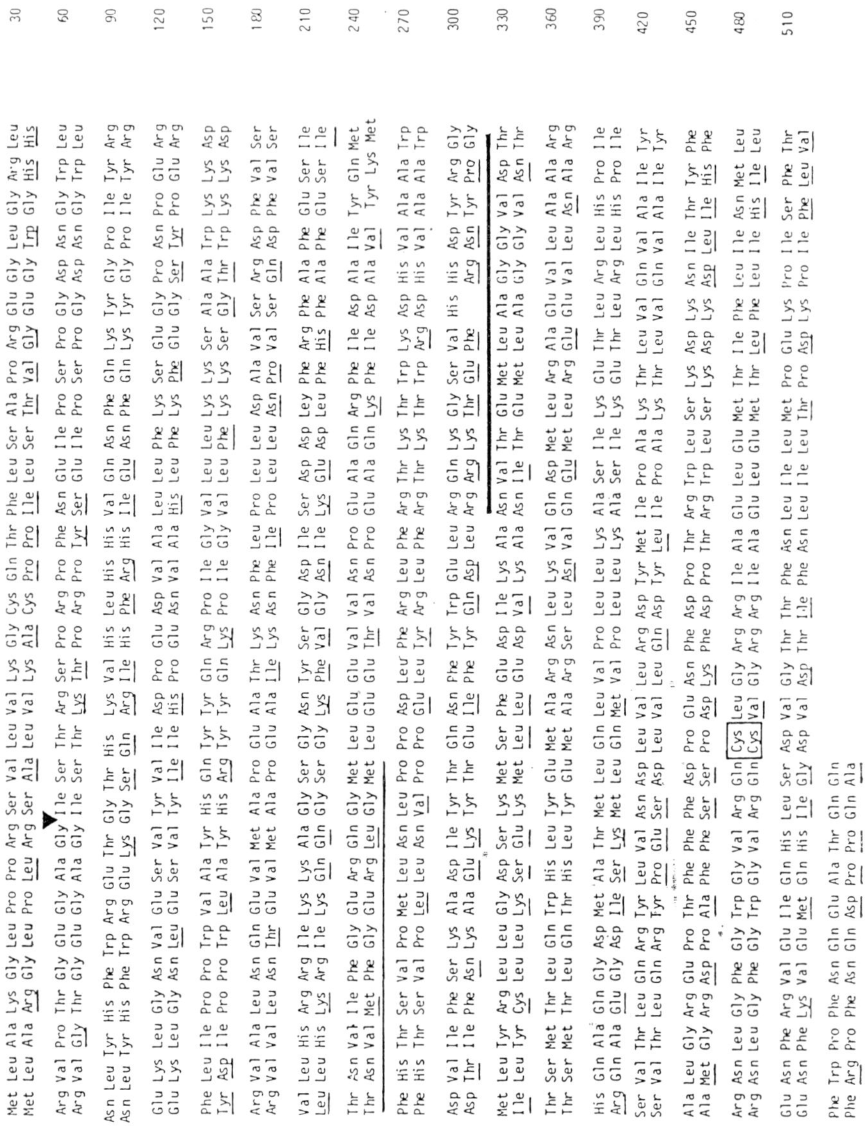

Fig. 2. Comparison of the primary structure of the human placental (CHUNG et al., 1986) and bovine adrenocortical (MOROHASHI et al., 1984) cytochrome P-450scc. The substitutions in the primary structure are indicated. The cleavage point of the extrapeptide is shown by the arrowhead. The trypsin-sensitive hinge is underlined. The possible heme-binding peptide is also indicated.

tional domains of the cytochrome P-450scc molecule. Cytochrome P-450scc proteolytically modified to equimolar mixture of F_1 and F_2 or F_1 and F_3 fragments, retains the spectral and functional properties of the native hemeprotein. The domains of cytochrome P-450scc are tightly bound and could not be separated under non-denaturing conditions. The aspartylpropyl bonds, sulfhydryl groups, tyrosyl residues are distrubuted nonuniformly between fragment F_1 and F_2. The complete separation of fragments F_1 and F_2 has been achieved by

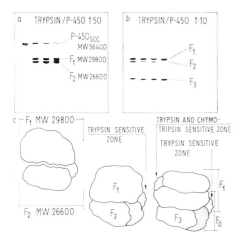

Fig. 3. Limited tryptic proteolysis of cytochrome P-450scc with trypsin. SDS PAGE of the time-course of limited proteolysis of cytochrome P-450scc with 1:50 (a) and 1:10 (b) trypsin/cytochrome P-450 ratio. Hypothetical model of limited tryptic proteolysis of cytochrome P-450scc (c).

chromatography under denaturing conditions on activated thiopropyl-Sepharose via selective covalent immobilization of fragment F_2, since unlike fragment F_1 this fragment contains sulfhydryl groups (Fig. 4).

Structural analysis of individual F_1 and F_2 fragments indicates that the N-terminal sequence of fragment F_1 coincides with the N-terminal sequence of cytochrome P-450scc, suggesting that this fragment represents the N-terminal part of cytochrome P-450scc. The C-terminal sequence of fragment F_2 is identical to that of cytochrome P-450scc indicating that this domain represents C-terminal sequence of cytochrome P-450scc. Both fragments connected with the short polypeptide hinge are limited by R(250) and N(257). The cleavage of this hinge results in the formation of F_1 and F_2 fragments. The second trypsin-sensitive site is R(399): cleavage of fragment F_2 at this place was shown to result in fragment F_3. Inactivation of cytochrome P-450scc resulting in cytochrome P-450 is followed by the appearance of a new polypeptide fragment at R(92) which is accessible to trypsin. Apparently, cytochrome P-450scc consists of at least two independently folded domains connected with polypeptide hinge that remain rigidly associated following trypsin treatment.

The localization of the heme group and catalytic center of cytochrome

P-450scc in the polypeptide chain has been the subject of much investigation. Recent advances in determining the primary structure of proteins have revealed the complete amino acid sequences of cytochrome P-450 isozymes. Unlike other cytochrome P-450 isozymes, cytochrome P-450scc contains only two cysteine residues which might be candidates for heme binding at the proximal site. One of the two cysteine residues is located near the trypsin

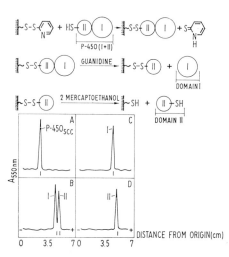

Fig. 4. Scheme for covalent thio-propyl chromatography of proteolytically modified cytochrome P-450scc. Gel-scanning of SDS PAGE of cytochrome P-450scc (A), proteolytically modified cytochrome P-450scc (B), fragment F_1 (C) and fragment F_2 (D).

sensitive site, while the other is located in the C-terminal sequence of the hemeprotein. As shown recently (USANOV et al., 1985; USANOV et al., 1987) of the two cysteine residues, only one (Cys 264) is accessible to modification with sulfhydryl reagents, while cysteine residue (Cys 422) may be modified only after cytochrome P-450scc inactivation and heme removal. Thus, it is evident that cysteine residue Cys 422 which is located in the C-terminal region of steroidogenic cytochrome P-450 enzymes is the proximal thiolate ligand to the cytochrome P-450scc heme iron. This sequence was shown to be highly conserved among steroidogenic cytochromes P-450 (Fig. 5).

Chemical modification experiments have shown that tyrosine residues of cytochrome P-450scc might be essential for heme environment from the distal side (USANOV et al., 1984). Additional studies indicate that the N-terminal tyrosines Tyr-93 and Tyr-94 essential for catalytic activity were selectively labeled with tetranitromethane (PIKULEVA et al., 1987).

Immunochemical methods proved to be useful in studies of the topology of cytochrome P-450scc in phospholipid membrane (USANOV et al., 1987). Since cytochrome P-450scc has a domain-like structure and contains at least F_1, F_2 and F_3 fragments, the monospecific antibodies against these fragments were shown to an efficient instrument to clarify the accessibility of the cytochrome

---

S. A. USANOV et al.

P-450scc (bovine)	Phe	Gly	Trp	Val	Arg	Gln		Cys	Val	Gly	Arg	Arg	Ile	Ala	Glu	Leu	Glu	Met	Thr	Leu	Phe
P-450scc (human)	Phe	Gly	Trp	Val	Arg	Gln		Cys	Leu	Gly	Arg	Arg	Ile	Ala	Glu	Leu	Glu	Met	Thr	Ile	Phe
$P-450_{11\beta}$ (bovine)	Phe	Gly	Phe	Val	Arg	Gln		Cys	Leu	Gly	Arg	Arg	Val	Ala	Glu	Val	Glu	Met	Leu	Leu	Leu
$P-450_{21}$ (bovine)	Phe	Gly	Cys	Ala	Arg	Val		Cys	Leu	Gly	Glu	Ser	Leu	Ala	Arg	Leu	Glu	Leu	Phe	Val	Val
$P-450_{21}$ (human)	Phe	Gly	Cys	Ala	Arg	Val		Cys	Leu	Gly	Glu	Pro	Leu	Ala	Arg	Leu	Glu	Leu	Phe	Val	Val
$P-450_{17\alpha}$ (human)	Phe	Gly	Ala	Gly	Pro	Arg	Ser	Cys	Ile	Gly	Glu	Ile	Leu	Ala	Arg	Gln	Glu	Leu	Phe	Leu	Ile
$P-450_{17\alpha}$ (human)	Phe	Gly	Ala	Gly	Pro	Arg	Ser	Cys	Val	Gly	Glu	Met	Leu	Ala	Arg	Gln	Glu	Leu	Phe	Leu	Phe

Fig. 5. Comparison of the sequence of the putative heme-binding site in various steroidogenic cytochromes P-450.

Pathways of Adrenal Steroidogenesis

P-450scc polypeptide chain to antibodies in the membrane-bound hemeprotein. Highly specific antibodies have also been useful in comparison with a proteolytic modification picture for soluble and membrane-bound cytochrome P-450scc.

Cytochrome P-450scc in mitoplast is not accessible to trypsin (USANOV, 1988). Ultrasonic disintegration of mitoplasts and preparation of outside vesicles makes cytochrome P-450scc accessible to trypsin and the products of limited proteolysis were found to be identical with the fragments F_1 and F_2. This was taken as an indication that proteolytic modification of cytochrome P-450scc in submitochondrial particles appears to be similar to that of soluble cytochrome P-450scc. Similarity of proteolytic modification for both membrane-bound and soluble cytochrome P-450scc was found to be an indication that polypeptide hinge accessible to trypsin is exposed on the surface of phospholipid membrane of the matrix side. Both fragments of cytochrome P-450scc were shown to be tightly bound with phospholipid membrane.

Cytochrome P-450scc proved to be accessible to antibodies against F_1 and F_2 fragments from the outer side of the mitoplast membrane. However, antibodies against fragment F_3 do not recognize cytochrome P-450scc from the outer side (USANOV et al., 1987). Efficient binding of antibodies against fragments F_1, F_2 and F_3 was observed after ultrasonic treatment of mitoplasts, i.e. with the matrix side of the inner mitochondrial membrane. These data allowed the conclusion that both N- and C-terminal sequences of cytochrome P-450scc span the phospholipid membrane and therefore cytochrome P-450scc is a typical transmembrane integral protein.

3.2. 11β-Hydroxylation

The final steps of corticosteroid hormone biosynthesis are realized in adrenocortical mitochondria by cytochrome P-450$_{11\beta}$. Cytochrome P-450$_{11\beta}$ according to physico-chemical, immunochemical, catalytic properties as well as with regard to its stability, is rather different from cytochrome P-450scc (TAKEMORI et al., 1975; WATANUKI et al., 1977; WATANUKI et al., 1978). Topological studies indicate that cytochrome P-450$_{11\beta}$ is much more deeply embedded in the phospholipid membrane of mitochondria than cytochrome P-450scc (CHURCHILL et al., 1978; CHURCHILL and KIMURA, 1979; WANG and KIMURA, 1978). Cytochromes P-450scc and P-450$_{11\beta}$ appear to be the two major cytochrome P-450 species present in a 2:1 stoichiometry in the inner mitochondrial membrane.

The purification of cytochrome P-450$_{11\beta}$ is based on hydrophobic chromatography (TAKEMORI et al., 1975; SATO et al., 1978). Important features of this procedure are using of a substrate as stabilizer of cytochrome P-450$_{11\beta}$ and differences in the solubility of cytochrome P-450scc and P-450$_{11\beta}$. Highly purified cytochrome P-450$_{11\beta}$ differs from cytochrome P-450scc immunologically (SUHARA et al., 1978), in amino acid composition, molecular weight (SUHARA et al., 1978; MARTZEV et al., 1982) and in EPR spectra (KOMINAMI et al., 1979).

Cytochrome P-450$_{11\beta}$ is responsible for regio- and stereo-selective 11β-hydroxylation of deoxycorticosterone, deoxycortisol, progesterone, 4-androstene-3,17-dione and testosterone resulting in respective 11β-hydroxyderivatives and concomitant hydroxylations at C_{18} and C_{19} positions. These activities were proved by experiments with a reconstituted system (WATANUKI et al., 1978; SATO, et al., 1978; SUHARA et al., 1978). Cytochrome P-450$_{11\beta}$ catalyzes hydroxylation of 4-androstene-3,17-dione at 11β- and 19-positions in a ratio of about 4:1 (SATO et al., 1978). The correlation between ability of steroids to modulate low to high spin conversion and their efficiency to serve as substrates were found for cytochrome P-450$_{11\beta}$ (KOMINAMI et al., 1979). This allows us to assess the importance of different elements of steroid molecule, which are necessary for interaction with cytochrome P-450$_{11\beta}$: cytochrome P-450$_{11\beta}$ is active with respect to 3-keto-Δ4-steroids in which additional hydroxy groups at 17α-, 21- and 18-position are not crucial for 11β-hydroxylation (KATAGIRI et al., 1980). Recently it was reported that cytochrome P-450$_{11\beta}$ catalyzes 19-hydroxylation of 18-hydroxy-11-deoxycorticosterone in addition to 11β-hydroxylation of the steroid (MOMOI et al., 1983). 19-Oxoandrostenedione, the product of 19-hydroxyandrostenedione by the oxidase activity of the purified cytochrome P-450$_{11\beta}$ was shown to be further oxidized and demethylated at the C_{10} position (SUHARA et al., 1988) to give the C_{18} steroids estrone (aromatase reaction) and 19-norandrostenedione (nonaromatizing 10-demethylase or $C_{10}-C_{19}$ lyase reaction). These reactions, together with the initial hydroxylation of androstenedione at C_{19} form a sequence of cytochrome P-450$_{11\beta}$ catalyzed C_{19} steroid 19-monooxygenase reactions. Thus, cytochrome P-450$_{11\beta}$ is somewhat similar to placental endoplasmic cytochrome P-450-dependent aromatase in some of its substrate specificity but appears P-450 very different to both cytochromes according to physico-chemical properties and primary structure.

Evidence is provided that a single cytochrome P-450$_{11\beta}$ molecule catalyzes both 11β- and 18-hydroxylations or 11β- and 19-hydroxylations (SATO et al., 1978; SUHARA et al., 1978; WATANUKI et al., 1978; MARTZEV et al., 1982). These results are based on the facts that: (i) different activities disappear at the same rate on storing; (ii) various inhibitors affect different reactions to the same degree; (iii) antibodies to cytochrome P-450$_{11\beta}$ inhibit different reactions to the same degree (WATANUKI et al., 1978). The nature of the dual activities of cytochrome P-450$_{11\beta}$ is not clear, although, in accordance with the multistep cholesterol hydroxylation by cytochrome P-450scc, one can propose the dynamic interaction of heme-bound active oxygen species with different positions of the steroid molecule. Both C_{18} and C_{19} angular methyl groups are in close proximity to the C_{11}-position so that also may be reactive with activated oxygen. However, the mode of regulation of different activities of cytochrome P-450$_{11\beta}$ is still unclear.

Much less is known about the structure-function relationships in cytochrome P-450$_{11\beta}$. The peptide pattern at limited proteolysis of cytochrome P-450$_{11\beta}$ is quite different from that of cytochrome P-450scc: treatment of cytochrome P-450$_{11\beta}$ with trypsin results in two fragments F_1 (32 kDa) and F_2 (14 kDa) (AKHREM et al., 1980). Recently, very similar findings were obtained, indicating that treatment of soluble cytochrome P-450$_{11\beta}$ with trypsin yielded transiently major peptides of 34, 30 and 17 kDa which thereafter almost disappeared with time (LOMBARDO et al., 1986). By contrast, proteolysis of the membrane-bound cytochrome P-450$_{11\beta}$ yielded preferably a 34 kDa fragment which represented about 80% of the products and was not further degraded. This data indicates that the 34 kDa fragment of cytochrome P-450$_{11}$ might represent the hydrophobic domain of the hemeprotein which is embedded in phospholipid membrane. Based on the inability of membrane bound cytochrome P-450$_{11\beta}$ to interact with nonpenetrating reagents it was concluded that the steroid binding site of cytochrome P-450$_{11\beta}$ is localized inside the mitochondrial inner membrane (RYDSTROM et al., 1976).

Recently, bovine and human adrenal cDNA clones encoding the cytochrome P-450 specific to 11β-hydroxylation have been isolated (CHUA et al., 1987; MOROHASHI et al., 1987). Sequence analysis revealed the primary structure of cytochrome P-450$_{11\beta}$ which consisted of 503 amino acids (Fig. 6). Cytochrome P-450$_{11\beta}$ has been predicted to contain 479 amino acid residues in the mature form, in addition to an amino-terminal mitochondrial signal sequence of 24-residues. The predicted amino acid sequence of cytochrome P-450$_{11\beta}$ is 40% homologous to the sequence of cytochrome P-450scc but only 22% homologous to the microsomal steroidogenic cytochrome P-450$_{21}$. Comparison of the primary structures of cytochrome P-450scc and cytochrome P-450$_{11\beta}$ revealed four highly conserved regions. It is thought that putative steroid binding site is in the C-1 and heme binding site in the C-4 regions (MOROHASHI et al., 1987).

3.3. Electron transfer components

The progress made in purification of electron transfer components of mitochondrial steroid hydroxylases allows us to understand the mechanism of electron transfer from the viewpoint of specific protein-protein interactions. With regard to the electron transfer pathway, the mitochondrial cytochromes P-450 are different from their microsomal counterparts in that the former receive electrons via an iron-sulfur protein, adrenodoxin, from NADPH-adrenodoxin reductase, whereas the latter receive electrons directly from a

Met Ala Leu Trp Ala Lys Ala Arg Val Arg Met Ala Gly Pro Trp Leu Ser Leu His Glu Ala Leu Leu Leu Gly Thr Arg Gly Ala Ala 30

Ala Pro Lys Ala Val Leu Pro Phe Glu Ala Met Pro. Arg Cys Pro Gly Asn Lys Trp Met Arg Met Leu Gln Ile Trp Lys Glu Gln Ser 60

Ser Glu Asn Met His Leu Asp Met His Gln Thr Phe Gln Glu Leu Gly Pro Ile Phe Arg Tyr Asp Val Gly Gly Arg His Met Val Phe 90

Val Met Leu Pro Glu Asp Val Glu Arg Leu Gln Gln Ala Asp Ser His His Pro Gln Arg Met Ile Leu Glu Pro Trp Leu Ala Tyr Arg 120

Gln Ala Arg Gly His Lys Cys Gly Val Phe Leu Leu Asn Gly Pro Gln Trp Arg Leu Asp Arg Leu Asn Pro Asp Val Leu Ser 150

Leu Pro Ala Leu Gln Lys Tyr Thr Pro Leu Val Ala Gly Val Ala Arg Asp Phe Ser Gln Thr Leu Lys Lys Ala Arg Val Leu Gln Asn Ala 180

Arg Gly Ser Leu Thr Arg Leu Gly His Ala Gln Leu Phe Arg Leu Gln Ala Ser Tyr Ile Glu Ala Ser Thr Leu Val Gly Glu Arg Leu Gly 210

Leu Leu Thr Gln Gln Pro Asn Pro Asp Ser Leu Asn Met Trp Arg Leu Ser His Ala Leu Glu Lys Ser Thr Val Gln Leu Met Phe Val 240

Pro Arg Arg Leu Ser Arg Trp Met Ser Thr Asn Met Trp Arg Glu His Phe Glu Ala Leu Trp Asp Tyr Ala Ala Asp Met Thr 270

Ile Gln Arg Ile Tyr Gln Glu Leu Ala Leu Gly His His Pro Trp His Tyr Ser Gly Ile Val Ala Glu Leu Leu Met Arg Ala Asp Met Thr 300

Leu Asp Thr Ile Lys Ala Asn Thr Ile Asp Leu Thr Ala Gly Ser Val Asp Thr Thr Ala Phe Pro Leu Leu Met Thr Leu Phe Glu Leu 330

Ala Arg Asn Pro Glu Glu Val Gln Gln Leu Gln Ala Val Arg Asp Glu Leu Ser Glu Ser Glu Gln Asn Pro Gln Arg Ala Ala Ile Thr 360

Glu Leu Pro Leu Leu Arg Ala Ala Leu Lys Glu Thr Lys Arg Leu Glu Lue Tyr Pro Val Gly Ile Pro Leu Arg Glu Glu Val Ser Asp Leu 390

Val Leu Gln Asn Tyr His Ile Pro Ala Gly Thr Leu Val Lys Val Lue Tyr Ser Ser Arg Phe Ala Asn Pro Ala Val Phe Ala Arg Pro 420

Glu Ser Tyr His Pro Gln Arg Trp Leu Asp Arg Gln Gly Ser Gly Ser Arg Phe Pro His Leu Ala Phe Gly Phe Gly Val Arg Gln Cys 450

Leu Gly Arg Arg Val Ala Glu Val Glu Met Leu Leu Leu Leu His His Val Leu Lys His Leu Val Glu Thr Leu Glu Gln Glu Asp 480

Ile Lys Met Val Tyr Arg Phe Ile Leu Met Pro Ser Thr Leu Pro Leu Phe Thr Phe Arg Ala Ile Gln

Fig. 6. The primary structure of bovine adrenal cytochrome P-450$_{11\beta}$ deduced from the sequencing of cDNA (MOROHASHI et al., 1987). The arrow indicates the processing point of the precursor of cytochrome P-450$_{11\beta}$. The putative steroid-binding site and conserved heme-binding site are underlined.

flavoprotein, NADPH-cytochrome P-450 reductase. Cytochrome P-450scc and cytochrome P-450$_{11\beta}$ are both embedded in the inner mitochondrial membrane and receive electrons from a common electron donor, adrenodoxin.

Electron transfer from NADPH to cytochrome P-450scc and cytochrome P-450$_{11\beta}$ may be envisaged as three sequential events: (i) interaction of NADPH-adrenodoxin reductase with NADPH and reduction of flavoprotein; (ii) electron transfer from adrenodoxin reductase to adrenodoxin; (iii) electron transfer from adrenodoxin to cytochrome P-450scc and cytochrome P-450$_{11\beta}$.

3.3.1. Adrenodoxin reductase

NADPH-adrenodoxin oxidoreductase (EC 1.18.1.2) is a flavoprotein localized in the inner mitochondrial membrane of adrenal cortex. The enzyme was purified by OMURA (OMURA et al., 1966). Later adrenodoxin (SUGIYAMA and YAMANO, 1975) was used as affinity matrix to isolate NADPH-adrenodoxin reductase. 2′,5′-ADP-Sepharose was also shown to be an efficient affinity matrix for adrenodoxin reductase purification (HIWATASHI et al., 1977). Adrenodoxin reductase contains a flavin as prosthetic group which is non-covalently bound to the protein. The flavin can be easily removed from flavoprotein by treatment with trichloroacetic acid and identified as FAD (FOSTER et al., 1975). Adrenodoxin reductase is characterized by an absorption spectrum typical of flavoproteins with maxima at 378 and 450 nm and an extinction coefficient 10.9—11.3 mM^{-1} cm^{-1} (HIWATASHI et al., 1976 a; HIWATASHI et al., 1976 b). Flavin group determines the presence of negative ellipticity at 450 and 475 nm in circular dichroism spectrum (HIWATASHI et al., 1976 a).

Pig adrenodoxin reductase, unlike bovine adrenodoxin reductase, which is a glycoprotein, is free of carbohydrate (HIWATASHI and ICHIKAWA, 1978). Although pig adrenodoxin reductase is free of carbohydrate, its physico-chemical properties are generally similar to those of bovine enzyme. The results of photooxidation and selective chemical modification with histidine and sulfhydryl-specific reagents indicate that histidyl and cysteinyl residues are essential for NADPH-binding by flavoprotein (HIWATASHI et al., 1978). Adrenodoxin reductase has essential arginyl residue which is crucial to the enzymatic activity as the recognition site for the negatively charged 2′-phosphate group of NADP^{+} (NONAKA et al., 1982).

Recently, cDNA clone for bovine adrenodoxin reductase has been isolated and the primary structure of flavoprotein precursor was deduced from the nucleotide sequence (NONAKA et al., 1987; SAGARA et al., 1987). The precursor form of adrenodoxin reductase consists of 492 amino acid residues, including an extrapeptide of 32 amino acids at N-terminal sequence which is hydrophobic and rich in arginine (Fig. 7). The deduced amino acid sequence (SAGARA

S. A. USANOV et al.

et al., 1987) of adrenodoxin reductase differs in 45 amino acid assignments from that reported by NONAKA (NONAKA et al., 1987). The amino acid sequence of adrenodoxin reductase in the N-terminal region from Gln 40 to Glu 70 was shown to be homologous to the conserved sequence of FAD- and NADPH-binding sites. This means that the FAD- and NADPH-binding sites of adrenodoxin reductase are in the N-terminal sequence. Chemical modification of adrenodoxin reductase with pyridoxal 5′-phosphate indicated that

Met Ala Pro Arg Cys Trp Arg Trp Trp Pro Trp Ser Ser Trp Thr Arg Thr Arg Leu Pro Pro Ser Arg Ser Ile Gln Asn Phe Gly Gln   30

His Phe Ser Thr Gln Glu Gln Thr Pro Gln Ile Cys Val Val Gly Ser Gly Pro Ala Gly Phe Tyr Thr Ala Gln His Leu Leu Lys His   60

His Ser Arg Ala His Val Asp Ile Tyr Glu Lys Gln Leu Val Pro Phe Arg Leu Val Arg Val Trp Leu Ala Leu Thr Thr Pro Arg Ser   90

Arg Met Leu Leu Asn Thr Phe Thr Gln Thr Ala Arg Ser Asp Arg Cys Ala Phe Tyr Gly Asn. Val Glu Val Gly Arg Asp Val Thr Val   120

Gln Glu Leu Arg Val Tyr Arg Leu Thr Ala Val Val Leu Ser Tyr Gly Ala Glu Asp His Gln Ala Leu Asp Ile Pro Gly Glu Glu Leu   150

Pro Gly Val Phe Ser Ala Arg Ala Phe Val Gly Trp Tyr Asn Gly Leu Pro Glu Asn Arg Glu Leu Ala Pro   Asp Leu Ser Cys Asp Thr   180

Ala Val Ile Leu Gly Gln Gly Asn Val Ala Leu Asp Val Ala Arg Ile Leu Leu Thr Pro Pro Asp His Leu Glu Lys Thr Asp Ile Thr   210

Glu Ala Ala Leu Gly Ala Leu Arg Gln Ser Arg Val Lys Thr Val Trp Ile Val Gly Arg Arg Gly Pro Leu Gln Val Ala Phe Thr Ile   240

Lys Glu Leu Arg Glu Met Ile Gln Leu Pro Gly Thr Arg Pro Met Leu Asp Pro Ala Asp Phe Leu Gly Leu Gln Asp Arg Ile Arg Glu   270

Ala Ala Arg Pro Arg Lys Arg Leu Met Glu Leu Leu Leu Arg Thr Ala Thr Glu Lys Pro Gly Val Glu Glu Ala Ala Arg Arg Ala Ser   300

Ala Ser Arg Ala Trp Gly Leu Arg Phe Phe Arg Ser Pro Gln Gln Val Leu Arg Leu Pro Asp Gly Arg Ala Arg Arg Ser Ala Trp Gln   330

Ser Pro Glu Leu Glu Gly Ile Gly Glu Ala His Pro Gly Ser Ala His Trp Gly Cys Gly Gly Pro Pro Cys Gly Leu Val Leu Ser Ser   360

Ile Gly Tyr Lys Ser Arg Pro Ile Asp Pro Ser Val Pro Phe Asp Pro Lys Leu Gly Val Val Pro Asn Met Glu Gly Arg Val Val Asp   390

Val Pro Gly Leu Tyr Cys Ser Gly Trp Val Lys Arg Gly Pro Thr Gly Val Ile Thr Thr Thr Met Thr Asp Ser Phe Leu Thr Gly Gln   420

Ile Leu Leu Gln Asp Leu Lys Ala Gly His Leu Pro Ser Gly Pro Arg Pro Gly Ser Ala Phe Ile Lys Ala Leu Leu Asp Ser Arg Gly   450

Val Trp Pro Val Ser Phe Ser Asp Trp Glu Lys Leu Asp Ala Glu Glu Val Ser Arg Gly Gln Ala Ser Gly Lys Pro Arg Glu Lys Leu   480

Leu Asp Pro Gln Glu Met Leu Arg Leu Leu Gly His

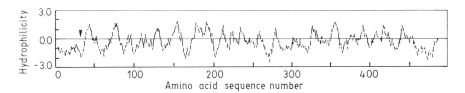

Fig. 7. The primary structure of adrenodoxin reductase from bovine mitochondria deduced from cDNA sequencing (SAGARA et al., 1987) and hydrophilicity/hydrophobicity analysis, calculated over every 9 residues. The arrowhead indicates the predicted processing point of the adrenodoxin reductase precursor. The underlined region in the N-terminal sequence is the putative FAD- or NADPH-binding site of adrenodoxin reductase. The second underlined sequence indicates the putative adrenodoxin-binding site with a marked lysine residue determined by chemical modification.

flavoprotein contains an essential lysine residue (Lys 276) which might be involved in interaction with adrenodoxin (HAMAMOTO and ICHIKAWA, 1984; HAMAMOTO et al., 1988).

Limited proteolysis of adrenodoxin reductase was shown to result in some fragments with 33, 29, 23, 19 and 13.5 kDa (AKHREM et al., 1980), the fragment with 29 kDa being the major polypeptide fragment. Of the limited proteolysis products, the only fragment with 23 kDa does not interact with immobilized adrenodoxin. Prolonged proteolysis of adrenodoxin reductase (50 min at 1:100 ratio) allows the obtaining of the 29 kDa fragment as a single major polypeptide which retains the spectral properties of the native adrenodoxin reductase as well as the reductase activity. These data are consistent with the conclusion that 29 kDa fragment contains adrenodoxin-, NADPH- and FAD-binding sites. The findings, which are very similar to these results, have been presented recently, indicating that adrenodoxin reductase can be selectively cleaved with trypsin to form two major peptides with 30.5 and 20.2 kDa (WARBURTON and SEYBERT, 1988), the flavin, NADPH and adrenodoxin binding sites being located in the 30.5 kDa fragment. These data might be indicative for the domain structure of adrenodoxin reductase.

The interaction of adrenodoxin reductase with NADPH results in reduction of flavoprotein followed by appearance of the distinctive long wavelength absorption from 505 nm to beyond 750 nm suggesting charge transfer complex, which could also be formed by anaerobic titration of the two electron reduced flavoprotein with $NADP^+$ (LAMBETH and KAMIN, 1976). The adrenodoxin reductase exhibits a preference for NADPH over NADH and by means of the intervention of the flavin semiquinone state can separate two reducing equivalents supplied in one step by pyridine nucleotide into two one-electron donation steps necessary to reduce one electron acceptors.

The rapid complex formation between NADPH and adrenodoxin reductase is followed by a slower hydride transfer (18 sec^{-1}) to generate the fully reduced flavoprotein (LAMBETH and KAMIN, 1976; LAMBETH et al., 1977). The complex between $NADP^+$ and reduced adrenodoxin reductase was shown to have the following structure: $AR(FADH_2)$-$NADP^+$. The spectrum of this complex is not typical for any known form of free flavin (oxidized, reduced or semiquinone) suggesting that the species represents a charge transfer complex (LAMBETH and KAMIN, 1976). This complex appears to exist as a multistep equilibrium between semiquinone and reduced forms of flavoprotein (KITAGAWA et al., 1982; SAKAMOTO et al., 1982; SUGIYAMA et al., 1979).

Oxidized adrenodoxin reductase binds $NADP^+$ much less effectively than do reduced flavoprotein, Kd 14 μM and 0.01 μM, respectively. Stopped-flow experiments indicate that reduction of adrenodoxin reductase with equimolar NADPH to form $AR(FADH_2)$-$NADP^+$ is a first order process with k = 28 s^{-1} (LAMBETH and KAMIN, 1976). When a large excess of NADPH is used, a second apparently first order process is observed with k = 4.25 s^{-1}

which is interpreted as a replacement of NADPH for $NADP^+$ in the $AR(FADH_2)\text{-}NADP^+$.

Adrenodoxin reductase was shown to be localized on the inner mitochondrial membrane with the NADPH-binding site being exposed to the matrix side of the inner membrane (CHURCHILL et al., 1978). Recently, at least two forms (less or more tightly bound to the membrane) of adrenodoxin reductase have been found in adrenocortical mitochondria (SUHARA et al., 1982), which was supported by immunochemical analysis after solubilization of adrenodoxin reductase by ultrasonic treatment (USANOV et al., 1987). Adrenodoxin reductase was shown to be accessible to antibodies either from the outer or the matrix side of the inner mitochondrial membrane. However, at present it is difficult to discriminate between either the existence of two pools of adrenodoxin reductase in the inner membrane or an indication of the transmembrane organization of flavoprotein.

3.3.2. Adrenodoxin

Adrenal iron-sulfur protein (adrenodoxin), an electron carrier in the mitochondrial steroid hydroxylases, is a low molecular weight protein with an iron-sulfur cluster of 2Fe-2S type which participates in electron transfer from NADPH to cytochromes P-450scc and $P\text{-}450_{11}$. The isolation and partial purification of adrenodoxin from bovine adrenocortical mitochondria was performed simultaneously by KIMURA (SUZUKI and KIMURA, 1965; KIMURA and SUZUKI, 1967) and OMURA (OMURA et al., 1966). Adrenodoxin has molecular weight 12000 and its 2Fe-2S chromophor imparts a reddish brown color to the protein (OMURA et al., 1966). It has unique spectral properties with a low 276 nm absorbance and maxima in the visible region of the spectrum at 414 nm and 454 nm (SUHARA et al., 1972). Although different physicochemical methods have been used to interpret the unique spectral properties of adrenodoxin, the question remains complex and unresolved. Adrenodoxin represents approximately 0.3% of the proteins of adrenal cortex and 2% of adrenal cortex mitochondria (ESTABROOK et al., 1973).

The iron-sulfur cluster of adrenodoxin is known to exhibit a binuclear tetrahedral symmetry. Each iron of the iron-sulfur cluster is bridged to two labile sulfurs (S) and two cysteinyl sulfurs of the polypeptide chain. In the reduced form adrenodoxin has an electron paramagnetic resonance signal at $g = 1.94$ (WATARI and KIMURA, 1966), where the two iron atoms are antiferromagnetically exchanged (KIMURA et al., 1970). Reduction of the iron sulfur center is accompanied by the uptake of protons by an adjacent histidine residue (LAMBETH et al., 1982). Despite the fact that oxidized adrenodoxin has two ferric ions, it mediates only one electron in the oxidation-reduction cycle.

Adrenodoxin represents a single polypeptide chain of 114 amino acid residues with single tyrosine residue at position 82 (TANAKA et al., 1973). Alignment of the primary structure of bovine adrenodoxin (TANAKA et al., 1973) with that of pig adrenodoxin (AKHREM et al., 1978) indicates the high degree of homology. Adrenodoxin is an acidic protein: it contains 21 acidic amino acids. The only other aromatic amino acids are 4 phenylalanine residues. The single tyrosine residue of adrenodoxin displays an anomalous peak at 331 nm which is opposite to a normal tyrosine emission at 305 nm (LIM and KIMURA, 1980). The anomalous fluorescence of tyrosine indicates that conformation of poly-peptide chain near tyrosine residue is closely related to this anomalous be-haviour. It is thought that the single tyrosine residue is located in the helix region of the polypeptide chain (LIM and KIMURA, 1981).

Recently, the amino acid sequence of mature adrenodoxin has been deduced from the nucleotide sequence of cDNA (OKAMURA et al., 1985). This sequence proved to be identical with that determined from protein sequencing except for three amide changes. The sequence of the adrenodoxin NH_2-terminal precursor segment (58 amino acids) contains several (10) basic residues, a characteristic feature of precursors of mitochondrial proteins. In addition, a 14 amino acid extension is present at the C-terminal sequence. The absence of this 14 amino acid C-terminal sequence in earlier protein sequencing studies (TANAKA et al., 1973; AKHREM et al., 1978) led to the speculation that either it was proteolytically cleaved during protein purification or removed upon insertion of the precursor molecule into mitochondria. Immunochemical ana-lysis with antibodies raised against C-terminal 14 amino acid fragment indicates that it is indeed an integral part of the adrenodoxin molecule and it is not processed upon maturation by mitochondria (BHASKER et al., 1987).

Adrenodoxin was shown to be selectively phosphorylated upon incubation with purified cAMP-dependent protein kinase (MONNIER et al., 1987). Phos-phorylation results in a changed electrophoretic mobility of adrenodoxin and makes it more resistant to mild trypsin degradation. The serine in position 88 is the most likely target for phosphorylation. Phosphorylation of adrenodoxin results in an increased affinity of ferredoxin to cytochrome P-450scc and cyto-chrome P-450$_{11\beta}$. The biological significance of the adrenodoxin phosphorylation by the cAMP-dependent protein kinase in intact adrenocortical cells remains to be established.

Earlier studies failed to detect the reactivity of adrenodoxin with plant $NADP^+$-ferredoxin reductase and that of ferredoxin with adrenodoxin reduc-tase (SUZUKI and KIMURA, 1965; KIMURA and OHNO, 1968). Recently, ferre-doxin and adrenodoxin were found to be able to form complexes with hetero-logous flavoproteins (JACQUOT et al., 1988). Adrenodoxin is also active in some other plant-specific electron transfer reactions. The heterologous ferre-doxins, however, are used at much higher molar ratio to flavoprotein. Thus, being different from plant ferredoxin in its structure and oxidation-reduction potential, adrenodoxin is still able to interact with electron transfer partners of ferredoxin indicating the presence of the same common mechanisms of protein-protein interaction.

Mitochondria from non steroidogenic tissues such as liver and kidney, revealed the presence of ferredoxin-dependent cytochrome P-450 monooxygenases. Recently, the primary structures of ferredoxins from liver mitochondria — hepatoredoxin (CHASHCHIN et al., 1986) and kidney mitochondria — renoredoxin (AKHREM et al., 1988) have been elucidated by protein sequencing. cDNA of human adrenodoxin (PICADO-LEONARD et al., 1988), human placental ferredoxin (MITTAL et al., 1988), and from chicken testis mitochondria (KAGIMOTO et al., 1988) have been cloned and sequenced. Alignment of the primary structure of ferredoxins indicates (Fig. 8) a high degree of homology indicating the conserve nature of ferredoxin molecules, which implies a similar molecular organization of the iron-sulfur cluster. The main question regarding the structural organization of ferredoxin molecule is the following: what cysteine is not involved in cluster fixation. Chemical modification experiments revealed that the cysteinyl residue at position 95 is accessible to modification in the native ferredoxin molecule and appears not to be involved in cluster binding (AKHREM et al., 1988). Based on results of chemical modification and theoretical analysis of the conformation of the polypeptide loop (Cys-52—Cys-95) in ferredoxins two different models (Fig. 8) for molecular organization of iron-sulfur cluster have been proposed (AKHREM et al., 1988). These models were shown to be applicable also to the putidaredoxin — ferredoxin of cytochrome P-450cam.

3.3.3. Principles of molecular organization of mitochondrial monooxygenases and the mechanism of electron transfer

Adrenodoxin reductase was shown to form a tightly associated complex with adrenodoxin as revealed by kinetic and gel-filtration experiments (CHU and KIMURA, 1973). This complex has a 1:1 stoichiometry and dissociation constant at low ionic strength is 10^{-9} M. The interaction of adrenodoxin reductase with adrenodoxin was further proved by using immobilized adrenodoxin for affinity purification of the flavoprotein (SUGIYAMA et al., 1975). Cross-linking experiments proved to be effective in showing complexation of the flavoprotein with ferredoxin (LAMBETH et al., 1984; USANOV et al., 1985). At present, the complex formation between adrenodoxin reductase and adrenodoxin has been proved by using different approaches.

The quenching of the fluorescence of FAD and tryptophan of adrenodoxin reductase upon interaction with adrenodoxin suggests that adrenodoxin bind in the neighborhood of a tryptophanyl residue near the prosthetic group of adrenodoxin reductase and that the chromophor of adrenodoxin is not in-

```
                           20
A:  S S S Q D K I - T Y - H F I N R - - - - - - D G E T L T T K G K I G D S L L D Y Y Y Q N N
H:  S S S E D K I - T Y - H F I N R - - - - - - D G E T L T T K G K I G D S L L D Y Y Y Q N N
R:  S S S E D K I - T Y - H F I X R - - - - - - D G E T L T T K G K I G X S L X D Y Y Y X X X
P:  - - S - - K Y Y Y Y S H N G T R R Q L D Y A D G Y S L M Q - - - - - - - - - - A A Y - S N

       40                                60
A:  - L D I D G F G A C E G T L A C S T C H L I F E N H I F - E K L E A I T N E - E N N M L D
H:  - L D I D G F G A C E G T L A C S T C H L I F E Q H I F - E K L E A I T D E - E N D M L D
R:  - X X X X X X X X X G T L A C S X X X X I F E X H I F - E K L E A I T X X - E X X M L D
P:  G I Y - D I Y G D C G G S A S C A T C H - Y Y Y N E A F T D K Y P A - A N E R E I G M L E

       80                              100
A:  - L A Y G L T D R S R L G C Q I C L T K A M D - N M T Y R Y P D A Y S D A
H:  - L A Y G L T D R S R L G C Q I C L T K A M D - N M T Y R Y P D A Y S D A R E S
R:  - L A Y G L T D R S R L G C X I C L X K A M D - X M T Y R Y P D A Y S X A R E S I D M G M N S S K I E
P:  C Y T A E L K P N S R L C C Q I I M T P Q L D G I Y - Y D Y P D - - - - - R Q W
```

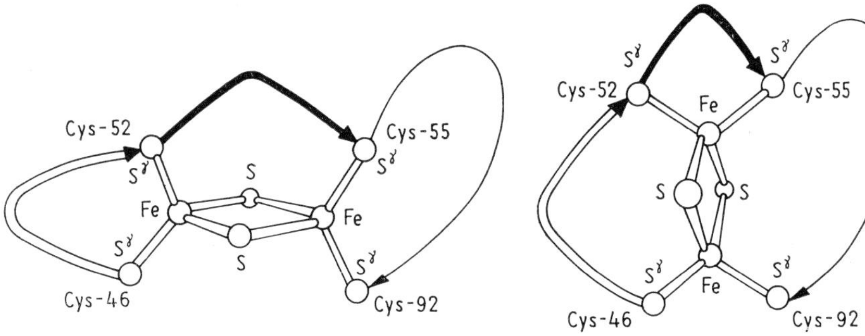

Fig. 8. The alignment of the primary structure of adrenodoxin (A), hepatoredoxin (H), renoredoxin (R) and putidaredoxin (P) and the models for the molecular organization of the iron-sulfur cluster of ferredoxins. Indicated are cysteinyl residues of ferredoxins. Underlined is the C-terminal sequence of renoredoxin determined by direct protein sequencing previously not detected in adrenodoxin and deduced from sequencing of cDNA complementary to adrenodoxin.

volved in the process of interaction (HIWATASHI et al., 1976; HIWATASHI et al., 1978). Treatment of adrenodoxin reductase with neuraminidase decreases its affinity to adrenodoxin. This was interpreted as involvement of the sugar component of adrenodoxin reductase in complex formation with adrenodoxin (HIWATASHI et al., 1982). In this connection it is not clear why pig adrenodoxin reductase does not contain carbohydrate (HIWATASHI et al., 1978) but still interacts with adrenodoxin. A lysine residue of adrenodoxin reductase spatially close to the sugar component of flavoprotein proved important for interaction with adrenodoxin, indicating that charge ion pairing is involved in complex fixation (HAMAMOTO and ICHIKAWA, 1984). This essential lysine residue was identified as Lys-276 (HAMAMOTO et al., 1988).

Based on chemical modification studies, it has been proposed that the single tyrosine residue of adrenodoxin is involved in interaction with adrenodoxin reductase (TANIGUCHI and KIMURA, 1975; TANIGUCHI and KIMURA, 1976). Chemical modification of lysine residues of adrenodoxin has practically no effect on the kinetic parameters of interaction with adrenodoxin reductase (AKHREM et al., 1977; GEREN et al., 1984; TULS et al., 1987). Modification of free carboxyl groups of adrenodoxin results in an increased Kd for complex formation with adrenodoxin reductase (AKHREM et al., 1977; LAMBETH et al., 1984). Thus, carboxyl groups of aspartate and glutamate were suggested as essential for complex formation with adrenodoxin reductase, despite the fact that affinity chromatography experiments indicate involvement of hydrophobic interactions to stabilize the complex. The secondary structure of adrenodoxin has been predicted to consist of two helical segments (residues 65—85) separated by a β-turn at residues 72—75 (LIM and KIMURA, 1981). This is one of the most negatively charged sequences of adrenodoxin which contains 6 acidic amino acid residues and appears to be involved in interaction with adrenodoxin reductase. The chemical modification of adrenodoxin with radioactive water-soluble carbodiimide revealed modified carboxyl groups at Glu-74, Asp-79, Asp-86, which are located in the sequence with a high negative charge density (GEREN et al., 1984).

An excellent model of the electron transfer from reduced adrenodoxin reductase to adrenodoxin has been derived from the cytochrome c reduction. Although cytochrome c reduction results from electron transfer through reduced adrenodoxin, the rate is limited by other electron transfer processes within the adrenodoxin reductase — adrenodoxin complex. The kinetic mechanism of cytochrome c reduction by the complex of adrenodoxin reductase and adrenodoxin has been determined using stopped-flow spectrophotometry and the nature of electron containing intermediates was identified (LAMBETH and KAMIN, 1977; LAMBETH and KAMIN, 1979; LAMBETH et al., 1979). Cytochrome c reduction is assumed to be catalyzed by both proteins in a complex with 1:1 stoichiometry cycling between 1- and 3-electron reduced (via a 2-electron reoxidation intermediate) states (LAMBETH et al., 1976; LAMBETH and KAMIN, 1977). This cycle was originally proposed to be independent of requiring dissociation of the protein-protein complex. However, later kinetic studies showed rapid dissociation-association of this complex (LAMBETH et al., 1981). The flavin-to-iron sulfur electron transfer appears to be the rate-limiting step in cytochrome c reduction (LAMBETH and KAMIN, 1979).

During adrenodoxin reduction by adrenodoxin reductase, the oxidation-reduction potential of ferredoxin shifts by -50 to -100 mV (LAMBETH et al., 1976) indicating that the reduced form of adrenodoxin binds more weakly to flavoprotein. Thus reduction of adrenodoxin promotes dissociation of the flavoprotein-ferredoxin complex. This data led to the proposal of a model where electron transfer performs via adrenodoxin, acting as a mobile electron "shuttle", binding sequentially first with adrenodoxin reductase and then with cytochrome c (LAMBETH et al., 1979; HANUKOGLU et al., 1980).

The interaction of adrenodoxin with cytochrome P-450scc is of special interest since reduced adrenodoxin serves as electron donor for cytochrome P-450scc. The first evidence for an interaction of cytochrome P-450scc with adrenodoxin was obtained by KATAGIRI (KATAGIRI et al., 1977), who found that interaction of adrenodoxin with cytochrome P-450scc takes place in the presence of cholesterol. Complex formation was further confirmed by spectro-photometric, gel-filtration and gradient centrifugation experiments (KATAGIRI et al., 1977). The complex is characterized by a dissociation constant of $Kd = 0.16 \mu M$. Affinity chromatography of cytochrome P-450scc on im-mobilized adrenodoxin proved to be the most effective procedure for cyto-chrome P-450scc purification (SUGIYAMA et al., 1976). The affinity of adreno-doxin to cytochrome P-450scc is highly dependent on substrate, detergents, phospholipids and pH (JEFCOATE et al., 1982; RADYUK et al., 1982; LAMBETH and PEMBER, 1983). Later it was shown that substrate-free cytochrome P-450scc also interacts with adrenodoxin (HANUKOGLU et al., 1981). The binding of cholesterol and adrenodoxin to cytochrome P-450scc are not independent: the affinity of adrenodoxin to cytochrome P-450scc is increased by a factor of 20 at saturating concentration of substrate (LAMBETH et al., 1980; HANU-KOGLU et al., 1981).

Interaction of cytochrome P-450scc with adrenodoxin is a function of detergent and phospholipid concentration. The effect of phospholipids and detergents is complex and not completely understood. Nevertheless, the dissociation constant of the cytochrome P-450scc-adrenodoxin complex is $10^{-7}-10^{-8}$ M, which is larger than that of the adrenodoxin reductase-adreno-doxin complex by one order of magnitude. Upon interaction of cytochrome P-450scc with adrenodoxin the oxidation-reduction potential of ferredoxin decreases (-273 to -291 mV), indicating the preferential binding of reduced adrenodoxin to oxidized cytochrome P-450scc (LAMBETH and PEMBER, 1983). The Kd for the complex between oxidized adrenodoxin and reduced cyto-chrome P-450scc (120 nM) is higher than that of the complex between reduced adrenodoxin and oxidized cytochrome P-450scc (60 nM) and at least 5 times higher than the Kd for the complex of oxidized ferredoxin with reduced adrenodoxin reductase (25 nM).

This fact together with data indicating a $50-100$ mV shift in the midpoint potential of adrenodoxin upon complex formation with adrenodoxin reductase as well as a decrease in the affinity of ferredoxin to reduced cytochrome

P-50scc led to the model of electron transfer consistent with the "shuttle" mechanism (LAMBETH et al., 1979). According to this mechanism, adrenodoxin first interacts and accepts one electron from adrenodoxin reductase, then dissociates, and interacts and transfers the electron to cytochrome P-450scc; the oxidized iron-sulfur protein moves back to flavoprotein and the cycle is repeated. Both the oxidation-reduction state of ferredoxin and substrate interaction with cytochrome P-450scc have been shown to regulate the protein-protein interactions which comprise the "shuttle" mechanism (LAMBETH et al., 1980; LAMBETH and PEMBER, 1983). The differential binding of oxidized and reduced adrenodoxin with flavoprotein and hemeprotein respectively and the dependence of activity on the concentration of free reduced adrenodoxin (HANUKOGLU and JEFCOATE, 1980) are the most important characteristics of the "shuttle" mechanism. The important conclusion from this model is that binding sites on the surface of adrenodoxin responsible for interaction with adrenodoxin reductase and cytochrome P-450scc appear to be nearly the same. However, comparing the ionic strength dependence of the complexes of adrenodoxin with adrenodoxin reductase and cytochrome P-450scc some differences can be observed. The cytochrome P-450scc-adrenodoxin complex is more salt-sensitive indicating that adrenodoxin under physiological conditions is considerably more tightly bound to adrenodoxin reductase. The specificity of the ions affecting the complexes of adrenodoxin with flavoprotein and hemeprotein is not identical (LAMBETH and KRIENGSIRI, 1985). This data might indicate that different types of interactions are involved in protein-protein interaction.

Adrenodoxin reductase, adrenodoxin and cytochrome P-450scc was also found to form a stable complex as detected by difference spectroscopy and gel-filtration (KIDO and KIMURA, 1979). Immobilized adrenodoxin proved to be a very effective approach to understanding the mechanism of electron transfer. The fact that immobilized adrenodoxin retains the ability to interact with adrenodoxin reductase and cytochrome P-450scc, affinity being much more high with respect to hemeprotein, allowed the reconstitution of enzymatic activity in heterologous systems (AKHREM et al., 1978; AKHREM et al., 1979). Adrenodoxin-Sepharose was useful for proving the interaction of both sub-strate-free and substrate-bound cytochrome P-450scc with adrenodoxin. The complex of cytochrome P-450scc with immobilized adrenodoxin is destabilized by non-ionic detergents that may be compensated by exogeneous cholesterol (RADYUK et al., 1982). Pregnenolone does not affect the affinity of cytochrome P-450scc with immobilized adrenodoxin (RADYUK et al., 1983). Reduction of cytochrome P-450scc with a chemical reductant does not result in dissociation of the complex with immobilized adrenodoxin (RADYUK et al., 1983).

Adrenodoxin reductase and cytochrome P-450scc immobilized on adreno-doxin-Sepharose are able to enzymatically reduce cytochrome P-450scc in the presence of low NADPH concentrations without desorption of proteins from the column (RADYUK et al., 1983). However, when adrenodoxin-Sepharose is saturated with cytochrome P-450scc and adrenodoxin reductase in the presence of high NADPH concentration, a slow protein dissociation in the course of cholesterol side chain cleavage and 11β-hydroxylation takes place.

Chemical cross-linking proved to be a useful approach in studying the mechanism of electron transfer in steroidogenic enzyme systems (CHASHCHIN et al., 1985; USANOV et al., 1985). A complex of cytochrome P-450scc with adrenodoxin has been covalently fixed by using different cross-linking reagents. Covalently cross-linked with dimethyl-3,3'-diothiobispropionimidate a complex of adrenodoxin and cytochrome P-450scc was found to be active in electron transfer and cholesterol oxidation in the presence of exogeneous NADPH and adrenodoxin reductase (CHASHCHIN et al., 1985). The catalytic competence of a cross-linked complex of adrenodoxin with cytochrome P-450scc (CHASHCHIN et al., 1985) and adrenodoxin reductase (USANOV et al., 1985) indicates the existence of two different binding sites on the surface of adrenodoxin responsible for interaction with flavoprotein and hemeprotein, respectively. Chemical cross-linking of cytochrome P-450scc with adrenodoxin by water-soluble carbodiimide (EDC) and subsequent structural analysis indicates that polypeptides of cytochrome P-450scc Leu-88—Trp-108; Leu-368—Trp-400; Leu-401—Trp-

```
                                                            50
ISTKTPRPYSEIPSPGDDGWLNLYHFWREKGSQRIHFRHIENFQKYGPIY
                                                            100
REKLGNLESVYIIHPENYAHLF(K)FEGSYPERYDIPPWLAYTRYYQKPIGV
                                                            150
LF(K)(K)SGTWKKDRYYLNTEYMAPEAI(K)NFIPLLNPYSQDFYSLLH(K)RI(K)QQ
                                                            200
GSG(K)FYGNIKEDLFHFAFESITNYMFGERLGMLEETYNPEAQKFIDAYYK
                                                            250
MFHTSYPLLNYPPELYRLFRTKTWRDHYAAWDTIFNKAEKYTEIFYQDLR
                                                            300
RKTEFRNYPGILYCLL(K)SE(K)MLLEDYKANITEMLAGGYNTTSMTLQYHLY
                                                            350
EMARSLNYQEMLREEYLNARRQAEGDISKMLQMYPLL(K)ASI(K)ETLRLHPI
                                                            400
SYTLQRYPESDLYLQDYLIPAKTLYQVAIYAMGRDPAFFSSPDKFDPTRW
                                                            450
LSKDKDLIHFRNLGFGWGYRQCYGRRIAELEMTLFLIHILENFKYEMQHI
GDYDTIFNLILTPDKPIFLYFRPFNQDPPQA
```

Fig. 9. Primary structure of cytochrome P-450scc. Underlined are the polypeptides which are involved in the interaction with adrenodoxin as determined by chemical cross-linking. Also indicated are the lysine residues which might be involved in the interaction with adrenodoxin as determined by differential chemical modification of cytochrome P-450scc with succinic anhydride.

S. A. USANOV et al.

417 (Fig. 9) might be involved in interaction with adrenodoxin (TURKO et al., 1988; TURKO et al., 1988).

Modification of cytochrome P-450scc with succinic anhydride was found to result in the loss of its ability to interact with adrenodoxin indicating the involvement of lysine residues of hemeprotein in interaction with ferredoxin (USANOV et al., 1988; ADAMOVICH et al., 1989). Structural analysis of peptides labelled with radioactive reagent showed 11 of 33 lysine residues of cytochrome P-450scc to be in these peptides. Most radioactivity was found in Lys-73, Lys-109, Lys-110, Lys-126, Lys-148, Lys-154 representing the N-terminal sequence of cytochrome P-450scc and Lys-264, Lys-270, Lys-338, Lys-342 from C-terminal sequence (Fig. 9). These data indicate that both fragments of cytochrome P-450scc are involved in the very complex interaction with adrenodoxin involving multipoint protein-protein interaction.

Figure 10 shows structure-function models of the electron transfer in a mitochondrial cholesterol side chain cleavage system. Reconstitution of cholesterol side chain cleavage activity by interaction of immobilized adrenodoxin with adrenodoxin reductase and cytochrome P-450scc presumes electron transfer in an organized complex of electron transfer proteins (Fig. 10a). This model also requires the presence of different binding sites on the surface of ferredoxin responsible for interaction with flavoprotein and hemeprotein respectively. The model of the molecular organization is modified by the same elemental steps of monooxygenase catalysis such as interaction with substrate, 1-electron reduction, oxygen binding but presuming formation of protein-protein complexes of different half-lifes (Fig. 10b). The possible regulatory role of cholesterol via protein-protein complexes is also shown in this model: in the presence of cholesterol, the oxidized proteins may be associated in the complex, but in the absence of substrate there exist preferably complexes of adrenodoxin reductase with adrenodoxin and free cytochrome P-450scc.

Figure 10c shows the dynamic model of electron transfer, the so-called "shuttle" mechanism (LAMBETH et al., 1982). This model assumes that the same or overlapping binding site(s) on the surface of adrenodoxin is (are) involved in the interaction with adrenodoxin reductase and cytochrome P-450scc. The electrons are transferred by sequential complex formation via adrenodoxin acting as a "shuttle".

Figure 10d shows the possible role of the sugar component of steroid hydroxylases in the electron transfer process (ICHIKAWA and HIWATASHI, 1982). Thus, the models presented in Figure 10 reflect some elemental steps of the complicated mechanism of electron transfer. There is experimental evidence supporting every model. Nevertheless, the final mechanism of electron transfer during monooxygenation catalyzed by cytochrome P-450scc has to be clarified. One should also keep in mind the specific role of phospholipid membrane in electron transfer.

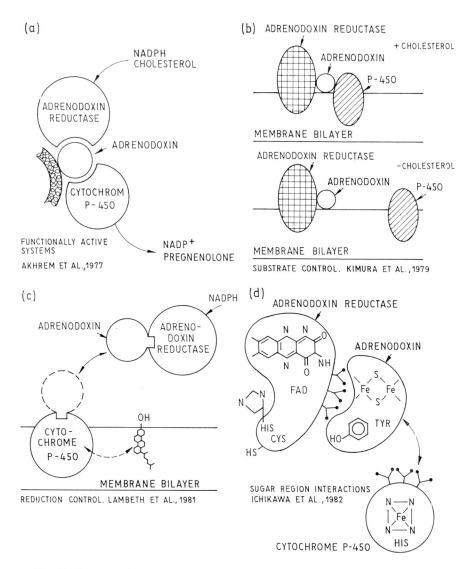

(a)

NADPH
CHOLESTEROL

ADRENODOXIN
REDUCTASE

ADRENODOXIN

CYTOCHROM
P-450

FUNCTIONALLY ACTIVE
SYSTEMS

AKHREM ET AL.,1977

NADP+
PREGNENOLONE

(b) ADRENODOXIN REDUCTASE

+CHOLESTEROL

ADRENODOXIN

P-450

MEMBRANE BILAYER

ADRENODOXIN REDUCTASE

-CHOLESTEROL

ADRENODOXIN

P-450

MEMBRANE BILAYER

SUBSTRATE CONTROL. KIMURA ET AL.,1979

(c)

ADRENODOXIN

NADPH (d)

ADRENO-
DOXIN
REDUCTASE

CYTO-
CHROME
P-450

OH

MEMBRANE BILAYER

REDUCTION CONTROL. LAMBETH ET AL.,1981

ADRENODOXIN REDUCTASE

N   N   O

N   NH
  O

FAD

ADRENODOXIN

S
Fe   Fe
S

N

HIS
CYS

HS

HO

TYR

SUGAR REGION INTERACTIONS
ICHIKAWA ET AL.,1982

N—N
Fe
N—N

CYTOCHROME P-450   HIS

Fig. 10. Structure-function models of the molecular organization of the mitochondrial steroid hydroxylases: a — ternary complex reconstituted on immobilized adrenodoxin; b — model presuming the regulatory role of cholesterol via the formation of binary and ternary complexes; c — model of a dynamic electron transfer; d — model reflecting the possible role of sugar regions in adrenodoxin reductase and cytochrome P-450scc in the interaction with adrenodoxin.

S. A. Usanov et al.

Monospecific antibodies against cytochrome P-450scc, adrenodoxin reductase and adrenodoxin proved to be efficient to study the stoichiometry of the components of mitochondrial electron transfer systems as well as their topology (Usanov et al., 1987). The stoichiometry of adrenodoxin reductase, adrenodoxin and cytochrome P-450scc as determined by radioimmunoassay was shown to be the following: 1 : 7.5 : 3.4. Figure 11 shows the proposed model for the molecular organization of cholesterol side chain cleavage system as revealed by immunochemical approaches based on accessibility of the electron transfer components to antibodies.

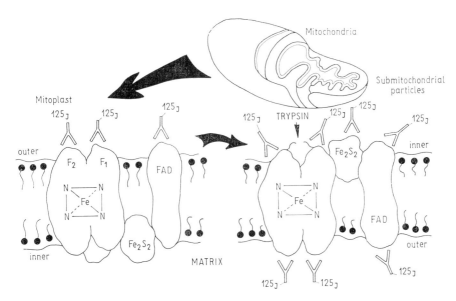

Fig. 11. Hypothetical model for the molecular organization of the cholesterol side chain cleavage system in the inner mitochondrial membrane as determined by limited proteolysis and accessibility to monospecific antibodies.

4. Microsomal steroid transforming enzymes

Further important steps in the biosynthesis of steroid hormones are hydroxylation at C_{21} position and cleavage of side chain of C_{21} steroids which are catalyzed by enzymes of endoplasmic reticulum membranes. These reactions are intermediate between the first (side chain cleavage of cholesterol) and final (11β-hydroxylation) step in the biosynthesis of glucocorticoids. Both reactions are catalyzed by two forms of steroidogenic cytochromes P-450: cytochrome P-450$_{21}$ and cytochrome P-450$_{17\alpha}$. The latter possess 17α-hydroxylase and 17α-hydroxylase and 20-lyase activities. The electrons for both reactions are supplied by NADPH via NADPH-cytochrome P-450 reductase. Unlike mitochondrial steroidogenic monooxygenases, microsomal enzymes require only flavoprotein, without an additional ferredoxin.

Important questions related to steroid hormones biosynthesis by microsomal steroidogenic enzymes are:

1. what hydroxylation at 17α- or 21-position precedes biosynthesis of cortisol?
2. is 17α-hydroxylation and 17,20-lyase activity catalyzed by the same cytochrome P-450?
3. what pathway Δ^4 or Δ^5 is favourable for steroidogenesis?

4.1. 21-Hydroxylase

The 21-hydroxylation of steroids in microsomes of different tissues is catalyzed by cytochrome P-450$_{21}$. This cytochrome P-450 was one of the first for which involvement in steroid hormones hydroxylation was proved by photochemical action spectrum (ESTABROOK et al., 1963). The main physiological role of cytochrome P-450$_{21}$ is to convert progesterone and 17α-hydroxyprogesterone to deoxycorticosterone and 11-deoxycortisol. Cytochrome P-450$_{21}$ reacts more actively with 17α-hydroxysteroids than with the corresponding 17α-deoxysteroids. Therefore, it is thought that normally 17α-hydroxylation precedes 21-hydroxylation. The 21-hydroxylation was also found in some extrasteroidogenic tissues including liver. However, substrate specificity of cytochrome P-450$_{21}$ in the liver was found to differ from that of adrenocortical microsomes (ICHIKAWA et al., 1984). The physiological role of extraadrenal cytochrome P-450$_{21}$ is not clearly understood.

While the nature of cytochrome P-450$_{21}$ was demonstrated in 1965 (COOPER et al., 1965), systematic efforts to purify cytochrome P-450$_{21}$ have been reported only recently (KOMINAMI et al., 1978; KOMINAMI et al., 1980; HIWATASHI and ICHIKAWA, 1981; BUMPUS and DUS, 1982). Later, cysteine-containing peptides of cytochrome P-450$_{21}$ have been identified by chemical modification and subsequent microsequencing (YUAN et al., 1983). The complete amino acid sequence of cytochrome P-450$_{21}$ was determined by protein sequencing and shown to contain 492 amino acid residues (HANIU et al., 1987). Recently, cytochrome P-450$_{21}$ from different animal species was cloned and the primary structure deduced from the cDNA (CHUNG et al., 1986; CHAPLIN et al., 1986; WHITE et al., 1986; HIGASHI et al., 1986). Alignment of the primary structure of cytochrome P-450$_{21}$ from bovine and human adrenocortical microsomes is presented in Figure 12. The structure of cytochrome P-450 with a particular substrate specificity does appear to be highly conserved in different species. Thus, alignment of bovine and porcine cytochrome

S. A. USANOV et al.

Met Val Leu Ala Gly Leu Leu Leu Leu Leu Thr Leu Leu Ala Gly Ala His Leu Leu Trp Gly Arg Trp Lys Leu Arg Asn Leu His Leu 30
Met Leu Leu Leu Gly Leu Leu Leu Leu - Pro Leu Leu Ala Gly Ala Arg Leu Leu Trp Asn Trp Trp Lys Leu Arg Ser Leu His Leu 29

Pro Pro Leu Val Pro Gly Phe Leu His Leu Leu Gln Pro Asn Leu Pro Ile His Leu Leu Ser Leu Thr Gln Lys Leu Gly Pro Val Tyr 60
Pro Pro Leu Ala Pro Gly Phe Leu His Leu Leu Gln Pro Asp Leu Pro Ile Tyr Leu Leu Gly Leu Thr Gln Lys Phe Gly Pro Ile Tyr 59

Arg Leu Arg Leu Gly Leu Gln Glu Val Val Val Leu Asn Ser Lys Arg Thr Ile Glu Glu Ala Met Ile Arg Lys Trp Val Asp Phe Ala 90
Arg Leu His Leu Gly Leu Gln Asp Val Val Val Leu Asn Ser Lys Arg Thr Ile Glu Glu Ala Met Val Lys Lys Trp Ala Asp Phe Ala 89

Gly Arg Pro Gln Ile Pro Ser Tyr Lys Leu Val Ser Gln Arg Cys Gln Asp Ile Ser Leu Gly Asp Tyr Ser Leu Leu Trp Lys Ala His 120
Gly Arg Pro Glu Pro Leu Thr Tyr Lys Leu Val Ser Lys Asn Tyr Pro Asp Leu Ser Leu Gly Asp Tyr Ser Leu Leu Trp Lys Ala His 119

Lys Lys Leu Thr Arg Ser Ala Leu Leu Leu Gly Thr Arg Ser Ser Met Glu Pro Trp Val Asp Gln Leu Thr Gln Glu Phe Cys Glu Arg 150
Lys Lys Leu Thr Arg Ser Ala Leu Leu Leu Gly Ile Arg Asp Ser Met Glu Pro Val Val Glu Gln Leu Thr Gln Glu Phe Cys Glu Arg 149

Met Arg Val Gln Ala Gly Ala Pro Val Thr Ile Gln Lys Glu Phe Ser Leu Leu Thr Cys Ser Ile Ile Cys Tyr Leu Thr Phe Gly Asn 180
Met Arg Ala Gln Pro Gly Thr Pro Val Ala Ile Gln Glu Glu Phe Ser Leu Leu Thr Cys Ser Ile Ile Cys Tyr Leu Thr Phe Gly Asp 179

Lys - Glu Asp - Thr Leu Val His Ala Phe His Asp Cys Val Gln Asp Leu Met Lys Thr Trp Asp His Trp Ser Ile Gln Ile Leu 208
Lys Ile Lys Asp Asp Asn Leu Met Pro Ala Tyr Tyr Lys Cys Val Gln Glu Val Leu Lys Thr Trp Ser His Trp Ser Ile Gln Ile Val 209

Asp Met Val Pro Phe Leu Arg Phe Phe Pro Asn Pro Gly Leu Trp Arg Leu Lys Gln Ala Ile Glu Asn Arg Asp His Met Val Glu Lys 238
Asp Val Ile Pro Phe Leu Arg Phe Phe Pro Asn Pro Gly Leu Arg Arg Leu Lys Gln Ala Ile Glu Lys Arg Asp His Ile Val Glu Met 239

Gln Leu Thr Arg His Lys Glu Ser Met Val Ala Gly Gln Trp Arg Asp Met Thr Asp Tyr Met Leu Gln Gly Val Gly Arg Gln Arg Val 268
Gln Leu Arg Gln His Lys Glu Ser Leu Val Ala Gly Gln Trp Arg Asp Met Met Asp Tyr Met Leu Gln Gly Val Ala Gln Pro Ser Met 269

Glu Glu Gly Pro Gly Gln Leu Leu Glu Gly His Val His Met Ser Val Val Asp Leu Phe Ile Gly Gly Thr Glu Thr Thr Ala Ser Thr 298
Glu Gln Gly Ser Gly Gln Leu Leu Glu Gly His Val His Met Ala Ala Val Asp Leu Leu Ile Gly Gly Thr Glu Thr Thr Ala Asn Thr 299

Leu Ser Trp Ala Val Ala Phe Leu Leu His His Pro Glu Ile Gln Arg Arg Leu Gln Glu Glu Leu Asp Arg Glu Leu Gly Pro Gly Ala 328
Leu Ser Trp Ala Val Val Phe Leu Leu His His Pro Glu Ile Gln Gln Arg Leu Gln Glu Glu Leu Asp His Glu Leu Gly Pro Gly Ala 329

Ser Cys Ser Arg Val Thr Tyr Lys Asp Arg Ala Arg Leu Pro Leu Leu Asn Ala Thr Ile Ala Glu Val Leu Arg Leu Arg Pro Val Val 358
Ser Ser Ser Arg Val Pro Tyr Lys Asp Arg Ala Arg Leu Pro Leu Leu Asn Ala Thr Ile Ala Glu Val Leu Arg Leu Arg Pro Val Val 359

Pro Leu Ala Leu Pro His Arg Thr Thr Arg Pro Ser Ser Ile Phe Gly Tyr Asp Ile Pro Glu Gly Met Val Val Ile Pro Asn Leu Gln 388
Pro Leu Ala Leu Pro His Arg Thr Thr Arg Pro Ser Ser Ile Ser Gly Tyr Asp Ile Pro Glu Gly Thr Val Ile Ile Pro Asn Leu Gln 389

Gly Ala His Leu Asp Glu Thr Val Trp Glu Gln Pro His Glu Phe Arg Pro Asp Arg Phe Leu Glu Pro Gly Ala Asn Pro Ser Ala Leu 418
Gly Ala His Leu Asp Glu Thr Val Trp Glu Arg Pro His Glu Phe Trp Pro Asp Arg Phe Leu Glu Pro Gly Lys Asn Ser Arg Ala Leu 419

Ala Phe Gly Cys Gly Ala Arg Val Cys Leu Gly Glu Ser Leu Ala Arg Leu Glu Leu Phe Val Val Leu Leu Arg Leu Leu Gln Ala Phe 448
Ala Phe Gly Cys Gly Ala Arg Val Cys Leu Gly Glu Pro Leu Ala Arg Leu Glu Leu Phe Val Val Leu The Arg Leu Leu Gln Ala Phe 449

Thr Leu Leu Pro Pro Pro Val Gly Ala Leu Pro Ser Leu Gln Pro Asp Pro Tyr Cys Gly Val Asn Leu Lys Val Gln Pro Phe Gln Val 478
Thr Leu Leu Pro Ser Gly Asp - Ala Leu Pro Ser Leu Gln Pro Leu Pro His Cys Ser Val Ile Leu Lys Met Gln Pro Phe Gln Val 478

Arg Leu Gln Pro Arg Gly Val Glu Ala Gly Ala Trp Glu Ser Ala Ser Ala Gln
Arg Leu Gln Pro Arg Gly Met Gly Ala His Ser Pro Gly Gln Asn - - Gln

Fig. 12. Primary structure of bovine (Chung et al., 1986) and human (Higashi et al., 1986) cytochrome P-450_{21} deduced from sequencing of cDNA. Underlined are the putative steroid-binding site and the conserved hemebinding site, respectively.

P-450$_{21}$ reveals 90% homology of amino acid sequences except for the C-terminal sequence (HANIU et al., 1987). Unlike porcine cytochrome P-450$_{21}$, bovine cytochrome P-450$_{21}$ contains 496 amino acid residues (CHUNG et al., 1986). The most likely candidate for the steroid-binding site is the sequence 173—184, I-C-C-L-T-F-G-D-K-E-D-T, which contains three homologous residues as compared to the steroid binding site of cytochrome P-450scc (GOTOH et al., 1985).

A wide variety of gene conversions, point mutations, crossovers and polymorphism were shown to account for most cases of 21-hydroxylase deficiency resulting in congenital adrenal hyperplasia (MILLER, 1987).

Cytochrome P-450$_{21}$ purified from bovine adrenocortical microsomes was found to be low spin hemeprotein (KOMINAMI et al., 1980). The extinction coefficients of substrate-free and 17α-hydroxyprogesterone-bound forms of cytochrome P-450$_{21}$ were found to be 122 mM^{-1} cm^{-1} at 419 nm and 100 mM^{-1}cm^{-1} at 396 nm. The difference extinction coefficient of 113 mM^{-1} cm^{-1} at 389 nm versus 422 nm was found to be for the complex of cytochrome P-450$_{21}$ with 17α-hydroxyprogesterone. Cytochrome P-450$_{21}$ seems to contain two different binding sites for steroids (KOMINAMI et al., 1980). Site I is characterized by a higher affinity to steroids with a type I spectral response than to steroids causing modified type I spectra. Site II has a higher affinity to modified type I steroids. Binding of steroids to site I causes low- to high spin changes whereas binding to site II results in shifting the spin equilibrium to the low spin form. The apparent substrate dissociation constant was shown to depend on cytochrome P-450$_{21}$ concentration, indicating that binding reaction does not follow a simple two- component mass-action equilibrium (NARASHIMHULU et al., 1985). This data is in accordance with the model in which cytochrome P-450$_{21}$ exists in a monomer-dimer equilibrium and the dimer does not interact with substrate. The intrinsic dissociation constant (K_1) and the dissociation constant for dimerization reaction (K_2) were found to be independent of the enzyme concentration. Obviously mutual interaction between substrate binding and self association reactions of cytochrome P-450$_{21}$ exists (NARASHIMHULU et al., 1985).

The substrate-binding site of cytochrome P-450$_{21}$ is in the hydrophobic pocket containing at least one of the tryptophan residue (NARASIMHULU, 1988) and was shown to face the membrane lipid phase (KOMINAMI et al., 1986). The interaction of membrane-bound cytochrome P-450$_{21}$ with steroids depends on phosphatidylcholine concentration in the system, showing the substrate to be partitioned between the aqueous and lipid phase. The apparent rate of substrate binding to cytochrome P-450$_{21}$ was found independent of substrate partitioning, indicating partitioning to occur much more quickly than substrate binding (KOMINAMI et al., 1986). Thus, it is possible to conclude that the substrate binding affinity of cytochrome P-450$_{21}$ as well as other steroidogenic cytochromes P-450, is mainly controlled by hydrophobic

S. A. USANOV et al.

interactions of the steroid nucleus and the hydrophobic pocket of the respective cytochrome P-450. The rate-limiting step of hydroxylation might be determined by the phospholipid membrane.

4.2. 17α-Hydroxylase/17,20-lyase

Our present understanding of steroid hormones biosynthesis has been hampered by several uncertainties connected with 17α-hydroxylase/17,20-lyase activity. The questions which have to be solved are:

1. mediate one or different forms of cytochrome P-450$_{17α}$ 17α-hydroxylation and 17,20-lyase activities?
2. whether $Δ^4$ or $Δ^5$ steroids are substrates for 17α-hydroxylase/17,20-lyase?
3. is 17α-hydroxyprogesterone or 17α-hydroxypregnenolone a real intermediate in the lyase reaction?

The 17α-hydroxylation reaction is of special interest since it occurs at the branching point in the metabolic pathway leading to cortisol or corticosterone. The side chain cleavage reaction between C-17 and C-20 of C_{21} steroids is an inevitable reaction in the production of sex steroid hormones. Thus, cytochrome P-450 catalyzing 17α-hydroxylation and the 17,20-lyase reaction is unique among other steroidogenic enzymes in that it apparently catalyzes two distinct catalytic activities and represents a branching between mineralocorticoids or corticosterone on the one hand and between cortisol and sex hormones on the other.

Recently, cytochrome P-450 that catalyzes both 17α-hydroxylation and the 17,20-lyase reaction (cytochrome P-450$_{17α}$) has been purified from neonatal pig testis (NAKAJIN and HALL, 1981; NAKAJIN et al., 1983), adrenocortical microsomes (KOMINAMI et al., 1982; BUMPUS and DUS, 1982; NAKAJIN, et al., 1984) and pig testis microsomes (SUHARA et al., 1984). By comparing the physico-chemical properties of purified cytochrome P-450$_{17α}$, the authors came to the conclusion that only one single cytochrome P-450 mediates both 17α-hydroxylase and 17,20-lyase activities both in testis or in adrenal. No evidence has been obtained for the possible existence of a steroid 17α-hydroxylating cytochrome P-450 without lyase activity. Nevertheless, in adrenal cells this cytochrome P-450 possesses 17α-hydroxylase activity providing C_{21} steroids with a functional group. This intermediate (17α-hydroxyprogesterone) then undergoes 21-hydroxylation to form 21-deoxycortisol.

Immunochemical studies have led to the conclusion that the testis cytochrome P-450$_{17α}$ and corresponding species of adrenocortical microsomes are immunochemically the same or very similar (NAKAJIN et al., 1983). Cytochrome P-450$_{17α}$ was shown to differ immunochemically from microsomal cytochrome P-450$_{21}$ (KOMINAMI et al., 1983).

Met Trp Glu Leu Val Ala Leu Leu Leu Leu Thr Leu Ala Tyr Leu Phe Trp Pro Lys Arg Arg Cys Pro Gly Ala Lys Tyr Pro Lys Ser 30
Met Trp Leu Leu Leu Ala Val Phe Leu Leu Thr Leu Ala Tyr Leu Phe Trp Pro Lys Thr Arg His Ser Gly Ala Lys Tyr Pro Arg Ser

Leu Leu Ser Leu Pro Leu Val Gly Ser Leu Pro Phe Leu Pro Arg His Gly His Met His Asn Asn Phe Phe Lys Leu Gln Lys Lys Tyr 60
Leu Pro Ser Leu Pro Leu Val Gly Ser Leu Pro Phe Leu Pro Arg Arg Gly Gln Gln His Lys Asn Phe Phe Lys Leu Gln Glu Lys Tyr

Gly Pro Ile Tyr Ser Val Arg Met Gly Thr Lys Thr Thr Val Ile Val Gly His His Gln Leu Ala Lys Glu Val Leu Ile Lys Lys Gly 90
Gly Pro Ile Tyr Ser Phe Arg Leu Gly Ser Lys Thr Thr Val Met Ile Gly His His Gln Leu Ala Arg Glu Val Leu Leu Lys Lys Gly

Lys Asp Phe Ser Gly Arg Pro Gln Met Ala Thr Leu Asp Ile Ala Ser Asn Asn Arg Lys Gly Ile Ala Phe Ala Asp Ser Gly Ala His 120
Lys Glu Phe Ser Gly Arg Pro Lys Val Ala Thr Leu Asp Ile Leu Ser Asp Asn Gln Lys Gly Ile Ala Phe Ala Asp His Gly Ala His

Trp Gln Leu His Arg Arg Leu Ala Met Ala Thr Phe Ala Leu Phe Lys Asp Gly Asp Gln Lys Leu Glu Lys Ile Ile Cys Gln Glu Ile 150
Trp Gln Leu His Arg Lys Leu Ala Leu Asn Ala Phe Ala Leu Phe Lys Asp Gly Asn Leu Lys Leu Glu Lys Ile Ile Asn Gln Glu Ala

Ser Thr Leu Cys Asp Met Leu Ala Thr His Asn Gly Gln Ser Ile Asp Ile Ser Phe Pro Val Phe Val Ala Val Thr Asn Val Ile Ser 180
Asn Val Leu Cys Asp Phe Leu Ala Thr Gln His Gly Glu Ala Ile Asp Leu Ser Glu Pro Leu Ser Leu Ala Val Thr Asn Ile Ile Ser

Leu Ile Cys Phe Asn Thr Ser Tyr Lys Asn Gly Asp Pro Glu Leu Asn Val Ile Gln Asn Tyr Asn Glu Gly Ile Ile Asp Asn Leu Ser 210
Phe Ile Cys Phe Asn Phe Ser Phe Lys Asn Glu Asp Pro Ala Leu Lys Ala Ile Gln Asn Val Asn Asp Gly Ile Leu Glu Val Leu Ser

Lys Asp Ser Leu Val Asp Leu Val Pro Trp Leu Lys Ile Phe Pro Asn Lys Thr Leu Gly Lys Leu Lys Ser His Val Lys Ile Arg Asn 240
Lys Glu Val Leu Leu Asp Ile Phe Pro Val Leu Lys Ile Phe Pro Ser Lys Ala Met Gly Lys Met Lys Gly Cys Val Gln Thr Arg Asn

Asp Leu Leu Asn Lys Ile Leu Glu Asn Tyr Lys Glu Lys Phe Arg Ser Asp Ser Ile Thr Asn Met Leu Asp Thr Leu Met Gln Ala Lys 270
Glu Leu Leu Asn Glu Ile Leu Glu Lys Cys Gln Glu Asn Phe Ser Ser Asp Ser Ile Thr Asn Leu Leu His Ile Leu Ile Gln Ala Lys

Met Asn Ser Asp Asn Gly Asn Ala Gly Pro Asp Gln Asp Ser Glu Leu Leu Ser Asp Asn His Ile Leu Thr Thr Ile Gly Asp Ile Phe 300
Val Asn Ala Asp Asn Asn Asn Ala Gly Pro Asp Gln Asp Ser Lys Leu Leu Ser Asn Arg His Met Leu Ala Thr Ile Gly Asp Ile Phe

Gly Ala Gly Val Glu Thr Thr Thr Ser Val Val Lys Trp Thr Leu Ala Phe Leu Leu His Asn Pro Gln Val Lys Lys Lys Leu Tyr Glu 330
Gly Ala Gly Val Glu Thr Thr Thr Ser Val Ile Lys Thr Ile Leu Ala Phe Leu Leu His Asn Pro Ser Leu Lys Lys Arg Ile Gln Asp

Glu Ile Asp Gln Asn Val Gly Phe Ser Arg Thr Pro Thr Ile Ser Asp Arg Asn Arg Leu Leu Leu Leu Glu Ala Thr Ile Arg Glu Val 360
Asp Ile Asp Gln Ile Ile Gly Phe Asn Arg Thr Pro Thr Ile Ser Asp Arg Asn Arg Leu Val Leu Leu Glu Ala Thr Ile Arg Glu Val

Leu Arg Leu Arg Pro Val Ala Pro Met Leu Ile Pro His Lys Ala Asn Val Asp Ser Ser Ile Gly Glu Phe Ala Val Asp Lys Gly Thr 390
Leu Arg Ile Arg Pro Val Ala Pro Thr Leu Ile Pro His Lys Ala Val Ile Asp Ser Ser Ile Gly Asp Leu Thr Ile Asp Lys Gly Thr

Glu Val Ile Ile Asn Leu Trp Ala Leu His His Asn Glu Lys Glu Trp His Gln Pro Asp Gln Phe Met Pro Glu Arg Phe Leu Asn Pro 420
Asp Val Val Val Asn Leu Trp Ala Leu His His Ser Glu Lys Glu Trp Gln His Pro Asp Leu Phe Met Pro Glu Arg Phe Leu Asp Pro

Ala Gly Thr Gln Leu Ile Ser Pro Ser Val Ser Tyr Leu Pro Phe Gly Ala Gly Pro Arg Ser Cys Ile Gly Glu Ile Leu Ala Arg Gln 450
Thr Gly Thr Gln Leu Ile Ser Pro Ser Leu Ser Tyr Leu Pro Phe Gly Ala Gly Pro Arg Ser Cys Val Gly Glu Met Leu Ala Arg Gln

Glu Leu Phe Leu Ile Met Ala Trp Leu Leu Gln Arg Phe Asp Leu Glu Val Pro Asp Asp Gly Gln Leu Pro Ser Leu Glu Gly Ile Pro 480
Glu Leu Phe Leu Phe Met Ser Arg Leu Leu Gln Arg Phe Asn Leu Glu Ile Pro Asp Asp Gly Lys Leu Pro Ser Leu Glu Gly His Ala

Lys Val Val Phe Leu Ile Asp Ser Phe Lys Val Lys Ile Lys Val Arg Gln Ala Trp Arg Glu Ala Gln Ala Glu Gly Ser Thr
Ser Leu Val Leu Gln Ile Lys Pro Phe Lys Val Lys Ile Glu Val Arg Gln Ala Trp Lys Glu Ala Gln Ala Glu Gly Ser Thr Pro

Fig. 13. Primary structure of human adrenocortical (Chung et al., 1987) and bovine adrenocortical (Zuber et al., 1986) cytochrome P-$450_{17\alpha}$. Underlined are the putative steroid-binding and heme-binding sites, respectively.

The cDNA coding for bovine adrenocortical (ZUBER et al., 1986; CHUNG et al., 1987) and human adrenocortical as well as testis cytochrome P-450$_{17\alpha}$ have been isolated and sequenced. Figure 13 shows alignment of the deduced primary structures of cytochrome P-450$_{17\alpha}$ from bovine and human adrenocortical cells. Bovine cytochrome P-450$_{17\alpha}$ contains 509 amino acid residues whereas human cytochrome P-450$_{17\alpha}$ consists of 508. Alignment of bovine and human amino acid sequences indicates 70.5% homology (358 of 508 amino acid residues). Human testis cytochrome P-450$_{17\alpha}$ is identical to that in adrenal cells. The amino acid sequence of cytochrome P-450$_{17\alpha}$ is more similar to cytochrome P-450$_{21}$ (28.9%) than to that of cytochrome P-450scc (12.3%). Comparison of the deduced human and bovine amino acid sequences reveals a hypervariable region in cytochrome P-450$_{17\alpha}$ from position 160 to 268.

Binding studies with highly purified cytochrome P-450$_{17\alpha}$ showed that Δ^5 steroids are better substrates than Δ^4 steroids (NAKAJIN et al., 1981). The preference for the Δ^5 substrates revealed by spectral studies was confirmed by kinetic studies using progesterone and pregnenolone as substrates: pregnenolone was found to be more rapidly consumed than progesterone (NAKAJIN et al., 1981). Evidence that 17α-hydroxypregnenolone is a true intermediate of lyase activity was obtained by using a reconstituted system and [^{14}C]-pregnenolone as substrate: (i) the disappearance of substrate could be entirely accounted for by the appearance of the product — dehydroepiandrosterone plus 17α-hydroxypregnenolone; (ii) the incorporation of [^{14}C] into the product was decreased by addition of 17α-hydroxypregnenolone as the result of exchange between 17α-hydroxy-[^{14}C] pregnenolone generated by enzyme and 17α-hydroxy-[^{14}C] pregnenolone generated by enzyme and 17α-hydroxypregnenolone added.

The photochemical action spectra are identical for both 17α-hydroxylase and 17,20-lyase activities, confirming that both steps require cytochrome P-450 as catalyst (HALL, 1980).

Cytochrome P-450$_{17\alpha}$ is similar to mitochondrial cytochrome P-450scc with respect to its ability to bind to adrenodoxin-Sepharose, proteins being clearly separated from each other during chromatography (BUMPUS and DUS, 1982). Cytochrome P-450$_{17\alpha}$ in the presence of adrenodoxin reductase and adrenodoxin possesses cholesterol side chain cleavage activity, which is about 60% of that of cytochrome P-450scc.

Besides steroid transformation activities, cytochrome P-450$_{17\alpha}$ reveals considerable specificities towards various xenobiotics, suggesting that cytochrome P-450$_{17\alpha}$ and microsomal cytochrome P-450 are basically similar with regard to enzymatic activities. This fact is consistent with the widely accepted concept that some livermicrosomal cytochromes P-450 exhibit a high degree of regio- and stereoselectivity at the steroid binding site and on the other hand are able to loosely bind a variety of xenobiotics not strictly competitive with steroid substrates.

The interaction of cytochrome P-450$_{17\alpha}$ with pregnenolone, progesterone and 17α-hydroxyprogesterone was shown to result in a nearly complete conversion of hemeprotein to the high spin state. Progesterone competitively inhibits the lyase activity of 17α-hydroxyprogesterone, confirming that the binding of steroids occur at the same site (NAKAJIN et al., 1981). In addition to clear preference for Δ^5-3β-hydroxy, as opposed to Δ^4-ketosteroids, it is obvious that C$_{21}$ steroids produce greater spectral shifts than C$_{19}$ steroids. Evidently, a 20-ketone group is not required for substrate binding.

Affinity chemical modification of cytochrome P-450$_{17\alpha}$ with 17-(bromoacetoxy) progesterone proved to be useful in the identification of peptides which were reactive with reagent. One of these peptides was shown to be at the C-terminal sequence in conserved region (HR2) which is involved in heme binding (ONODA et al., 1987). The other cysteine residue of cytochrome P-450$_{17\alpha}$ which is protected from inactivation by the substrate was attributed to the HR1-region. These results suggest the localization of a substrate binding site very close to the heme binding site of cytochrome P-450$_{17\alpha}$.

Metabolic intermediates in the biosynthetic pathways of steroid hormones are very hydrophilic and favor partitioning into the lipid phase of the membrane. Previous studies on steroidogenic cytochromes P-450 led to suggesting that the substrate binding site might face the phospholipid membrane this being experimentally proved for cytochrome P-450$_{17\alpha}$, recently (KOMINAMI, et al., 1988).

4.3. NADPH-cytochrome P-450 reductase

NADPH-cytochrome P-450 reductase plays a very important role in the steroidogenesis in adrenal microsomes where two different cytochromes P-450 are involved in stereo- and regioselective hydroxylations. The flavoprotein transfers electrons from NADPH to cytochrome P-450. Since the relative activities of cytochrome P-450$_{21}$ and cytochrome P-450$_{17\alpha}$ are crucial for the regulation of the biosynthesis of steroid hormones, these activities might be controlled by the affinity to NADPH-cytochrome P-450 reductase.

In contrast to flavoprotein of mitochondrial monooxygenases, NADPH-cytochrome P-450 reductase of adrenocortical microsomes contains besides FAD an additional flavin-FMN. This makes it rather similar to microsomal flavoprotein of liver which has been intensively studied (MASTERS, 1978).

NADPH-cytochrome P-450 reductase from adrenocortical microsomes has been extensively purified by detergent solubilization and affinity chromatography on 2′,5′-ADP-Sepharose (HIWATASHI and ICHIKAWA, 1979). The spectral properties of NADPH-cytochrome P-450 reductase from bovine adrenocortical microsomes are very similar to those of the liver microsomal enzyme: the absorption peaks are at 274, 380 and 455 nm with shoulders at 290, 360 and 480 nm. The molecular weight of flavoprotein is near 80000. NADPH-cytochrome P-450 reductase was found to contain FAD and FMN at 1:1 stoichiometry (HIWATASHI and ICHIKAWA, 1979). The comparison of amino acid residues of NADPH-cytochrome P-450 reductase from bovine adrenocortical microsomes and that of liver microsomal enzyme was found to be similar. The latter was extensively studied. It was cloned and amino acid composition deduced from sequencing of cDNA (PORTER and KASPER, 1985; KATAGIRI et al., 1986). The primary structure of hydrophilic domain (VOGEL and LUMPER, 1986) and total flavoprotein (HANIU et al., 1986) from porcine hepatic microsomes have been elucidated by protein sequencing. Recently, liver microsomal NADPH-cytochrome P-450 reductase has been expressed in *E. coli* (PORTER et al., 1987).

Yeast NADPH-cytochrome P-450 reductase (YABUSAKI et al., 1988) has been cloned recently and its primary structure deduced from the cDNA sequence. Figure 14 shows the alignment of the primary structures of NADPH-cytochrome P-450 reductase from different sources. Since the primary structure of bovine adrenocortical NADPH-cytochrome P-450 reductase is unknown, this data might be useful in further studies of the is flavoprotein. The Km values of NADPH-cytochrome P-450 reductase from bovine adrenocortical microsomes were found to be 5.3 µM for NADPH, 1.1 mM for NADH and 9—24 µM for cytochrome c. Chemical modification of the NADPH-cytochrome P-450 reductase indicates that histidyl and cysteinyl residues are essential for interaction with NADPH (HIWATASHI and ICHIKAWA, 1979).

4.4. 3β-Hydroxysteroid dehydrogenase/Δ^5 — Δ^4 isomerase

The conversion of pregnenolone to progesterone is the key step in the biosynthesis of steroid hormones. This transformation involves 3β-hydroxysteroid dehydrogenase and Δ^5-Δ^4 3-ketosteroid isomerase activities. The enzymes responsible for these activities are NAD$^+$-dependent 3β-hydroxysteroid dehydrogenase, and isomerase which does not require any cofactors. Both activities are of microsomal origin and the former appear to be the rate-limiting step. These are not cytochrome P-450-dependent enzymes. There has been considerable debate as to whether the 3β-hydroxysteroid dehydrogenase and isomerase activities are connected with the same protein and whether several substrate specific 3β-hydroxydehydrogenase/Δ^5-Δ^4 isomerase complexes exist or only one multisubstrate complex occurs.

By solubilization of microsomal fraction with Triton X-100 and subsequent ion-exchange chromatography, 3β-hydroxysteroid dehydrogenase/Δ^5-Δ^4 isomerase has been purified to an apparent homogeneity (FORD and ENGEL, 1974). Based on substrate specificity, electrophoretic properties and N-terminal sequence analysis it was suggested that both dehydrogenase and isomerase

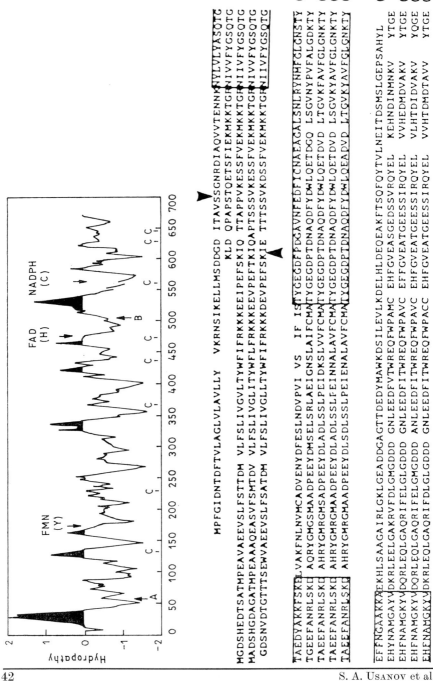

MPFGIDNTDFTVLAGLVLAVLLY VKRNSIKELLMSDDGD ITAVSSGNRDIAQVVTENNNYLVLVYASQTG (70)
  KLD QPAPSTQETSFIEKMKKTGRIVVFYGSQTG (89)
MGDSHEDTSATHPEAVAEEVSLFSTTDM VLFSLIVGLTYWFIFRKKKEEIPEFSKIQ TTAPPVKESSFVEKMKKTGRIVVFYGSQTG
MADSHGDAGATHPEAAAQEASVFSMTDV VLFSLIVGLITYWFLFRKKKEEVPEFTKIQAPTSSSVKESSFVEKMKKTGRIVVFYGSQTG (90)
GDSNVDTGTTTSEWVAEEVSLFSATDM VLFSLIVGLLTYWFIFRKKKDEVPEFSKIE TTTSSVKDSSFVEKMKKTGRIIVFYGSQTG (88)

TAEDYAKKFSKEL VAKFNLNVMCADVENYDFESLNDVPVI VS IF ISTYGEGDFPDGAVNFEDFICNAEAGALSNLRYNHFGLGNSTY (157)
TGEEFANRLSKE AQRYGMGSMAADPEEYDMSELSRLAEIGNSLAIFCMATYGEGDPTDNAQDFYDWLQETDGQ LSGVNYPVFALGDKTY
TAEEFANRLSKE AHRYGMRGMSADPEEYDLADLSSLPEIDKSLVVFCMATYGEGDPTDNAQDFYDWLQETDVD LTGVKFAVFGLGNKTY (178)
TAEEFANRLSKE AHRYGMRGMAADPEEYDLADLSSLFEINNALAVFCMATYGEGDPTDNAQDFYDWLQETDVD LSGVKYAVFGLGNKTY (179)
TAEEFANRLSKE AHRYGMRGMAADPEEYDLSDLSSLPEIENALAVFCMATYGEGDPTDNAQDFYDWLQEADVD LTGVKYAVFGLGNKTY (177)

EFFNGAAKKAEKHLSAAGAIRLGKLGEADGAGTTDEDYMAWKDSILEVLKDELHLDEQEAKFTSQFQYTVLNEITDSMSLGEPSAHYL (246)
EHYNAMGAYVDKRLEELGAKRVFDLGMGDDD GNLEEDFVTWREQFWPAMC EHFGVEASGEDSSVRQYEL KEHNDINMNKV YTGE (262)
EHFNAMGKYVDQRLEQLGAQRIFELGLGDDD GNLEEDFITWREQFWPAVC EFFGVEATGEESSIRQYEL VVHEDMDVAKV YTGE (263)
EHFNAMGKYVDQRLEQLGAQRIFELGMGDDD ANLEEDFITWREQFWPAVC EHFGVEATGEESSIRQYEL VLHTDIDVAKV YQGE (261)
EHFNAMGKYVDKRLEQLGAQRIFDLGLGDDD GNLEEDFITWREQFWPACC EHFGVEATGEESSIRQYEL VVHTDMDTAVV YTGE

```
                                                                                                                    (337)
LGR L KSFET QKPPFDAKNPFLAPVTVNR..NKAGELHKMHLEVDITGSKIRYESGDHVAVYPT.NTVIVNRLGQILGVDLDSVISLNN
MGR L KSYEN QKPPFDAKNPFLAAVTANR..NQGTERHLMHLELDISDSKIRYESGDHVAVYPAN.SALVNQIGEILGADLVIMSLNN  (350)
MGR L KSYEN QKPPFDAKNPFLAAVTANR..NQGTERHLMHLELDISDSKIRYESGDHVAVYPAN.SALVNQLGEILGADLVVMSLNN  (351)
MGR L KSYEN QKPPFDAKNPFLATVTTNR..NQGTERHLMHLELDISDSKIRYESGDHVAVYPAN.DSALVNQLGEILGTDLDIVMSLNN (349)


LD    PTVKVPFPTTIGAAIKHYLELITGPVSRQLFSSLIQFAPNADVKEKL  TLLSKDKDQFAVEITSKYFNIADALKYLSDGAKWDNVP (425)
LDEESNKKHPFPCPTTYRTALTHYLDIIHPPRTNVLYELAQYATDLKDQENTDSNASSAPEGKALYOSFVLEDNRNILAILEDLPSL RPP (440)
LDEESNKKHPFPCPTTYRTALTYYLDITNPPRTNVLYELAQYASEPSEQEHLHKMASSSGEGKELYLSWVVEARRHILAILQDYPSL RPP (441)
LDEESNKKHPFPCPTSYRTALTYYLDITNPPRTNVLYELAQYAADPAEQEOLRKMASSSGEGKELYLSWVVEARRHILAILQDYPSL RPP (439)
LDEESNKRHPFPCPTTYRTALTYYLDITNPPRTNVLYELAQYASEPSEQEOLRKMASSSGEGKELYLSWVVEARRHILAILQDYPSL RPP


MQFLVESVPQMT[RYYSISSSSLSEKQTVHVTSI VE]NFPNPELPDAPPGVGVTTNLLRNIQLAQNNVNIAETNLPVHYDLNGPRKLFANY (515)
IDHLCELMPRLQA[RYYSIASSSKVHPNSIHICAVLVE]Y -TK -- ---GVATTWLK  --  |  --  ------
IDHLCELLPRLQA[RYYSIASSSKVHPNSVHICAVAVE]Y EAK SG RVNKGVATSWLR  AK  E  PAG E NGGRAL         (508)
IDHLCELLPRLQA[RYYSIASSSKVHPNSVHICAVAVE]Y ETK AG RLNKGVATSWLR  AK  E  PAG E NGGRAL         (509)
IDHLCERLPRLQA[RYYSIASSSKVHPNSVHICAVVVE]Y ETK SG RVNKGVATSWLR  AK  E  PAG E NGRRAL         (507)


KLPVHVRRSNFRLPSNPSTH[VIMIGPGTGVAPFRGFIREM]VAFLESQKKGGNNVSLGKHILFYGSRNTD DFLYQDEWPEYAKKLDGSFEM (605)
---YIRKSQFRLPFKASNF[VIMVGPGTGIAPFMGFIGER]GWLKES G KEV  GETVLYCGCRHKEEDYLYQEELEQAHKKG ALTKL
VPMFVRKSQFRLPFKSTTF[VIMVGPGTGIAPFMGFIQER]AWLREQ G KEV  GETLLYGCRRSDEDYLYREELARFHKDG ALTQL  (591)
VPMFVRKSQFRLPFKATTF[VIMVGPGTGVAPFIGFIQER]AWLRQQ G KEV  GETLLYGCRRAAEDYLYREELAGFQKDG TLSQL  (592)
VPMFVRKSQFRLPFKATTF[VMVGRGTGVAPFEGFEIQEE]AWLOEQ G KEV  GETLLYYGCRRSDEDYLYREELAQFHAKG ALTRL  (590)


VVAHSRLPNTKKVYVYVQDKLKDYEDQVFEMINNG AFIYVCGDAKGMAKGVSTALVGILSRGKSITTDEATELIKMLKTSGRYQEDVWS (691)
NVAFSR EQDQKVYVQHLLRKNKVDLWRQIHEDYAHIYICGDARNMARDVQTAFYEIAEELGGMTRTQATDYIKKLMTKGRYSQDVWS    (691)
NVAFSR EQAHKVYVQHLLKRDREHLWKLIHEGGAHIYVCGDARNMAKDVQNTFYDIVAEFGPMEHTQAVDYVKKLMTKGRYSLDVWS    (678)
NVAFSR EQAQKVYVQHLLRRDKEHLWRLIHEGGAHIYVCGDARNMARDVQNTFYDIVAELGAMEHAQAVDYVKKLMTKGRYSLDVWS    (679)
SVAFSR EQPQKVYVQHLLKRDKEHLWKLIHDGGAHIYICGDARNMARDVQNTFCDIVAEQGPMEHAQAVDYVKKLMTKGRYSLDVWS    (677)
```

Fig. 14. Alignment of the yeast NADPH-cytochrome P-450 reductase (YABUSAKI et al., 1988) with that of trout (URENJAK et al., 1987), rat (PORTER and KASPER, 1985), rabbit (KATAGARI et al., 1986) and pig (HANIU et al., 1986) flavoproteins. The arrowhead indicates the sites of proteolytic modification. The boxed sequences are putative functional domains of NADPH-cytochrome P-450 reductase responsible for FMN, FAD and NADPH-binding. Hydropathy plot of porcine NADPH-cytochrome P-450 reductase calculated using a window of 12 amino acid residues (HANIU et al., 1986).

activities are associated with the same polypeptide chain. At the same time, dehydrogenase and isomerase activities in *Ps. testosteroni* were shown to be separate proteins having different molecular weights (KAWAHARA et al., 1962). However, in disagreement with these results, GALLAY et al. (GALLAY et al., 1978) reported separation of dehydrogenase and isomerase activities from bovine adrenocortical microsomes. Recently, these results were proved by showing that homogeneous 3β-hydroxysteroid dehydrogenase fail to catalyze $\varDelta^5$-$\varDelta^4$ conversion (HIWATASHI et al., 1985). The molecular weight of the dehydrogenase was found to be 41 kDa. The Km values of dehydrogenase proved to be 6.2 μM for NAD$^+$, 4.9 mM for NADP$^+$, 2.0 μM for pregnenolone and 5.3 μM for 17α-hydroxypregnenolone. Histidyl and cysteinyl residues might be present in the NAD$^+$-binding site of 3β-hydroxysteroid dehydrogenase (HIWATASHI et al., 1985).

Based on kinetic experiments, evidence has been provided for the existence of multiple forms of dehydrogenase and isomerase: isomerisation of C_{21} and C_{19}-$\varDelta^5$- 3-ketosteroids is catalyzed by the same enzyme. On the other hand, dehydrogenase reaction of corresponding $\varDelta^5$-3β-hydroxysteroids needs two different enzymes (NEVILLE and ENGEL, 1968; YATES and DESPHANDE, 1975). In this connection, the question of the molecular organization of the dehydrogenase/isomerase complex in adrenocortical microsomes and relations between the two activities is still open.

In the first step of transformation of $\varDelta^5$-3β-hydroxysteroids to $\varDelta^4$-3-ketosteroids NAD$^+$-dependent dehydrogenase catalyzes oxidation of hydroxyl group at 3β-position of steroid. The hydrogen is accepted by 4β-position of the nicotinamide ring (WARREN and CHEATUM, 1968). The kinetic data of the 3β-hydroxysteroid dehydrogenase indicates a compulsory-sequence substrate-binding mechanism and the formation of a ternary complex (HIWATASHI et al., 1985).

The mechanism of isomerase reaction has been studied in more detail for enzyme purified from *Ps. testosteroni* and appears to be the same for isomerase from adrenocortical microsomes (TALALAY and BENSON, 1972). This mechanism suggests the abstraction of a hydrogen atom at 4-position with concomitant enolization; $\varDelta^{3'}\varDelta^5$-dienol formed further undergoes ketonization with axial reprotonization at 6-position. The hydrogen atom transfer from 4- to 6-position is an example of highly specific intramolecular rearrangement.

4.5. Molecular organization and mechanism of electron transfer in microsomal monooxygenases

To realize steroid hormones biosynthesis in endoplasmic reticulum membranes the cytochrome P-450-dependent monooxygenases (cytochrome P-450$_{21}$, cytochrome P-450$_{17a}$) should interact with NADPH-cytochrome P-450 reductase as well as with the other components, including 3β-hydroxysteroid dehydrogenase and isomerase to regulate electron transfer and steroidogenesis. The electrons for 21-hydroxylation and 17α-hydroxylation as well as for 17/20-lyase activities are provided by NADPH via NADPH-cytochrome P-450 reductase. The reduction of cytochrome P-450 by NADPH-cytochrome P-450 reductase has been extensively used as a model to study the interaction of both proteins and the mechanism of electron transfer. Studies on the interaction of cytochrome P-450 with NADPH-cytochrome P-450 reductase are of special interest since flavoprotein is present in adrenocortical microsomes in limiting quantities. The ratio of cytochrome P-450 and NADPH-cytochrome P-450 reductase in bovine adrenocortical microsomes was shown to be about 10 (KOMINAMI et al., 1983). Therefore, steroidogenesis might be controlled by the cytochrome P-450 species which are in close contact with flavoprotein.

The stable complex between cytochrome P-450$_{21}$ and NADPH-cytochrome P-450 reductase was proved by gel-filtration and steady state kinetic experiments (KOMINAMI et al., 1984). Stopped-flow studies indicate that the time course of cytochrome P-450$_{21}$ reduction by NADPH-cytochrome P-450 reductase follows biphasic kinetics composed of fast and slow phases with first order rate constants 1.0 s^{-1} and 0.05 s^{-1}. Careful analysis of kinetics indicates that the fast phase corresponds to the electron transfer within the complex and the slow phase to the electron transfer due to random collision between cytochrome P-450$_{21}$ and NADPH-cytochrome P-450 reductase. The reduction of cytochrome P-450$_{21}$ is not affected by the high spin content of substrate-bound cytochrome P-450$_{21}$ but is very slow in the absence of substrate (KOMINAMI and TAKEMORI, 1982). There are no significant differences in the rate and the amount of the fast phase of reduction between 17α-hydroxyprogesterone- and progesterone-bound cytochrome P-450$_{21}$. This data made possible the conclusion that the spin state of cytochrome P-450$_{21}$ does not dramatically affect electron transfer from NADPH to cytochrome P-450$_{21}$, via NADPH-cytochrome P-450 reductase but substrate is essential for efficient electron transfer.

At present it is difficult to discriminate between two models of molecular organization of cytochrome P-450$_{21}$ and NADPH-cytochrome P-450 reductase in the membrane: whether it is the cluster or the random diffusion model is still to be clarified. Nevertheless, based on the data of KOMINAMI et al. (KOMINAMI et al., 1984), the cluster model is considered more appropriate.

The interaction of cytochrome P-450$_{17α}$ with NADPH-cytochrome P-450 reductase is much less studied. Furthermore studies on this problem are complicated by the findings that cytochrome b$_5$ may also participate in reactions catalyzed by cytochrome P-450$_{17α}$ (SHINZAWA et al., 1985). Stimulation of C$_{21}$ steroid side chain cleavage activities catalyzed by testicular cytochrome P-450$_{17α}$ and NADPH-cytochrome P-450 reductase have also been reported

(KATAGIRI et al., 1982; ONODA and HALL, 1982). The physiological role of cytochrome b_5 in reactions catalyzed by cytochrome P-450$_{17\alpha}$ in microsomes was interpreted to be either as participation in electron transfer or as regulation of dual activity of cytochrome P-450$_{17\alpha}$. Removal of cytochrome b_5 decreases lyase activity of cytochrome P-450$_{17\alpha}$ (SHINZAWA et al., 1985). A very speculative model (Fig. 15) of the molecular organization of microsomal steroidogenic enzymes has been recently proposed (HALL, 1984). How-

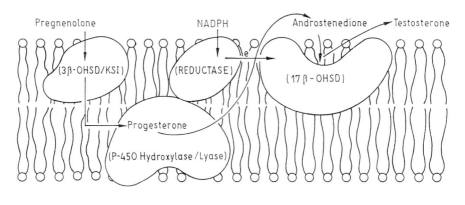

——▶ Flow of steroid intermediates

Fig. 15. Hypothetical model for the molecular organization of steroidogenic enzymes of adrenocortical endoplasmic reticulum membranes (HALL, 1984).
3β-OHSD = 3β-hydroxysteroid dehydrogenase; 3-KSI = 3-ketosteroid isomerase; 17β-OHSD = 17β-hydroxysteroid dehydrogenase.

ever, this model neither considers the role of cytochrome P-450$_{21}$ and cytochrome b_5 nor their mutual relations. But it might be useful for an understanding of the steroidogenesis in microsomes.

5. Perspectives of the application of steroid transforming enzymes

The high level of recent fundamental research on cytochrome P-450 dependent enzymes participating in the steroid hormones biosynthesis pushes the search for practical application of these enzymes in biotechnology and medicine. There are at least three important applications of cytochrome P-450 dependent enzymes:

1. preparation of physiologically active compounds by cytochrome P-450 catalyzed stereo- and regio-selective insertion of hydroxy groups into organic molecules: the development of biotechnological procedures;

2. medical therapy: low dose glucocorticoid replacement therapy of patients suffering from disorders in the steroid hormones biosynthesis, this therapy serves to replace the deficient cortisol and to suppress ACTH overproduction;

3. development of approaches to a gene therapy of diseases connected with disturbances in the steroid hormones biosynthesis.

The technology of biotransformation of the natural products with either the isolated enzymes or cells and microorganisms to realize specific chemical transformations enables many reactions in a highly regio- and stereo-specific manner with high yield to be carried out. The microbial transformation of steroids has been extensively applied in the production of steroids in the commercial scale.

Nearly 50 different steroid hormones and their intermediates occur in the organism. Some of them such as cortisol, corticosterone, aldosterone, progesterone as well as their derivatives (prednisolone) proved to be efficient in medical therapy to replace the deficiency of cortisol. Nevertheless, the production of physiologically active steroids is of special interest. The development of chemical methods to introduce the hydroxy group into the steroid nucleus by direct conversion of carbon-hydrogen bond to the carbon-hydroxyl with defined regio- and stereo-selectivity has attracted much attention. However, the production of physiologically active steroids, by chemical methods involves great problems due to expensive multistep chemical conversions.

There are currently two main biotechnological applications of microorganisms to produce steroids: (i) the use of microorganisms to produce appropriate intermediate products from raw material for a general steroid production; (ii) the use of microorganisms for specific transformations of intermediate steroids to the desired product. Since the discovery that the side chain of various naturally occurring steroids (cholesterol, β-sitosterol, campesterol) may be selectively degraded by microorganisms, this conversion method has attracted most attention and several useful processes have been developed. However, microbial side chain degradation of β-sitosterol results in only two C_{17} ketosteroids which are further used to synthesize the sex steroid hormones. In contrast to cholesterol side chain cleavage in mammalian systems, where C_{17} ketosteroids are formed via cleavage of the $C_{20}-C_{22}$ bond followed by cleavage of the $C_{17}-C_{20}$ bond, microorganisms shorten the side chain of cholesterol or β-sitosterol by a mechanism similar to that of the β-oxidation of fatty acids (SIH and WHITLOCK, 1968).

Side chain cleavage of cholesterol as catalyzed by cytochrome P-450scc in adrenal mitochondria resulting in pregnenolone — the pivotal precursor of main steroid hormones is of special interest in biotechnology. Whereas the same subsequent reactions of pregnenolone transformation are realized in some microbial processes, the initial side chain cleavage reaction was shown not to occur in procaryots. Therefore, the conversion of natural sterols (cholesterol and β-sitosterol) to intermediates to be used in the synthesis of medically important steroids is the most valid biotechnological implication of cytochrome P-450scc. Since the gene responsible for the biosynthesis of cytochrome P-450scc, has been cloned and expressed (MOROHASHI et al., 1984; CHUNG et al., 1986; MOROHASHI et al., 1987) the nearest task in this direction is the expression of cytochrome P-450scc in commercially favourable microorganisms.

The second biotechnological application of cytochrome P-450 dependent hydroxylases is the development of chemico-enzymatical methods for synthesizing physiologically active compounds. The idea of this approach is to use highly regio- and stereospecific reactions catalyzed by cytochrome P-450 to synthesize unique physiologically active compounds with very high yield. The technological approach is to use enzyme reactors containing immobilized substrate-specific cytochrome P-450. Further progress in this field may be connected with the development of simple chemical models of cytochrome P-450 catalyzing the same reactions as cytochrome P-450 with a high degree of regio- and stereoselectivity. An alternative approach is to develop the cyclic electrochemical reduction of cytochrome P-450 to simplify the electron transfer chain by substituting flavoprotein and ferredoxin.

Congenital adrenal hyperplasia due to disorder of adrenocortical functions originates from an enzymatic defect at one specific step in the conversion of cholesterol to cortisol and represents one of the most common inborn errors (NEW, 1987; WHITE, 1987). The most prevalent cause of congenital hyperplasia is a defective 21-hydroxylation, characterized by overproduction of androgens and, in a significant percentage of cases, also by deficient synthesis of aldosterone. There are at least two genes encoding cytochrome P-450$_{21}$ (21-OH "A" and 21-OH "B"), adjacent to one of the two genes for the fourth component of complement (C4A and C4B). The 21-OH "B" gene was shown to be the active gene, while the 21-OH "A" is pseudogene (WHITE et al., 1984). Congenital adrenal hyperplasia due to deficient 21-hydroxylase activity is caused by a large number of different lesions in the human cytochrome P-450$_{21}$ "B" gene. The understanding of the molecular mechanisms leading to the congenital adrenal hyperplasia is of special interest during lowdose glucocorticoid replacement therapy which is usually employed to treat the patients with classical and nonclassical symptoms. The oligonucleotide probes corresponding to different mutations in cytochrome P-450$_{21}$ gene might be useful in diagnostics of disorders in steroid metabolism. It should be possible to prenatally diagnose 21-hydroxylase deficiency by chorionic villus sampling, using a small number of oligonucleotide probes (WHITE et al., 1987). This would also be of special interest in the light of a possible prenatal gene therapy to prevent abnormal sexual differentiation.

Despite the fact that at present there are few commercial processes utilizing the unique catalytic properties of this enzyme, cytochrome P-450 is one of the most promising enzymes for future biotechnology. The importance of genetic engineering of microorganisms that would synthesize steroid intermediates from the raw material is further obvious. Another major advance would be the creation of microorganisms with definite specific hydroxylase activity that would further offer cost reduction in established hydroxylation fermentations or permit hydroxylations with immobilized cells or protein complexes being more efficient.

6. Concluding remarks

In the present review we have attempted to summarize the recently acquired knowledge of cytochrome P-450 dependent processes involved in steroid hormones biosynthesis starting from cholesterol. In this presentation we tried too to make understandable how definite function is realized in the terms of elements of the primary structure of cytochrome P-450 dependent enzymes. More importantly, this approach focuses on the idea that the cytochrome P-450 dependent steroidogenesis may be more complicated and more fascinating than have been thought before. The elucidation of the primary structure of cytochrome P-450 isozymes led to better understanding of the complicated multistep nature of cholesterol conversion to pregnenolone catalyzed by cytochrome P-450scc, dual activities of cytochrome $P-450_{11\beta}$ and cytochrome $P-450_{17\alpha}$ which are a manifestation of the polyfunctional catalysis which is an intrinsic property of cytochrome P-450. This phenomenon characterizes the interaction of cytochrome P-450 with steroid as a dynamic model for the interaction of active site bound steroid with heme bound active oxygen species. This might provide a reasonable explanation for alternative pathways of steroid hormones biosynthesis dependent on functional states of the organism. Further analysis of the data on catalytic activity, substrate binding and structural studies of cytochrome P-450 enzymes in terms of functionally important domains of the hemeprotein will contribute to understanding the nature of regio- and stereospecificity.

Some fundamental questions concerning the molecular organization of mitochondrial and microsomal steroidogenic enzymes still remain open: (i) the mode of interaction of electron transfer components, the "cluster" or diffusion model of organization in the phospholipid membrane; (ii) the relations between different cytochrome P-450 enzymes that need common electron service proteins; (iii) the topology of cytochrome P-450 dependent enzymes in membrane; (iv) relations between mitochondrial and microsomal monooxygenases.

The mechanism of the regulation of activity and the level of cytochrome P-450 dependent enzymes is acknowledged to be the most important factor in steroidogenesis. In this case, cytochromes P-450, as well as genes responsible for their synthesis, provide a unique possibility for understanding the mechanism of regulation control at the level of steroid hormones biosynthesis. Elucidation of the structure of genes encoding steroidogenic enzymes as well as their chromosomal localization will allow us to understand the molecular aspects of human diseases related to steroid hormones. These approaches will also lead to the understanding of the developmental regulation of the cytochrome P-450 dependent steroid hydroxylases gene expression as well as molecular mechanisms of tissue-specific gene expression. The regulation and control of the activity and the gene expression of cytochrome P-450 enzymes will be the most important fundamental question to be solved in the near future.

The development of methods for the construction of recombinant cDNA will lead to the engineering of protein by using techniques of site-directed mutagenesis to solve fundamental problems of cytochrome P-450 dependent monooxygenase catalysis and the nature of the high regio- and stereospecificity of oxidation. The construction of recombinant cDNA molecules of some substrate specific forms of cytochrome P-450 will lead also to the development of biotechnological procedures for the preparation of physiologically active compounds.

7. References

ADAMOVICH, T. B., I. A. PIKULEVA, S. A. USANOV, V. L. CHASHCHIN, (1989), Biokhymia 54.

AKHREM, A. A., V. M. SHKUMATOV, V. L. CHASHCHIN, (1977), Bioorg. Khymia 3, 1064—1069.

AKHREM, A. A., V. M. SHKUMATOV, V. L. CHASHCHIN, (1978), Bioorg. Khymia 4, 688—693.

AKHREM, A. A., A. G. LAPKO, V. N. LAPKO, L. A. MOROZOVA, V. A. REPIN, I. A. TISHCHENKO, (1978), Bioorg. Khymia 4, 462—475.

AKHREM, A. A., V. N. LAPKO, A. G. LAPKO, V. M. SHKUMATOV, V. L. CHASHCHIN, (1979), Acta biol. et med. germ. 38, 257—274.

AKHREM, A. A., V. I. VASILEVSKY, T. B. ADAMOVICH, A. G. LAPKO, V. M. SHKUMATOV, V. L. CHASHCHIN, (1980), in: J. A. GUSTAFSSON et al., (eds.), Biochemistry, Biophysics and Regulation of Cytochrome P-450, p. 57, Elsevier North Holland, Amsterdam.

AKHREM, A. A., V. I. VASILEVSKY, V. M. SHKUMATOV, V. L. CHASHCHIN, (1980), in: M. J. COON et al., (eds.), Microsomes, Drug Oxidations and Chemical Carcinogenesis, p. 77, Academic Press, New York.

AKHREM, A. A., V. I. VASILEVSKY, V. G. RADYUK, V. M. SHKUMATOV, V. L. CHASHCHIN, (1980), Bioorg. Khymia, 6, 285—295.

AKHREM, A. A., V. I. VASILEVSKY, S. P. MARTSEV, V. M. SHKUMATOV, V. L. CHASHCHIN, (1980), Doklady Akademii nauk USSR 252, 751—754.

AKHREM, A. A., I. I. BOVDEY, I. N. MOROZ, V. A. REPIN, V. M. SHKUMATOV, V. L. CHASHCHIN, (1980), Doklady Akademii nauk USSR, 250, 757—761.

AKHREM, A. A., T. B. ADAMOVICH, A. G. LAPKO, N. A. LOBANOV, S. A. USANOV, V. L. CHASHCHIN, (1988), 6th International Conference on Biochemistry, Biophysics of Cytochrome P-450, Abstracts, Vienna, p. 1.

ARTHUR, J. R., H. A. P. BLAIR, G. S. BOYD, I. MASON, K. E. SUCKLING, (1976), Biochem. J. 158, 47—51.

BHASTER, C. R., T. OKAMURA, E. R. SIMPSON, M. R. WATERMAN, (1987), Eur. J. Biochem. 164, 21—25.

BUMPUS, J. A., K. M. DUS, (1982), J. Biol. Chem. 257, 1269—1270.

BURSTEIN, S., M. GUT, (1976), Steroids 28, 115—131.

CHAPLIN, D. D., L. J. GALBRAITH, J. D. SEIDMAN, P. C. WHITE, K. L. PARKER, (1986), Proc. Natl. Acad. Sci. USA 83, 9601—9605.

CHASHCHIN, V. L., V. I. VASILEVSKY, V. M. SHKUMATOV, A. A. AKHREM, (1984), Biochim. Biophys. Acta 787, 27—28.

CHASHCHIN, V. L., V. I. VASILEVSKY, V. M. SHKUMATOV, V. N. LAPKO, T. B. ADAMOVICH, T. M. BERIKBAEVA, A. A. AKHREM, (1984), Biochim. Biophys. Acta 791, 375—383.

CHASHCHIN, V. L., I. V. TURKO, A. A. AKHREM, S. A. USANOV, (1985), Biochim. Biophys. Acta 828, 313—324.

CHASHCHIN, V. L., V. N. LAPKO, T..B. ADAMOVICH, V. M. KIRILLOVA, A. G. LAPKO, A. A. AKHREM, (1986), Bioorg. Khymia 12, 1286—1289.

CHASHCHIN, V. L., V. N. LAPKO, T. B. ADAMOVICH, A. G. LAPKO, N. S. KUPRINA, A. A. AKHREM, (1986), Biochim. Biophys. Acta 871, 217—223.

CHENG, B., D. HSU, T. KIMURA, (1985), Mol. and Cell. Endocrinology 40, 233—243.

CHU, J.-W., T. KIMURA, (1973), J. Biol Chem. 248, 5183—5187.

CHUA, S. C., P. SZABO, A. VITEK, K.-H. GRZESCHNIK, M. JOHN, P. WHITE, (1987), Proc. Natl Acad. Sci. 84, 7193—7197.

CHUNG, B.-C., K. MATTESON, W. L. MILLER, (1986), Proc. Natl. Acad. Sci. USA 83, 4243—4247.

CHUNG, B.-C., K. J. MATTESON, R. VOUTILAINEN, T. K. MOHANDAS, W. L. MILLER, (1986), Proc. Natl. Acad. Sci. 83, 8962—8966.

CHUNG, B.-C., J. PICADO-LEONARD, M. HANIU, M. BIENKOWSKI, P. HALL, J. E. SHIVELY, W. L. MILLER, (1987), Proc. Natl. Acad. Sci. 84, 407—411.

CHURCHILL, P. E., L. R. DE ALVARE, T. KIMURA, (1978), J. Biol. Chem. 253, 4924—4929.

CHURCHILL, P. E., T. KIMURA, (1979), J. Biol. Chem. 254, 10443—10448.

CONSTANTOPOULOS, G., T. T. TCHEN, (1961), J. Biol. Chem. 236, 65—67.

CONSTANTOPOULOS, G., A. CARPENTER, P. S. SATOH, T. T. TCHEN, (1966), Biochemistry 5, 1650—1652.

COOPER, D. Y., S. LEVIN, S. NARASHIMHULU, O. ROSENTHAL, R. W. ESTABROOK, (1965), Science 147, 400—402.

DUQUE, C., M. MORISAKI, N. IKEKAWA, M. SHIKITA, (1978), Biochem. Biophys. Res. Communs 82, 179—187.

ESTABROOK, R. W., D. Y. COOPER, O. ROSENTHAL, (1963), Biochem. Z. 338, 741—755.

ESTABROOK, R. W., K. SUZUKI, J. I. MASON, J. BARON, W. E. TAYLOR, E. S. SIMPSON, J. PURVIS, J. MCCARTHY, (1973), in: Ironsulfur proteins, p. 193—223, New York.

FORD, H. C., L. L. ENGEL, (1974), J. Biol. Chem. 249, 1363—1368.

FOSTER, R. P., L. D. WILSON, (1975), Biochemistry 14, 1477—1484.

GALLAY, J., M. VINCENT, C. DE PAILLERETS, A. ALFSEN, (1978), Biochim. Biophys. Acta 529, 79—87.

GEREN, L. M., P. O'BRIEN, J. STONCHUERNER, F. MILLET, (1984), 259, 2155—2160.

GOTOH, O., Y. TAGASHIRA, K. MOROHASHI, Y. FUJII-KURIYAMA, (1985), FEBS Letts 188, 8—10.

GREENGARD, P., S. PSYCHOYOS, H. H. TALLAN, D. Y. COOPER, O. ROSENTHAL, R. W. ESTABROOK, (1967), Arch. Biochem. Biophys. **121**, 298—303.

GROH, S. E., A. NAGAHISA, S. L. TAN, W. H. ORME-JOHNSON, (1983), J. Amer. Chem. Soc. **105**, 7445—7446.

HALKERSTON, I. D. K., J. EICHHORN, O. HECHTER, (1961), J. Biol. Chem. **236**, 374—380.

HALL, P. F., S. B. KORITZ, (1964), Biochim. Biophys. Acta **93**, 441—444.

HALL, P. F., M. WATANUKI, B. A. HAMKALO, (1979), J. Biol. Chem. **254**, 547—552.

HALL, P. F., (1980), in: J. A. GUSTAFSSON, J. CARLSTED-DUKE, A. MODE, J. RAFTER, (eds.), Biochemistry, Biophysics and Regulation of Cytochrome P-450, p. 461—475, Elsevier/North Holland, Amsterdam.

HALL, P. F., (1984), Intern. Rev. Cytology, **86**, 53—95.

HAMAMOTO, I., Y. ICHIKAWA, (1984), Biochim. Biophys. Acta **786**, 32—41.

HAMAMOTO, I., K. KUZUTAKA, S. TANAKA, Y. ICHIKAWA, (1988), Biochim. Biophys. Acta **953**, 207—213.

HANIU, M., T. IYANAGI, P. MILLER, T. D. LEE, J. E. SHIVELY, (1986), Biochemistry **25**, 7906—7911.

HANIU, M., K. YANAGIBASHI, P. F. HALL, J. E. SHIVELY, (1987), Arch. Biochem. Biophys. **254**, 380—384.

HANUCOGLU, I., C. R. JEFCOATE, (1980), J. Biol. Chem. **255**, 3057—3061.

HANUCOGLU, I., V. SPITSBERG, J. A. BUMPUS, K. M. DUS, C. R. JEFCOATE, (1981), J. Biol. Chem. **256**, 4321—4328.

HARDING, R. W., S. H. WONG, D. H. NELSON, (1964), Biochim. Biophys. Acta **92**, 415—417.

HASUMI, H., F. YAMAKURA, S. NAKAMURA, K. SUZUKI, T. KIMURA, (1984), Biochim. Biophys. Acta **787**, 152—157.

HECHTER, O., G. PINCUS, (1954), Physiol. Rev. **34**, 459—496.

HEYL, B. L., D. J. TYRRELL, J. D. LAMBETH, (1986), J. Biol. Chem. **261**, 12743—2749.

HIGASHI, Y., H. YOSHIOKA, M. YAMANE, O. GOTOH, Y. FUJII-KURIYAMA, (1986), Proc. Natl Acad. Sci. USA **83**, 2841—2845.

HIWATASHI, A., Y. ICHIKAWA, N. MARUYA, T. YAMANO, K. AKI, (1976 a), Biochemistry, **15**, 3082—3090.

HIWATASHI, A., Y. ICHIKAWA, T. YAMANO, N. MARUYA, (1976 b), Biochemistry, **15**, 3091—3097.

HIWATASHI, A., Y. ICHIKAWA, T. YAMANO, (1977), FEBS Letts **82**, 201—205.

HIWATASHI, A., Y. ICHIKAWA, (1978), J. Biochem. **84**, 1071—1086.

HIWATASHU, A., Y. ICHIKAWA, (1979), Biochim. Biophys. Acta **580**, 44—63.

HIWATASHI, A., Y. ICHIKAWA, (1981), Biochim. Biophys. Acta **664**, 33—48.

HIWATASHI, A., Y. ICHIKAWA, (1982), Biochim. Biophys. Acta **705**, 82—91.

HIWATASHI, A., I. HAMAMOTO, Y. ICHIKAWA, (1985), J. Biochem. **98**, 1519—1526.

HUME, R., R. W. KELLY, P. TAYLOR, G. S. BOYD, (1984), Eur. J. Biochem. **140**, 583 to 591.

ICHIKAWA, Y., A. HIWATASHI, (1982), Biochim. Biophys. Acta **705**, 82—91.

ICHIKAWA, Y., A. HIWATASHI, M. TSUBAKI, (1984), FEBS Letts **167**, 131—136.

JAQUOT, J.-P., A. SUZUKI, J.-B. PEYRE, R. PEYRONNET, M. MIGIVIAC-MASLOW, P. GADAL, (1988), Eur. J. Biochem. **174**, 629—635.

JEFCOATE, C. R., R. HUME, G. S. BOYD, (1970), FEBS Letts **9**, 41—44.

JEFCOATE, C. R., (1975), J. Biol. Chem. **250**, 4663—4670.

JEFCOATE, C. R., W. H. ORME-JOHNSON, (1975), J. Biol. Chem. **250**, 4671—4677.

JEFCOATE, C. R., W. H. ORME-JOHNSON, H. BEINERT, (1976), J. Biol. Chem. **251**, 3706 to 3715.

JEFCOATE, C. R., (1977), J. Biol. Chem. **252**, 8788—8796.
JEFCOATE, C. R., (1982), J. Biol. Chem. **257**, 4731—4737.
JEFCOATE, C. R., (1986), in: P. R. ORTIZ DE MOTELLANO, (eds.), Cytochrome P-450. Structure, Mechanism, and Biochemistry, p. 387—428, Plenum Press, New York and London.
KAGIMOTO, K., J. L. MCCARTHY, M. R. WATERMAN, M. KAGIMOTO, (1988), Biochem. Biophys. Res. Communs **155**, 379—383.
KATAGIRI, M., O. TAKIKAWA, H. SATO, K. SUHARA, (1977), Biochem. Biophys. Res. Communs **77**, 804—809.
KATAGIRI, M., K. SUHARA, (1980), in: J.-A. GUSTAFSSON, J. CARLSTED-DUKE, A. MODE, and J. RAFTER, (eds.), Biochemistry, Biophysics and Regulation of Cytochrome P-450, p. 97, Elsevier/North Holland Amsterdam.
KATAGIRI, M., K. SUHARA, M. SHIROO, Y. FUJISHIMA, (1982), Biochem. Biophys. Res. Communs **108**, 379—384.
KATAGIRI, M., H. MURAKAMI, Y. YABUSAKI, T. SUGIYAMA, M. OKAMOTO, T. YAMANO, H. OHKARA, (1986), J. Biochem. **100**, 945—954.
KAWAHARA, F. S., S. F. WANG, P. TALALAY, (1962), J. Biol. Chem. **237**, 1500—1508.
KIDO, T., T. KIMURA, (1979), J. Biol. Chem. **254**, 11806—11815.
KIDO, T., F. YAMAKURA, T. KIMURA, (1981), Biochim. Biophys. Acta **666**, 370—381.
KIMURA, T., K. SUZUKI, (1967), J. Biol. Chem. **242**, 485—491.
KIMURA, T., H. OHNO, (1968), J. Biochem. **63**, 717—724.
KIMURA, T., A. TASAKI, H. WATARI, (1970), J. Biol. Chem. **245**, 4450—4455.
KIMURA, T., (1981), Mol. and Cell. Biochem. **36**, 105—122.
KITAGAWA, T., H. SAKAMOTO, T. SUGIYAMA, T. YAMANO, (1982), J. Biol. Chem. **257**, 12075—12080.
KOMINAMI, S., S. MORI, S. TAKEMORI, (1978), FEBS Letts **89**, 215—218.
KOMINAMI, S., H. OCHI, S. TAKEMORI, (1979), Biochim. Biophys. Acta **577**, 170—176.
KOMINAMI, S., H. OCHI, Y. KOBAYASHI, S. TAKEMORI, (1980), J. Biol. Chem. **255**, 3386 to 3394.
KOMINAMI, S., K. SHINZAWA, S. TAKEMORI, (1982), Biochem. Biophys. Res. Communs, **109**, 916—921.
KOMINAMI, S., S. TAKEMORI, (1982), Biochim. Biophys. Acta, **709**, 147—153.
KOMINAMI, S., K. SHINZAWA, S. TAKEMORI, (1983), Biochem. Biophys. Acta, **755**, 163 to 169.
KOMINAMI, S., H. HARA, T. OGISHIMA, S. TAKEMORI, (1984), J. Biol. Chem. **259**, 2991 to 2999.
KOMINAMI, S., Y. ITOH, S. TAKEMORI, (1986), J. Biol. Chem. **261**, 2077—2083.
KOMINAMI, S., A. HIGUCHI, S. TAKEMORI, (1988), Biochim. Biophys. Acta **937**, 177—183.
KRAAIPOEL, R. J., H. J. DEGENHART, J. G. LEFERINCK, V. VAN BEEK, H. DE LEEUW-BOON, K. A. VISSER, (1975a), FEBS Letts **50**, 204—209.
KRAAIPOEL, R. J., H. J. DEGENHART, V. VAN BEEK, H. DE LEEUW-BOON, G. ABGEIN, H. K. A. VISSER, J. G. LEFERINK, (1975b), FEBS Letts **54**, 172—179.
LAMBETH, J. D., H. KAMIN, (1976), J. Biol. Chem. **251**, 4299—4306.
LAMBETH, J. D., D. R. MCCASLIN, H. KAMIN, (1976), J. Biol. Chem. **251**, 7545—7559.
LAMBETH, J. D., H. KAMIN, (1977), J. Biol. Chem. **252**, 2908—2917.
LAMBETH, J. D., H. KAMIN, (1979), J. Biol. Chem. **254**, 2766—2774.
LAMBETH, J. D., D. W. SEYBERT, H. KAMIN, (1979), J. Biol. Chem. **254**, 7255—7264.
LAMBETH, J. D., D. W. SEYBERT, H. KAMIN, (1980), J. Biol. Chem. **255**, 138—143.
LAMBETH, J. D., J. R. LANCASTER, H. KAMIN, (1981), J. Biol. Chem. **256**, 3674—3678.

LAMBETH, J. D., S. E. KITCHEN, A. A. FAROOQUI, R. TUCKEY, H. KAMIN, (1982), J. Biol. Chem. **257**, 1876—1884.

LAMBETH, J. D., D. W. SEYBERT, J. R. LANCASTER, J. C. SALERNO, H. KAMIN, (1982), Mol. and Cell. Biochem. **45**, 13—31.

LAMBETH, J. D., S. O. PEMBER, (1983), J. Biol. Chem. **258**, 5596—5602.

LAMBETH, J. D., L. M. GEREN, F. MILLET, (1984), J. Biol. Chem. **259**, 10025—10029.

LAMBETH, J. D., S. KRIENGSIRI, (1985), J. Biol. Chem. **260**, 8810—8816.

LAMBETH, J. D., (1986), Endocrine Research **12**, 371—392.

LANGE, R., L. MAURIN, C. LARROQUE, A. BIENVENUE, (1988), Eur. J. Biochem., **172**, 189—195.

LARROQUE, C., J. ROUSSEAU, J. E. VAN LIER, (1981), Biochemistry **20**, 925—929.

LIEBERMAN, S., N. J. GREENFIELD, A. WOLFSON, (1984), Endocrine Rev. **5**, 128—148.

LIM, B. T., T. KIMURA, (1980), J. Biol. Chem. **255**, 2440—2444.

LIM, B. T., T. KIMURA, (1981), J. Biol. Chem. **256**, 4400—4406.

LOMBARDO, A., M. LAINE, G. DERAFAYE, N. MONNIER, C. GUIDICELLI, E. M. CHAMBAZ, (1986), Biochim. Biophys. Acta **863**, 71—81.

MARTSEY, S. P., V. L. CHASHCHIN, A. A. AKHREM, (1982), Biokhimia, **47**, 1070—1083.

MASTERS, B. S. S., R. T. OKITA, (1978), Phar. Ther. **9**, 227—244.

MILLER, W. L., (1987), J. Steroid Biochem. **27**, 759—766.

MITHAL, S., Y.-Z. ZHU, L. E. VICKERY, (1988), Arch. Biochem. Biophys. **264**, 383—391.

MOMOI, K., M. OKAMOTO, T. YAMANO, (1983), J. Steroid Biochem. **22**, 267—271.

MONNIER, N., G. DEFAY, E. M. CHAMBAZ, (1987), Eur. J. Biochem. **169**, 147—153.

MORISAKI, M., K. BANNAI, M. SHIKITA, (1976), Biochem. Biophys. Res. Communs **69**, 481—488.

MORISAKI, M., C. DUQUE, N. IKEKAWA, M. SHIKITA, (1980), J. Steroid Biochem. **13**, 545—550.

MORISAKI, M., C. DUQUE, K. TAKANE, N. IKEKAWA, M. SHIKITA, (1982), J. Steroid Biochem. **16**, 101—106.

MOROHASHI, K., Y. FUJII-KURIYAMA, Y. OKADA, K. SOGAWA, T. HIROSE, S. INAYAMA, T. OMURA, (1984), Proc. Natl Acad. Sci. **81**, 4645—4653.

MOROHAHSHI, K., K. SOGAWA, T. OMURA, Y. FUJII-KURIYAMA, (1987), J. Biochem. **101**, 879—887.

MOROHASHI, K., H. YOSHIOKA, O. GOTOH, Y. OKADA, K. YAMAMOTO, T. MIYATA, K. SOGAWA, Y. FUJII-KURIYAMA, T. OMURA, (1987), J. Biochem. **102**, 559—568.

NABI, S., T. OMURA, (1980), Biochem. Biophys. Res. Communs **97**, 680—686.

NABI, S., S. KOMINAMI, S. TAKEMORI, T. OMURA, (1980), Biochem. Biophys. Res. Communs **97**, 687—693.

NAKAJIN, S., P. HALL, (1981), J. Biol. Chem. **256**, 3871—3876.

NAKAJIN, S., P. F. HALL, M. ONADA, (1981), J. Biol. Chem. **256**, 6134—6139.

NAKAJIN, S., M. SHINODA, P. HALL, (1983), Biochem. Biophys. Res. Communs, **112**, 512—517.

NAKAJIN, S., M. SHINODA, M. HANIU, J. E. SHIVELY, P. HALL, (1984), J. Biol. Chem. **259**, 3971—3976.

NARASIMHULU, S., C. R. EDDY, M. DIBARTOLOMEIS, R. KOWLURU, C. R. JEAFCOATE, (1985), Biochemistry **24**, 4287—4294.

NARASIMHULU, S., (1988), Biochemistry **27**, 1147—1153.

NEVILLE, A. M., L. L. ENGEL, (1968), Endocrinology **83**, 864—869.

NEW, M. I., (1987), J. Steroid Biochem. **27**, 1—7.

NONAKA, Y., T. SUGIYAMA, T. YAMANO, (1982), J. Biochem. **92**, 1693—1701.

NONAKA, Y., H. MURAKAMI, Y. YABUSAKI, S. KURAMITSU, H. KAGAMIYAMA, T. YAMANO, M. OKAMOTO, (1987), Biochem. Biophys. Res. Communs **145**, 1239—1247.

OKAMURA, T., M. E. JOHN, M. X. ZUBER, E. R. SIMPSON, M. R. WATERMAN, (1985), Proc. Natl Acad. Sci. **82**, 5705—5709.

OMURA, T., E. SANDERS, R. W. ESTABROOK, D. Y. COOPER, O. ROSENTHAL, (1966), Arch. Biochem. Biophys. **117**, 660—673.

ONODA, M., P. F. HALL, (1982), Biochem. Biophys. Res. Communs **108**, 454—460.

ONODA, M., M. HANIU, K. YANAGIBASHI, F. SWEET, J. E. SHIVELY, P. F. HALL, (1987), Biochemistry **26**, 657—662.

ORME-JOHNSON, N. R., D. R. LIGHT, R. W. STEVENS, W. H. ORME-JOHNSON, (1979), J. Biol. Chem. **254**, 2103—2111.

PICADO-LEONARD, J., R. VOUTILINEN, L.-C. KAO, B.-C. CHUNG, J. F. STRAUSS, W. L. MILLER, (1988), Biol. Chem. **263**, 3240—3244.

PIKULEVA, I. A., A. G. LAPKO, A. A. AKHREM, S. A. USANOV, V. L. CHASHCHIN, (1987), Bioorg. Khymia **13**, 739—747.

PORTER, T. D., C. B. KASPER, (1985), Proc. Natl Acad. Sci. USA, **82**, 973—977.

PORTER, T. D., T. E. WILSON, C. B. KASPER, (1987), Arch. Biochem. Biophys. **254**, 353—367.

RADYUK, V. G., V. M. SHKUMATOV, V. L. CHASHCHIN, A. A. AKHREM, (1982), Biokhymia **47**, 1700—1709.

RADYUK, V. G., V. M. SHKUMATOV, V. L. CHASHCHIN, A. A. AKHREM, (1982), Biokhymia **47**, 1792—1801.

RADYUK, V. G., V. M. SHKUMATOV, V. L. CHASHCHIN, A. A. AKHREM, (1983), Biokhymia **48**, 454—463.

RAMSEYER, J., B. W. HARDING, (1973), Biochim. Biophys. Acta **315**, 306—316.

ROSENTHAL, O., D. Y. COOPER, (1967), Meth. Enzymol. **10**, 616—629.

RYDSTROM, J., J.-A. GUSTAFSSON, M. INGELMAN-SUNDBERG, J. MONTELIUS, L. ERNSTER, (1976), Biochem. Biophys. Res. Communs **73**, 555—561.

SABA, N., O. HECHTER, D. STONE, (1954), J. Amer. Chem. Soc. **76**, 3862—3864.

SAGARA, Y., Y. TAKATA, T. MIYATA, T. HARA, T. HORIUCHI, (1987), J. Biochem. **102**, 1333—1336.

SAKAMOTO, H., M. OHTA, R. MIURA, T. SUGIYAMA, T. YAMANO, Y. MIYAKE, (1982), J. Biochem. **92**, 1941—1950.

SATO, H., N. ASHIDA, K. SUHARA, E. ITAGAKI, S. TAKEMORI, M. KATAGIRI, (1978), Arch. Biochem. Biophys. **190**, 307—317.

SCHLEYER, H., D. COOPER, O. ROSENTHAL, (1972), J. Biol. Chem. **247**, 6103—6109.

SEYBERT, D. W., J. R. LANCASTER, J. D. LAMBETH, H. KAMIN, (1979), J. Biol. Chem. **254**, 12088—12098.

SHEETS, J. J., L. E. VICKERY, (1982), Proc. Natl Acad. Sci. **79**, 5773—5777.

SHEETS, J. J., L. E. VICKERY, (1983), J. Biol. Chem. **258**, 1720—1725.

SHEETS, J. J., L. E. VICKERY, (1983), J. Biol. Chem. **258**, 11446—11452.

SHIMIZU, K., M. HAYANO, M. GUT, R. I. DORFMAN, (1961), J. Biol. Chem. **236**, 695—699.

SHIMIZU, K., M. GUT, R. I. DORFMAN, (1962), J. Biol. Chem. **237**, 699—702.

SHINZAWA, K., S. KOMINAMI, S. TAKEMORI, (1985), Biochim. Biophys. Acta **833**, 151 to 160.

SHIKITA, M., P. F. HALL, (1973), J. Biol. Chem. **248**, 5598—5604.

SHIKITA, M., P. F. HALL, (1973), J. Biol. Chem. **248**, 5605—5609.

SHKUMATOV, V. M., S. A. USANOV, V. L. CHASHCHIN, A. A. AKHREM, (1985), Pharmazie **40**, 757—766.

SIH, C. J., H. W. WHITLOCK, (1968), Annu. Rev. Biochem. **37**, 661—693.

SIMPSON, E. R., G. S. BOYD, (1966), Biochem. Biophys. Res. Communs **24**, 10—17.

SUGIYAMA, T., T. YAMANO, (1975), FEBS Letts **52**, 145—148.

SUGIYAMA, T., P. MIURA, T. YAMANO, (1976), in: Iron-Copper Proteins (K. T. YASUNOBU, H. F. MOVER, O. HAYAISHI, eds.), New York, London, p. 290—302.

SUGIYAMA, T., R. MIURA, T. YAMANO, (1979), J. Biochem. **86**, 213—223.
SUHARA, K., S. TAKEMORI, M. KATAGIRI, (1972), Biochim. Biophys. Acta **263**, 272—278.
SUHARA, K., T. GOMI, H. SATO, E. ITAGAKI, S. TAKEMORI, M. KATAGIRI, (1978), Arch. Biochem. Biophys. **190**, 290—299.
SUHARA, K., K. NAKAYAMA, O. TAKIKAWA, M. KATAGIRI, (1982), Eur. J. Biochem. **125**, 659—664.
SUHARA, K., Y. FUJIMURA, M. SHIROO, M. KATAGIRI, (1984), J. Biol. Chem. **259**, 8729 to 8736.
SUHARA, K., K. OHASHI, K. TAKAHASHI, M. KATAGIRI, (1988), Arch. Biochem. Biophys. **267**, 31—37.
SUZUKI, K., T. KIMURA, (1965), Biochem. Biophys. Res. Communs **18**, 340—345.
TAKEMORI, S., H. SATO, T. GOMI, K. SUHARA, M. KATAGIRI, (1975), Biochem. Biophys. Res. Communs **67**, 1151—1157.
TAKEMORI, S., K. SUHARA, S. HASHIMOTO, M. HASHIMOTO, H. SATO, T. GOMI, M. KATAGIRI, (1975), Biochem. Biophys. Res. Communs **63**, 588—593.
TALALAY, P., A. M. BENSON, (1972), Enzymes **6**, 591—595.
TANAKA, M., M. HANIU, K. YASUNOBU, T. KIMURA, (1973), J. Biol. Chem. **248**, 1141 to 1157.
TANIGUCHI, T., T. KIMURA, (1975), Biochemistry **14**, 5573—5578.
TANIGUCHI, T., T. KIMURA, (1976), Biochemistry **15**, 2849—2853.
TEICHER, B. A., N. KOIZUMI, M. KOREEDA, M. SHIKITA, P. TALALAY, (1978), Eur. J. Biochem. **91**, 11—19.
TULS, J., L. GEREN, J. D. LAMBETH, F. MILLER, (1987), J. Biol. Chem. **262**, 10020 to 10025.
TURKO, I. V., T. B. ADAMOVICH, N. M. KIRILLOVA, S. A. USANOV, V. L. CHASHCHIN, (1988), Biokhymia **53**, 1810—1816.
TURKO, I. A., S. A. USANOV, A. A. AKHREM, V. L. CHASHCHIN, (1988), Biokhymia **53**, 1352—1356.
URENJAK, J., D. LINDER, L. LUMPER, (1987), J. Chromatogr. **397**, 123—136.
USANOV, S. A., I. A. PIKULEVA, V. L. CHASHCHIN, A. A. AKHREM, (1984), Biochim. Biophys. Acta **790**, 259—267.
USANOV, S. A., I. V. TURKO, V. L. CHASHCHIN, A. A. AKHREM, (1985), Biochim. Biophys. Acta **832**, 288—296.
USANOV, S. A., I. A. PIKULEVA, I. V. TURKO, V. L. CHASHCHIN, A. A. AKHREM, (1985), in: L. VEREZKEY, K. MAGYAR, (eds.), Biochemistry, Biophysics and Induction of Cytochrome P-450, p. 25, Akademia Kiado, Budapest.
USANOV, S. A., I. A. PIKULEVA, A. A. AKHREM, V. L. CHASHCHIN, (1987), Bioorg. Khymia **13**, 725—738.
USANOV, S. A., A. A. CHERNOGOLOV, A. I. PETRASHIN, A. A. AKHREM, (1987), Biol. membrany **4**, 1102—1120.
USANOV, S. A., (1988), 6th International Conference on Biochemistry and Biophysics of Cytochrome P-450, Vienna, Austria, Abstracts, p. 85.
VAN LIER, J. F., L. L. SMITH, (1970), Ann N.Y. Acad. Sci. **212**, 276—289.
VAN LIER, J. F., J. ROUSSEAN, (1976), FEBS Letts **70**, 23—27.
VICKERY, L. E., J. T. KELLIS, (1983), J. Biol. Chem. **258**, 3832—3836.
VOGEL, F., L. LUMPER, (1986), Biochem. J. **236**, 871—876.
WANG, H.-P., T. KIMURA, (1978), Biochim. Biophys. Acta **542**, 115—127.
WARBURTON, R. J., D. W. SEYBERT, (1988), Biochem. Biophys. Res. Communs **152**, 177—183.
WARREN, J. G., S. G. CHEATUM, (1968), Biochim. Biophys. Acta **159**, 540—547.
WATANUKI, M., B. E. TILLEY, P. F. HALL, (1977), Biochim. Biophys. Acta **483**, 236 to 247.

WATANUKI, M., B. E. TILLEY, P. F. HALL, (1978), Biochemistry 17, 127—130.
WATARI, H., T. KIMURA, (1966), Biochem. Biophys. Res. Communs 24, 106—111.
WATERMAN, M. R., E. R. SIMPSON, (1985), Mol. and Cell. Endocrinology 39, 81—89.
WATERMAN, M. R., M. E. JOHN, E. R. SIMPSON, (1986), in: P. R. ORTIZ DE MONTELLANO,
    (eds.), Cytochrome P-450. Structure, Mechanism and Biochemistry, p. 345—385,
    Plenum Press, New York and London.
WHITE, P. V., (1987), Recent Prog. Horm. Res. 43, 305—336.
WHITE, P. C., M. I. NEW, B. DUPONT, (1984), Proc. Natl. Acad. Sci. USA 81, 7505—7509.
WHITE, P. C., M. I. NEW, B. DUPONT, (1986), Proc. Natl. Acad. Sci. USA 83, 5111—5115.
WHITE, P. C., B. DUPONT, M. NEW, (1987), U.S. Patent 4720454.
WILSON, L. D., D. H. NELSON, B. W. HARDING, (1965), Biochim. Biophys. Acta 99,
    391—393.
YABUSAKI, Y., H. MURAKAMI, H. OHKAWA, (1988), J. Biochem. 103, 1004—1010.
YAMAKURA, F., T. KIDO, T. KIMURA, (1981), Biochim. Biophys. Acta 649, 343—354.
YATES, J., N. DESPHANDE, (1975), J. Endocrinol. 64, 195—191.
YUAN, P. M., S. NAKAJIN, M. HANIU, M. SHINODA, P. F. HALL, J. E. SHIVELY, (1983),
    Biochemistry 22, 143—149.
ZUBER, M. X., M. E. JOHN, T. OKAMURA, E. R. SIMPSON, M. R. WATERMAN, (1986), J.
    Biol. Chem. 261, 2475—2482.

Chapter 2
Enzymology of Mitochondrial Side-Chain
Cleavage by Cytochrome P-450scc

J. D. Lambeth

1. Introduction

A large number of cytochromes P-450 occur in mammalian systems and may be categorized both according to their location (mitochondrial versus microsomal) and by the substrate or substrates upon which they act. The mitochondrial enzymes are distinguished from their microsomal counterparts not only by their inner membrane location (FARKASH et al., 1986; ISHIMURA et al., 1988), but also by the electron transfer systems which they utilize. Whereas the microsomal systems use a single, membrane-associated FAD/FMN-containing protein, NADPH-cytochrome P-450 reductase (plus in some cases NADH-cytochrome b_5 reductase, cytochrome b_5), the mitochondrial system utilizes an FAD protein (NADPH-adrenodoxin reductase) plus a ferredoxin (adrenodoxin), and, in this sense, more closely resembles bacterial P-450 sytems (KAMIN and LAMBETH, 1982). Cytochrome P-450$_{scc}$ (scc for side-chain cleavage) is an integral inner mitochondrial membrane enzyme (LAMBETH, 1984b; FARKASH et al., 1986; ISHIMURA et al., 1988), and is prototypical of the mitochondrial type of cytochrome. It represents one end of a spectrum in terms of substrate specificity (LAMBETH, 1986); unlike many of the microsomal cytochromes which show broad specificities, it is highly specific for its substrate cholesterol upon which it acts to produce pregnenolone via cleavage of the isocaproic side-chain.

The side-chain cleavage reaction is of particular biological interest, in that it is the first and rate-limiting step in steroid hormone biosynthesis in a variety of steroidogenic tissues (adrenal cortex, ovary, testis, placenta, and perhaps specific regions of the brain), and its activity therefore contributes to circulating levels of these hormones (including glucocorticoids, mineralocorticoids, progesterone, etc). The side-chain cleavage of cholesterol is acutely regulated (within several minutes) by tissue-specific peptide hormones [e.g., adrenocorticotropic hormone (ACTH) for the adrenal, luteinizing hormone (LH) for the corpus luteum]. The intracellular mechanisms of this rapid regulation remain enigmatic, but appear to involve the control of substrate delivery to the cytochrome rather than regulation of the activity or concentration of the cytochrome itself, although the latter undoubtedly contributes to the overall steroidogenic capacity of the adrenal following ACTH. An understanding of the enzymology of this system thus provides a backdrop for the study of the regulatory mechanisms; the regulation per se is beyond the scope of this review, and the reader is referred to recent reviews on this topic (PEDERSEN, 1985; LAMBETH and STEVENS, 1985).

The reaction is also of particular interest from a mechanistic point of view, given its complexity. As shown in Figure 1, the cleavage reaction proceeds in three stages, each of which utilizes 1 mol of NADPH as an electron donor plus 1 mol of molecular oxygen (HALL, 1976; HUME and BOYD, 1978; HUME et al., 1984; BURNSTEIN and GUT, 1976). Each stage is accomplished by more-or-less classical sequence of cytochrome P-450-catalyzed oxygen activation (substrate binding → initial electron transfer → oxygen binding → second electron transfer → oxygen activation → substrate oxidation). The first stage involves the stereospecific hydroxylation at the 22(R) position to yield 22(R)-hydroxy-cholesterol. A second hydroxylation then occurs at the 20(α) position, yielding

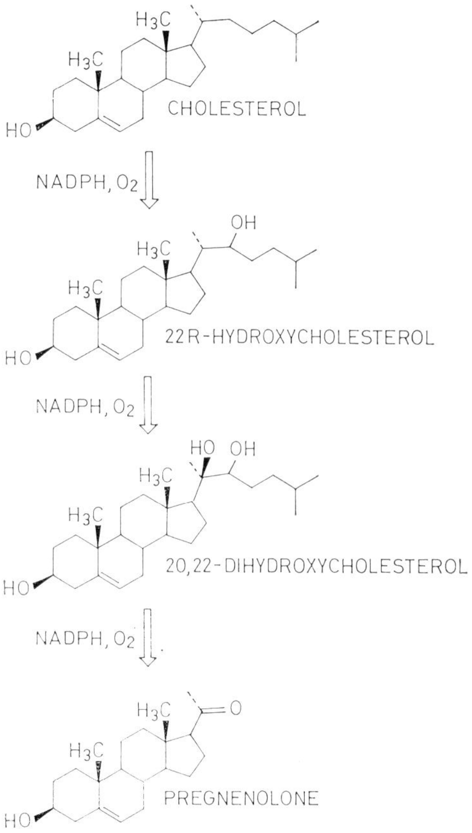

Fig. 1. Sequential reactions catalyzed by cytochrome P-450scc. Each NADPH of the three O_2-dependent steps is shown along with the hydroxysteroid intermediates.

the 20(α),22(R)-dihydroxycholesterol. A third oxidation results in the cleavage of the 20-22 carbon-carbon bond, yielding pregnenolone (the common precursor of the steroid hormones) plus isocaproaldehyde. These hydroxylations proceed to pregnenolone without the generation of significant quantities of free intermediates.

Several questions arise with regard to this mechanism. By what means does the enzyme prevent the accumulation of these hydroxylated steroid intermediates? How does the structure of the substrate binding/active site permit

---

and promote the sequential oxidations which comprise the side-chain cleavage reaction? What role does the membrane in general, and more specifically, the mitochondrial membrane play in the function of the enzyme? By what means does the system regulate the flow of electrons into the cytochrome, and how does electron transfer occur in such a way as to minimize the formation of toxic oxygen intermediates such as superoxide and hydrogen peroxide? What is the mechanism by which the enzyme catalyzes the cleavage of the 20-22 carbon-carbon bond? Studies which bear on these and other questions will be reviewed herein.

2. Isolation and initial characterization of the enzymatic components

A mitochondrial steroid hydroxylase system (the 11β-hydroxylase) was resolved into three components by ESTABROOK and colleagues (OMURA et al., 1966). Reconstitution of activity required two mitochondrial fractions: a membrane fraction which contained cytochrome P-450 plus a soluble fraction. The latter was resolved chromatographically into two protein components. The first, NADPH-adrenodoxin reductase contains FAD (flavin adenine dinucleotide) and has a single subunit with a molecular weight of 52,000 grams/mole (CHU and KIMURA, 1973b; FOSTER and WILSON, 1975). In the absence of other components, the flavoprotein catalyzed NADPH-dependent diaphorase activity (reduction of dichlorophenolindophenol, DCPIP), but did not directly reduce cytochrome c or cytochrome P-450, as illustrated in Figure 2. The second, adrenodoxin, was also isolated and characterized at about the same time by SUZUKI and KIMURA (SUZUKI and KIMURA, 1965). The highly acidic protein has a molecular weight of about 12.000, and contains two mols each of iron and labile sulfur which together comprise the single iron-sulfur center (KIMURA, 1968). In the presence of adrenodoxin, adrenodoxin reductase catalyzed the NADPH-dependent reduction of both cytochrome c and cytochrome P-450, and inclusion of the membrane fraction plus the two resolved electron transfer components reconstituted NADPH-dependent 11β-hydroxylation of deoxycorticosterone (OMURA et al., 1966). Inhibition of the reaction by carbon monoxide, and perturbation of the P-450 spectrum by steroid substrate and product (CAMMER and ESTABROOK, 1967) established that the active component in the membrane fraction was cytochrome P-450. The sequence of electron flow and hydroxylation, as originally proposed (OMURA et al., 1966), is shown in Fig. 2. It was subsequently shown (SHIKITA and HALL, 1973) that like 11β-hydroxylation, the cholesterol side-chain cleavage reaction is also catalyzed by a unique cytochrome P-450, which was later resolved from that which catalyzes the 11β-hydroxylation (TILLEY et al., 1977; TAKIKAWA et al.,

1978). The stoichiometry of the reaction is 3 mols each of molecular oxygen and NADPH for each mol of cholesterol converted to pregnenolone (SHIKITA and HALL, 1973).

The electron transport components are isolated from the supernatant fraction from sonicated beef adrenal mitochondria. We have also noted that significant quantities of adrenodoxin can be recovered from the post-mitochondrial supernatant by batch absorption on DEAE cellulose, apparently due to leakage of this component from damaged mitochondria. The initial steps in purification take advantage of the tight binding of adrenodoxin to DEAE-cellulose and the binding of the reductase to an affinity matrix of

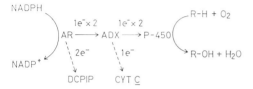

Fig. 2. Scheme for electron flow from NADPH to cytochrome P-450. Abbreviations are AR, adrenodoxin reductase; ADX, adrenodoxin; DCPIP, dichlorophenolindophenol; and cyt $\mathbf{c}$, cytochrome c. Two electrons are sequentially delivered to the cytochrome P-450 to support each hydroxylation of substrate (R—H) to product (R—OH).

$2'5'$-ADP Sepharose. We have developed a rapid and convenient single-step method (LAMBETH and KAMIN, 1979) for the initial resolution of these components in which the effluent from the initial DEAE column is flowed directly into the ADP-Sepharose column. By passage of the supernatant from sonicated mitochondria (adjusted to an appropriate ionic strength) through both columns, adrenodoxin is adsorbed as a tight brown band on the first, while adrenodoxin reductase adsorbs as a yellow band on the second. The columns are then disconnected and components eluted individually. Subsequent chromatographic steps allow the further purification to homogeneity of the resolved components.

Several laboratories (TAKIKAWA et al., 1978; SUGIYAMA et al., 1976; SUHARA et al., 1978; TAKEMORI et al., 1975b; TAKEMORI et al., 1975a; SATO et al., 1978) reported methods for the purification of cytochrome P-450$_{scc}$ and its resolution from cytochrome P-450$_{11\beta}$. Our purification (SEYBERT et al., 1979), which is similar to several other published methods, involves the solubilization of both cytochromes from beef adrenal mitochondrial membranes using sodium cholate, followed by ammonium sulfate fractionation which resolves the two cytochromes. Both are then purified further by "hydrophobic chromatography", using a hexyl-Agarose column, followed by elution with detergent (cholate for cytochrome P-450$_{scc}$ and cholate plus Tween 20 for the 11β-

hydroxylase). Typically, electrophoretically homogeneous preparations of the side-chain cleavage enzyme contain about $10-12$ nmols of heme per mg protein, while an occasional fraction may be somewhat higher (up to 16 nmol/mg). The apparent molecular mass by SDS gel electrophoresis was about 50,000 (actually 56,387 according to the predicted sequence, see below). Since the theoretical heme content is about 18, the preparation appears to contain significant quantities of apoprotein, a finding which is not unusual for a variety of mammalian cytochrome P-450 preparations. The heme is primarily in the high-spin optical form, suggesting substrate association, and analysis confirms the presence of $1-2$ mols of cholesterol per mol of heme (SEYBERT et al., 1979; ORME-JOHNSON et al., 1979). When it is necessary to prepare substrate-free enzyme, incubation in the presence of the electron transfer components and an electron source allows essentially complete metabolism of cholesterol, but the substrate-free form is less stable and must be stored with glycerol (LAMBETH et al., 1980a). In most cases, it is not necessary to utilize the substrate-free enzyme, since reconstitution into a lipid environment (phospholipid vesicles or micelles, see below) allows a many-fold dilution of the substrate into the lipid so that its effective concentration becomes negligible (LAMBETH et al., 1980). The preparation, after dialysis, is relatively free of residual cholate (3 mol/mol heme) or phospholipid (3 mol phosphate/mol heme) (SEYBERT et al., 1979), thus providing a good starting point for membrane reconstitution studies.

3. Primary and higher order structure of the components

The primary structures of cytochrome P-450$_{scc}$, cytochrome P-450$_{11\beta}$, adrenodoxin reductase, and adrenodoxin have been predicted from their cDNA clones (NONAKA et al., 1987; MOROHASHI et al., 1984; MOROHAHSI et al., 1987), and determined directly by complete protein sequencing of adrenodoxin and by partial sequencing of P-450$_{scc}$ (AKHREM et al., 1980a; TANAKA et al., 1970). All are synthesized with homologous N-terminal extension peptides of around 30 residues which participate in the mitochondrial uptake of these cytoplasmically synthesized, nuclear-coded proteins (NONAKA et al., 1987; DuBOIS et al., 1981; KRAMER et al., 1982).

Comparisons among the primary structures of various cytochromes P-450 (MOROHASHI et al., 1984) reveal that the two mitochondrial forms are significantly different from known microsomal forms (about 20% homology), and from the bacterial cytochrome P-450$_{cam}$ (14% homology). The homology between cytochromes P-450$_{11\beta}$ and P-450$_{scc}$ is higher, 36% overall, with 4 highly conserved regions showing $58-70\%$ homology. One of these regions (C-4, near the C terminus) is the putative heme binding region which contains the cysteinyl 5th ligand to the heme. Another region, C-1, is homologous in both cytochromes to a prostatic steroid binding protein (GOTOH et al., 1985), and has therefore been suggested to function as a steroid recognition site in both proteins. Presumably, the other conserved regions may participate in the interactions with adrenodoxin or membrane components such as cardiolipin (see below), but as yet, there is no information on this point. Interestingly, there are no extended hydrophobic regions which might represent membrane attachment regions, suggesting a mode of membrane attachment which is more complex than that seen with, for example, cytochrome b_5 which has a single membrane-spanning anchor domain. Both cytochrome P-450$_{scc}$ and adrenodoxin reductase have trypsin sensitive sites (CHASHCHIN et al., 1984; WARBURTON and SEYBERT, 1988) which allow cleavage into segments of approximately equal size, and separate functional domains have been suggested. Crystal structures are not available for any of these proteins. The determination of the secondary and higher order structures of the cytochrome (as well as the electron transport proteins) and of the relationship of structure to function remain as challenging areas for future studies.

4. Membrane association and reconstitution of steroidogenic enzymes

Cytochrome P-450$_{scc}$ is considered an integral membrane protein, since membrane disruption by detergents is required for its solubilization. In contrast, adrenodoxin and most of the adrenodoxin reductase are released readily following sonication, suggesting either that they are matrix proteins or that they are weakly associated with the inner membrane. A possible membrane association of the reductase has been suggested by studies showing an ionic strength-sensitive association with artificial phospholipid vesicles (SEYBERT et al., 1979) and by the finding that approximately 1/3 of the reductase is retained with the membrane fraction of sonicated mitochondria. The membrane-associated form has been isolated, and was essentially identical to the soluble form in catalytic function, amino acid composition, and immuno-chemical properties, but differed in carbohydrate composition (HIWATASHI and ICHIKAWA, 1982). The significance of membrane association of part of the reductase is not clear, since the fully soluble system has more than ample catalytic activity to support mitochondrial hydroxylations.

Removal of cytochrome P-450$_{scc}$ from its membrane environment results in a significant loss in activity (TAKIKAWA et al., 1978) which is presumably due to several factors, including both the insolubility of its substrate, cholesterol, and aggregation of the enzyme. As with other purified membrane proteins in the absence of their membrane environments, cytochrome P-450$_{scc}$ is self-associating. It exists in aqueous solution as a large molecular weight aggregate of approximately 8 to 16 monomers (SHIKITA and HALL, 1973), and has a tendency to precipitate with prolonged storage at 4 °C. Nevertheless, if frozen rapidly in aliquots, the cytochrome is maintained in a quasi-soluble state which can be used for subsequent membrane reconstitution studies.

In the presence of a nonionic detergent, Tween 80, the apparent molecular weight of the cytochrome decreases to approximately 115,000, a size which is consistent with a monomer embedded in a detergent micelle (AKHREM et al., 1980b). With added detergent, activation is seen (TAKIKAWA et al., 1978; NAKAJIN et al., 1979) and kinetic parameters for some substrates are altered (LAMBETH et al., 1982). For example, the K_m for hydroxylated intermediates of cholesterol decreases with increasing detergent, suggesting that the aggregation state of the enzyme affects its function. The state of aggregation in the mitochondrial membrane is not clear. While the possibility of functional multimeric forms of the enzyme has received precedence from studies of microsomal cytochromes P-450 (DEAN and GRAY, 1982), the mitochondrial enzyme is maximally active under conditions where the monomer is predicted as the predominant form.

While reconstitution of the enzyme into a membrane-like environment (either phospholipid vesicles or detergent micelles) results in considerable activation, the magnitude depends upon both the method of reconstitution and on the composition of the lipophilic phase. Two methods of reconstitution of cytochrome P-450$_{scc}$ into phospholipid vesicles have been reported. A cholate dialysis method was used to reconstitute the enzyme into vesicles comprised of both natural and synthetic phospholipids, and a greater than 10-fold increase in activity was seen compared with the solubilized cytochrome (HALL et al., 1979). We have utilized an incorporation method (SEYBERT et al., 1979) in which the solubilized cytochrome is added directly to phospholipid vesicles which were preformed by sonication. The incorporation occurs rapidly and is accompanied by activation. The reconstituted enzyme is unidirectionally oriented in the membrane: At low ionic strength, adrenodoxin interacts strongly with the cytochrome to form a 1:1 complex. Titration of the reconstituted enzyme with the adrenodoxin reveals that all of the adrenodoxin binding sites are exposed to the external aqueous environment (SEYBERT et al., 1979).

The rate of insertion was estimated from studies monitoring the spin state of the cytochrome (SEYBERT et al., 1979; TUCKEY and KAMIN, 1982a). As the cytochrome becomes associated with the membrane, the cholesterol dilutes from its binding site into the lipid environment with a corresponding shift from the high- to the low-spin state. Using this method, the rate of incorporation was second order, and depended upon the concentrations of both the cytochrome and the phospholipid vesicles. In addition, vesicles size affected the incorporation rate, as did vesicle composition and fluidity. In vesicles made of dimyristoylphosphatidylcholine, the rate of insertion showed a break at the

J. D. LAMBETH

transition temperature, with more rapid insertion in the more fluid vesicles. Further, the rate of incorporation increased with increasing unsaturation of the lipid (which causes increased fluidity) and decreased with membranes containing a higher content of cholesterol or stearic acid (which causes decreased fluidity). This reconstitution method was most successful when the cytochrome was isolated rapidly from fresh tissue. For unknown reasons, the cytochrome prepared from frozen tissue or mitochondria appeared to be more aggregated as judged by its slight turbidity. In our experience, reconstitution of this material into phospholipid vesicles was sometimes unsuccessful. To overcome some of these problems, we developed another reconstitution method (PEMBER et al., 1983) in which the isolated cytochrome is added to detergent micelles or to mixed micelles prepared from detergent plus phospholipid. Enzyme prepared from both fresh and frozen tissue was readily incorporated, and we have used this system for many of our subsequent studies. The mixed micelle system also has advantages for spectroscopic studies, since the smaller particle size results in less turbidity.

The phospholipid vesicle-reconstituted form of the enzyme appears to be functionally and topologically indistinguishable from the enzyme in its native membrane. As described above, the enzyme regains activity upon reconstitution. Using electron spin resonance, the orientation of the enzyme's heme with respect to the plane of the membrane was studied in oriented membrane multilayers (KAMIN et al., 1985). The heme was found to lie more-or-less parallel to the membrane surface both in the native mitochondrial membranes and in the reconstituted material. In both native and reconstituted membranes the heme shows equal accessibility to water-soluble complexes of paramagnetic ions which perturb the ESR signal of the heme. Thus, by physical and kinetic criteria, the reconstituted enzyme resembles that in native membranes.

5. Membrane-dissolved cholesterol serves as substrate for the cytochrome P-450scc

Cholesterol binding occurs via a steroid interaction site which is in communication with the hydrophobic membrane milieu rather than the aqueous environment. Cholesterol itself is poorly soluble in aqueous solution, with a critical micellar concentration in the nanomolar range, and an absolute solubility limit of about 1 µM (TANFORD, 1980). However, it is highly soluble in a phospholipid membrane environment, where it can attain concentrations up to equimolar with phospholipid. Thus, in the presence of a phospholipid membrane, cholesterol partitions almost exclusively into the membrane. A membrane-faced substrate binding site is therefore undoubtedly functionally necessary for this system.

This orientation of the substrate-binding site is supported by several lines of evidence. First, the **apparent** K_d for cholesterol binding (expressed with the total **aqueous** volume as the denominator in the concentration term) increases with increasing membrane (phospholipid) concentration. However, when the K_d is expressed in terms of the content of cholesterol in the **membrane**[1], its value is invariant with increasing phospholipid, and is independent of the total aqueous volume in which the lipid is dispersed. Such behaviour is consistent with an enzyme which "senses" the concentration in the membrane rather than the aqueous environment.

In another approach (SEYBERT et al., 1979), the cytochrome was first reconstituted into preformed phospholipid vesicles, the membranes of which contained either radioactive or unlabelled cholesterol. After incorporation of the enzyme, each preparation was then mixed with the other (enzyme-free) vesicle preparation. After addition of the electron transport components and a reducing system, the appearance of radiolabelled pregnenolone was monitored. If cholesterol must first partition out of the membrane to interact with an aqueous-faced binding site, then both mixtures should show appreciable activity. However, we observed radioactive pregnenolone production only when the cytochrome and the radiolabelled substrate were present in the same membrane. Thus, the cytochrome must utilize cholesterol in its own membrane environment, consistent with a membrane-communicating sterol binding site.

The implication from these studies is that inner mitochondrial membrane cholesterol is the substrate pool which is utilized by the cytochrome (LAMBETH and STEVENS, 1985; SEYBERT et al., 1979). The content of cholesterol in the inner membrane of steroidogenic tissues is reportedly 2.6 mol % for beef adrenal (SUZUKI and KIMURA, 1965). The K_d for cholesterol binding to cytochrome P-450 in reconstituted systems was about 2 mol % in the reconstitution studies (TUCKEY and STEVENSON, 1985; LAMBETH et al., 1982). In recent studies (PRIVALLE et al., 1987), the cholesterol contents in isolated mitochondrial preparations were adjusted by in vivo pretreatment of rats in various ways (e.g., ACTH-injection, aminoglutethimide injection, etc.). The cholesterol content in separated inner membranes was then related to the spin-state of the cytochrome in intact mitochondria to obtain the K_d for cholesterol binding to the cytochrome. Using this method (which did not require membrane reconstitution), a K_d of 2.5 mol% cholesterol was obtained, which was similar to values seen in the reconstituted systems. Thus, K_d values in both reconstituted and intact systems are in agreement, and indicate that cytochrome P-450$_{scc}$ functions in vivo well below saturating levels of cholesterol. This finding is consistent with the emerging concept (see LAMBETH

---

[1] We find it most convenient to express the membrane cholesterol concentration in terms of mols of cholesterol per mol of phospholipid, rather than per volume of membrane, since the latter is difficult to ascertain in the essentially two-dimensional environment of the membrane.

and STEVENS, 1985) that cholesterol movement to the inner mitochondrial membrane wherein cytochrome P-450$_{scc}$ is embedded determines the overall steroidogenic activity of adrenal mitochondria.

6. Effect of phospholipid composition on the function of cytochrome P-450scc: cardiolipin as an endogenous effector lipid

In early studies, we noted that the side-chain cleavage activity of cytochrome P-450$_{scc}$ reconstituted into phospholipid vesicles was sensitive to the fatty acyl side-chain composition of the phospholipid vesicles (LAMBETH et al., 1980). The effect was due not to differential incorporation, but to a direct effect of the lipid on the function of the enzyme. The mechanism of activation was complex in nature, since activity did not correlate with known bulk phase physical properties of the lipid (e.g., fluidity, transition temperature, chain length, etc.), or with the ability of a particular phospholipid to associate with cholesterol. At a constant membrane content of cholesterol, the variation in activity with different lipids correlated positively with the % of cytochrome in the high spin form. Thus, the effect of phospholipids was on the binding of cholesterol, a finding which was confirmed by measurements of the K_m and K_d values for cholesterol, both of which were modulated by the phospholipid composition.

In studies in which the content of an activator lipid was varied (keeping the total phospholipid content constant), activation showed a sigmoidal concentration dependence with respect to the activating lipid (LAMBETH, 1981a). The data showed an excellent fit to a kinetic model in which two molecules of activator lipid were bound to effector sites on the enzyme, suggesting that two mols of phosphatidylcholine participated in activation.

This observation provided the first clue that the mitochondrial phospholipid cardiolipin could be the physiological effector lipid. We reasoned that the two bound activator phospholipids might be arranged adjacent to one another in such a way as to resemble a single molecule of cardiolipin (which is essentially a "double" phospholipid, as it contains four rather than two fatty acyl side-chains). When cardiolipin was tested in this system, it was found to be a considerably more potent activator than any of the other phosphatidylcholines tested (LAMBETH, 1981a). In subsequent studies using mitochondrial inner membrane lipids, exclusion of cardiolipin significantly reduced activity (TUCKEY and STEVENSON, 1985). In these studies, no other mitochondrial phospholipid was implicated as an effector lipid. The mechanism of activation was the same as that for phosphatidylcholines: an increase in the association of cholesterol with the enzyme, as judged by both spectral binding and K_m

measurements (LAMBETH, 1981a). The half-maximal stimulation occurred at about 5—10 mol% cardiolipin. In isolated inner membranes from steroidogenic tissue, the cardiolipin content is reported to be 18% for beef adrenal cortex (CHENG and KIMURA, 1983) or 9% for bovine corpus luteum (TUCKEY and STEVENSON, 1985). Thus, cytochrome P-450$_{scc}$ is likely to be nearly saturated with cardiolipin under physiological conditions. Since the cholesterol content in the inner membrane is low [reportedly 3 mol% (CHENG and KIMURA, 1983)], it is likely that cardiolipin functions physiologically in this cholesterol-deficient environment to enhance the substrate interaction with the enzyme.

Cardiolipin interacts with a low stoichiometry (probably 1 mol/mol) binding site on cytochrome P-450$_{scc}$ (PEMBER et al., 1983). We have studied the enzyme-cardiolipin interaction in the mixed micelle system consisting of Tween 80 plus phospholipid. Using a spin-labelled derivative of cardiolipin, the cytochrome at equimolar concentration caused pronounced immobilization of the lipid, consistent with binding. We have also demonstrated light-stimulated covalent cross-linking of a photoreactive cardiolipin derivative to the cytochrome (LAMBETH et al., 1981b). At low detergent, increasing contents of cardiolipin resulted in an increased fraction of cytochrome in the high-spin optical form, due to cardiolipin-enhanced cholesterol binding. Under these conditions, it was possible to titrate the enzyme with the lipid and an endpoint was seen at approximately 2 mols of cardiolipin per mol of heme. Taking into account the content of apoprotein, the calculated stoichiometry was 1:1. Cardiolipin binding was weaker than that seen with some other mitochondrial enzymes (e.g. cytochrome c oxidase) for which cardiolipin functions as an effector, since in preparations similar to ours there was no enrichment of cardiolipin (or any other lipid) compared with the total mitochondrial lipid composition. (For cytochrome oxidase and some other mitochondrial enzymes, cardiolipin is bound sufficiently tightly that it copurifies with the enzyme (VIK and CAPALDI, 1977; YU et al., 1975)). Thus, in its purified form, cytochrome P-450$_{scc}$ is relatively free of cardiolipin, a finding which may partially account for its low activity.

The properties of the cardiolipin binding site were further investigated (PEMBER et al., 1983) using several methods. First, using a series of phospholipids with different headgroups, the ability to activate was related to the structural resemblance to cardiolipin and did not relate to either bulk phase physical properties or net charge. For example, phosphatidylglycerol which resembles cardiolipin was a relatively effective activator, whereas phosphatidylinositol phosphate which has the same charge as cardiolipin was ineffective. In addition, the cardiolipin headgroup analog α-glycerol phosphate, (but not glycerol or inorganic phosphate alone), reversed the effector action of cardiolipin, presumably by binding to the headgroup binding site. Calculations of the magnitude of inhibition by the soluble headgroup analog revealed that most of the binding energy for cardiolipin is provided by the interaction of the enzyme with the headgroup of cardiolipin (presumably the phosphates). This result was further confirmed using mono- and di-lyso cardiolipin (in which either one or two of the fatty acyl side-chains are absent). The binding of these lipids to enzyme was essentially the same as that of native cardiolipin. A careful thermodynamic analysis of cardiolipin and adrenodoxin binding data revealed that although the binding of both cardiolipin and adrenodoxin to cytochrome P-450$_{scc}$ both enhance the cholesterol binding, they do so by different mechanisms. Thus, the binding of one does not directly influence the binding of the other. We have suggested that while adrenodoxin binding to the cytochrome stabilizes a general steroid binding conformation, the fatty acyl side-chains of cardiolipin participate directly in the cholesterol binding site. The latter proposal is supported by the finding that the cardiolipin-induced, cholesterol-dependent increase in the high-spin form of the cytochrome is diminished by removal of two of the four fatty acyl side-chains of cardiolipin. Also consistent with this interpretation is the finding that although adrenodoxin increases the binding of side-chain hydroxycholesterols, cardiolipin does not. Thus, the presence of side-chain hydroxyls on cholesterol is expected to disrupt VAN DER WAAL's interactions with the fatty acyl chains of the lipid, accounting for the lack of effect. The data, taken together suggest that the major interaction of the enzyme with cardiolipin occurs at the membrane surface where both the phospholipids headgroups and the 3β-hydroxyl of cholesterol are located (i.e., in the interfacial region between the aqueous-exposed and membrane-associated domains of the enzyme).

In summary, cardiolipin binds to cytochrome P-450$_{scc}$ with low stoichiometry and relatively high affinity and specificity. Binding activates side-chain cleavage by promoting the binding of cholesterol to its active site. The effect is likely to be physiologically relevant, based upon the concentrations of cholesterol and cardiolipin present in adrenal mitochondria.

7. Experimental considerations and methods for assessing the interaction of steroids with the substrate binding site on cytochrome P-450scc

Before considering the properties of the steroid binding site per se, it is necessary to discuss the various methods which may be used to monitor binding and differences in methodologies which have led to considerable diversity in the values of published binding parameters. Several special problems are encountered with regard to binding studies of cytochrome P-450$_{scc}$. The first has to do with the aqueous-insolubility of the substrate, cholesterol, and its analogs, alluded to above. Aggregation and precipitation of substrate at concentrations above nanomolar render such methods as equilibrium dialysis difficult to quantitate, although it has been possible to determine binding stoichiometry with this method (ORME-JOHNSON et al., 1979). In addition, when isolated enzyme is used in the absence of a membrane environment, this method would detect not only binding to the true binding site, but also interactions with hydrophobic membrane-association regions which may have become exposed upon solubilization of the enzyme from its membrane. An alternative method, which we have used extensively, is to monitor optically the perturbation in the spin-state of the heme upon substrate binding. Since this method detects only substrate which is in close proximity to the heme, it is insensitive to interactions with artifactually created hydrophobic sites. Its disadvantage is that it may not detect steroid binding to possible allosteric sites. Despite this theoretical shortcoming, detection of spin-state changes using UV-visible absorbance spectroscopy is quite sensitive when the Soret peak of the heme is monitored.

To quantify binding, an entire binding curve must be obtained using sterol concentrations which span the K_d value. Many substrates differ in their binding-induced extinction coefficients, so that quantifying a sterol-induced absorbance change using a single sterol concentration is not sufficient to allow estimation of binding strength. For example, although cholesterol and 20(α),22(R)-dihydroxycholesterol convert the enzyme almost entirely to the high-spin form, 25-hydroxycholesterol produces only about 1/3 the extinction change, and 20(α)- and 22(R)-hydroxycholesterol are low-spin-inducing sterols (LAMBETH et al., 1982). Binding of the latter, however, can be readily monitored, since it produces a 6 nm red-shift in the low-spin peak (HEYL et al., 1986). When 20(α)-hydroxycholesterol is bound, however, the low-spin spectrum is indistinguishable from that of substrate-free cytochrome. For this substrate, it has therefore been necessary to extrapolate the binding constant from a series of apparent binding constants determined in the presence of varying concentrations of a high-spin-inducing substrate such as cholesterol (LAMBETH et al., 1982). Competition for binding reverts the spin-state towards low-spin, and an apparent K_d can be derived. Extrapolation of apparent K_d values to zero cholesterol then reveals the true K_d. Thus, spectral dissociation constants can be determined for both low- and high-spin inducing substrates.

Another consideration in binding studies is the choice of an experimental system. Binding and activity measurements have been carried out using whole and sonicated mitochondria, acetone powders, and purified enzyme, the latter either alone or reconstituted into a membrane-like environment. In addition, published methods of substrate delivery differ considerably. In some cases, cholesterol is added from an organic solvent such as acetone or ethanol, while in others, it is dispersed in and added from a detergent solution. In others, the cholesterol is pre-incorporated into phospholipid vesicles or mixed micelles, and the cytochrome is then reconstituted into these preparations. Since the lipid content and experimental conditions vary considerably, it comes as no surprise that reported binding constants and K_m values differ greatly.

Considerable care must be exercised in the choice of an appropriate experimental system which detects the true sterol binding characteristics of the enzyme rather than parameters which are dominated by other properties of

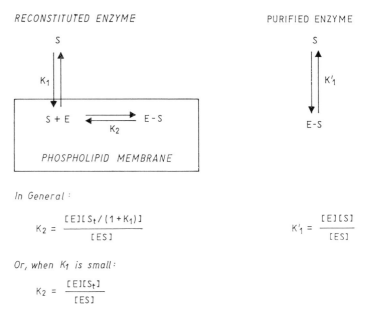

RECONSTITUTED ENZYME

PURIFIED ENZYME

In General :

$$K_2 = \frac{[E][S_t/(1+K_1)]}{[ES]}$$

$$K'_1 = \frac{[E][S]}{[ES]}$$

Or, when K_1 is small :

$$K_2 = \frac{[E][S_t]}{[ES]}$$

Fig. 3. Sterol binding equilibria in the presence or absence of a phospholipid membrane. The scheme at left shows the equilibria and binding equation for interaction of steroid with enzyme when a phospholipid membrane is present. That at the right shows the binding equilibrium when a membrane phase is absent.

the system, such as the aqueous versus hydrophobic solubility of the sterols. This is made evident by a consideration of sterol binding either in the presence or absence of a phospholipid membrane, as diagrammed in Figure 3. In the case of the enzyme in its native mitochondrial membrane, or enzyme reconstituted into phospholipid vesicles, the binding is dictated by the equilibria at the left. Partition of the substrate into the hydrophobic environment is governed by the constant K_1, which is defined as the equilibrium concentration of steroid in aqueous solution divided by that dissolved in the membrane. As such, K_1 is governed by the relative solubilities of a given substrate in the two environments. Because binding occurs from the hydrophobic membrane phase, the interaction of the membrane-dissolved substrate with the enzyme is dictated by a second constant, K_2, which is simply the equlibrium constant for binding from the membrane phase. Overall, binding is dictated by the equation shown at left in Fig. 3, which contains not only the K_2 term, but also a K_1 term. However, when the substrate partitions largely into the hydrophobic membrane as does cholesterol and most other substrates of the cytochrome, K_1 is a very small number and the equation simplifies to that shown at the bottom left in Fig. 1. Thus, in a native or reconstituted system, binding is dictated largely by the relative partitioning of membrane-associated substrate between the membrane and the active site and partitioning into the aqueous environment can be neglected. (As described above, the steroid concentration and binding constants must be expressed in terms of a cholesterol-to-phospholipid molar ratio, rather than cholesterol in the total aqueous environment.)

In contrast, when no membrane is present, as is the case in acetone powders or using the purified enzyme alone, the binding is dominated by the relative partitioning between the aqueous environment and the enzyme active site, as described by the diagram and equation for K_1' on the right side of Fig. 3. Since the steroid binding site is hydrophobic, K_1' is expected to be similar to K_1, and is dominated by the relative hydrophilic versus lipophilic properties of the substrate. Thus, experiments carried out under these two conditions yield considerably different binding constants. For example, for 22(R)- versus 22(S)-hydroxycholesterol, when no membrane phase is present, both bind tightly to the enzyme active site (ORME-JOHNSON et al., 1979). However, when these substrates are tested in a membrane-reconstituted system, the active site discriminates by a factor of 250 (HEYL et al., 1986). Thus, in the former case, the hydrophobic effect dominates, driving the substrate binding, while in the latter case, additional features of the binding site become apparent. It is possible that in the absence of a membrane-like environment, such hydrophobically-driven binding may produce artifactual orientations of the substrate. The diagram in Fig. 3 also serves as a caution against addition of the detergent-dissolved substrate to the isolated enzyme, since in this case not only the substrate concentration but also the concentration of the membrane-like environment will vary, resulting in a hybrid binding situation which is a mixture of both of the limiting cases shown in Figure 3.

An additional consideration in any purified system is the removal of one or more effectors during the isolation. As was mentioned above, cardiolipin appears to participate as a specific effector of cholesterol binding. As such, we now standardly include $15-20 \, \mathrm{mol}\%$ cardiolipin in phospholipid vesicle and detergent mixed micelle system for binding studies. Inclusion of this lipid has the additional advantage of promoting optical clarity in both systems.

8.   The steroid binding site of cytochrome P-450scc

8.1.   General considerations

The reader is referred to a recent review (LAMBETH, 1986) of the substrate specificity of cytochrome P-450$_{scc}$ and of adrenal mitochondria for additional detail and historical perspectives not provided herein. The present review will summarize some of the pertinent literature, with a focus primarily on some recent studies which explore the binding site using analogs, inhibitors and physical methods. Following a discussion of some general features of the binding sites in cytochromes P-450, specific regions of the cholesterol molecule will be considered individually. The structure and numbering of the cholesterol molecule is shown for reference in Figure 4.

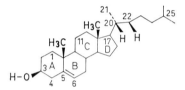

Fig. 4. Cholesterol structure with relevant positions numbered.

The most detailed knowledge of substrate binding to a cytochrome P-450 has been obtained from the crystal structure of the bacterial enzyme which catalyzes the hydroxylation of camphor (POULOS et al., 1985). Although it is uncertain how closely this soluble enzyme resembles mammalian membrane-bound cytochromes (there is very little sequence homology except in the heme binding domain), this structure can serve as a starting point for comparisons. In the bacterial enzyme, the heme and the camphor binding site are buried deeply at approximately the center of the protein structure. There is no obvious channel or cleft by which the substrate might gain access to its binding site, and significant conformational changes (e.g., the opening of protein "flaps") are required to account for substrate access. Using proton NMR, solvent protons can approach to within 2.5 Å of the heme iron. In the presence of substrate, solvent protons are excluded from the active site of cytochrome P-450$_{cam}$.

Proton NMR studies (JACOBS et al., 1987) of cytochrome P-450$_{scc}$ in the substrate-free form showed that like cytochrome P-450$_{cam}$, solvent protons have a close approach to the heme iron (about 2.5 Å). When cholesterol is bound, the distance is shifted to about 4 Å, indicating that although there is some obstruction by substrate, solvent continued to have access to the active site. For bound hydroxysterol intermediates, the solvent protons can again approach within 2.5 Å of the heme. Thus, solvent has considerably more access to the active site of cytochrome P-450$_{scc}$ than to that of cytochrome P-450$_{cam}$. Such access may be important to supply the protons required for the multiple oxidations catalyzed by the side-chain cleavage enzyme.

8.2. The 3β-hydroxyl group of cholesterol

The 3β-hydroxyl group of cholesterol (see Fig. 4) is critical for the initial interaction of cholesterol with cytochrome P-450$_{scc}$. Using various 3β-modified cholesterols as competitive inhibitors of [^{3}H]-cholesterol metabolism in a variety of experimental settings, native cholesterol is a better competitive inhibitor than either epicholesterol (3α-OH-cholesterol), 3-ketocholesterol, or 3β-fluorocholesterol (ARINGER et al., 1979; DEGENHART et al., 1984; BURSTEIN et al., 1976; KOBAYASHI and ICHII, 1969; RAGGATT and WHITEHOUSE, 1966). In the purified, micelle-reconstituted system (HEYL et al., 1986), binding was observed for native cholesterol, but could not be detected using 3-ketocholesterol, epicholesterol, 5-cholestene (no 3-substituent), or several 3β-substituted analogs. This was not due to a failure of the analogs to produce the high-spin state, since when cholesterol was included near its K$_d$ to partially convert the enzyme to the high-spin state there was no further spectral perturbation by any of the analogs. Thus, prior to the first hydroxylation, the 3β-hydroxyl is undoubtedly important for the initial enzyme-substrate interaction.

However, once the 22(R) position has been hydroxylated (see Fig. 1), the steroid becomes considerably more tightly bound to the enzyme (HEYL et al., 1986; LAMBETH et al., 1982) (see also below), and the 3β-hydroxyl becomes less dominant in the overall energetics of binding. This is supported by recent studies (DEGENHART et al., 1984) which show that while mitochondria failed to metabolize 4-cholestene-3-one, side-chain cleavage of the 3-keto steroid occurred readily when a 22(R)-hydroxyl was introduced.

The enzyme interaction with the 3β group occurs at the oxygen rather than the hydrogen of the hydroxyl, presumably via hydrogen bonding of an amino acid residue to the 3β-oxygen. Evidence for this conclusion comes from binding studies using 3β-ethers of cholesterol in the purified, micelle-reconstituted system (HEYL et al., 1986). The 3-methyl and ethyl ethers bind relatively strongly to the enzyme — in fact, about twice as strongly as cholesterol — indicating that the 3β-hydrogen is not crucial for binding. The available space beyond the 3β-hydroxyl is limited to about the size of a two carbon chain since ethers of 3 or more carbons do not bind. In contrast to the ethers, cholesterol esters did not bind strongly. Only cholesterol formate was observed to bind, and this interaction was considerably weaker than with cholesterol. Likewise, thiocholesterol (which is a poor acceptor for hydrogen bonding) failed to bind. In general, the binding correlated with the electron density at the 3β oxygen, consistent with hydrogen bonding occurring at this position.

J. D. LAMBETH

The above conclusions are in apparent contrast to reports that in mitochondrial acetone powders, cholesterol sulfate and cholesterol acetate undergo side-chain cleavage at rates comparable to cholesterol (ROBERTS et al., 1967; RAGGATT and WHITEHOUSE, 1966; WOLFSON and LIEBERMAN, 1979). Cleavage of cholesterol sulfate has also been noted in apparently homogenous preparations of isolated cytochrome P-450 (GREENFIELD and PARSONS, 1986; GREENFIELD et al., 1985), although experimental conditions in both systems may have emphasized artifactual enzyme-substrate interactions, as described above. Nevertheless, it is clear that cholesterol sulfate added to intact or sonicated mitochondria can be converted directly to pregnenolone sulfate. It is possible that either a weak interaction with the enzyme which is not detectable by spectral methods is sufficient to allow an appreciable cleavage, or that other factors may modulate the enzyme substrate interaction. LIEBERMAN and colleagues (WOLFSON and LIEBERMAN, 1979; GASPARINI et al., 1979) provided kinetic, inhibition, and stability data consistent with the presence of two side-chain cleavage systems, one of which was specific for cholesterol and one for its esters. Whereas both cholesterol and cholesterol sulfate were cleaved using a partially purified enzyme preparation, when both substrates were present at saturating concentrations their conversion was additive (YOUNG and HALL, 1969). Both of these approaches are consistent with two types of catalytic units with different specificities. It is not clear at this time whether there are distinct enzymes arising as different gene products, different post-translational modifications, or associated specificity modulation factors. It should be noted in this context that several forms of cytochrome P-450$_{scc}$ were recently isolated from bovine adrenal mitochondria (TSUBAKI et al., 1987c). Nevertheless, only a single gene which codes for the cytochrome has been detected (MILLER, 1987).

Substituents at the 3β position affect electronic properties of the heme iron such as its oxidation-reduction potential (HEYL et al., 1986). Data indicated that the strength of interaction with the 3β position increases upon reduction of the enzyme. As will be described below, the 22 position is located close to the heme iron. Using this as an "anchor point" in model heme-cholesterol systems, it is not possible to orient the cholesterol so that the 3β position comes nearer than 13 Å from the heme iron. As such, it cannot interact directly with any portion of the heme itself. Thus, the mutual interaction of the heme iron and the 3β-hydroxyl must be transmitted through the protein structure.

8.3. The ring system of cholesterol

Studies on the specificity of the side-chain cleavage enzyme for the ring system have apparently been limited by the paucity of ring-modified cholesterol analogs which would allow such studies. From available data, it seems clear that the binding site has a stringent requirement for the structure near the 5—6 double bond. Upon reduction of this bond, the resulting cholestanol is not a substrate (ARINGER et al., 1979; MORISAKI et al., 1982; KOBAYASHI and ICHII, 1969; TROUT and ARNETT, 1971). Using 22-amino derivatives of cholesterol and cholestanol, the absence of the Δ-5,6 double bond elevates K_i and K_d values by at least two orders of magnitude (SHEETS and VICKERY, 1983b). However, under some circumstances, Δ-4,5 cholesterol analogs are accepted as substrates (DEGENHART et al., 1984). Thus, the most important structural feature of this region is probably the planarity of the A-B ring juncture (at the 5 position).

Other ring-system regions for which modification results in decreased binding are the 4 and 17 positions of the A and D rings, respectively. Methyl groups at the 4 position diminish the ability of analogs to function as competitive inhibitors (KOBAYASHI and ICHII, 1969), as does the presence of a 17 hydroxyl group (BURSTEIN et al., 1976). Thus, insofar as has been tested, the binding site of cytochrome P-450$_{scc}$ exhibits a high degree of structural specificity for the ring system of cholesterol.

8.4. The side-chain of cholesterol

With the exception of the 20—22 region, cytochrome P-450$_{scc}$ does not have a great deal of specificity for the isocaproic side chain of cholesterol. In mitochondrial acetone powders and in isolated mitochondria, cholesterol derivatives with various side-chain lengths were converted equally well to pregnenolone (ARTHUR et al., 1976; ARINGER et al., 1979). In a purified enzyme system, a modest increase in side-chain cleavage activity was seen with increasing chain length up to the normal chain length of 27 carbons, with decreasing activity to 29 carbons and no activity using longer chain-length derivatives (MORISAKI et al., 1980). Thus, interactions with the carbons in the side-chain appear to provide a modest binding energy, but the binding cleft is limited in size to slightly longer than the native side-chain.

The side-chain region can accommodate a fair amount of width, since side-chain methyl and ethyl derivatives branched at the 24 position are readily bound (Morisaki et al., 1980). In addition, derivatives in which the side-chain is replaced by a series of 20-phenyl substituents bind tightly to the enzyme, and are good competitive inhibitors (UZGIRIS et al., 1977; VICKERY and KELLIS, 1983; MORISAKI et al., 1976).

The absence of strict structural specificity for the side-chain has allowed the use of a 25-spin labelled analog of cholesterol (25-doxyl-27-norcholesterol) to detect enzyme-substrate interactions (LANGE et al., 1988). Binding perturbed the low-temperature low-spin ESR spectrum of the cytochrome, and from this the authors estimated that the doxyl ring was approximately 6 to 10 Å from the heme iron, consistent with a close approach of the 22 position to the iron, as is described below.

In contrast to the more haphazard binding of the remainder of the side-chain, there is a high degree of structural specificity near the 22 position (HEYL et al., 1986). 22(R)-Hydroxycholesterol (the normal isomer which is produced by the first hydroxylation) binds 20-fold more tightly to the enzyme than cholesterol. In contrast, 22(S)-hydroxycholesterol binds more than 10-fold more weakly than cholesterol. Thus, the stereochemical orientation of the 22-hydroxyl results in a more than 200-fold binding differential.[1] Data are consistent with a steric exclusion of the 22(S)-hydroxyl from this region of the binding site.

The presence of a side-chain beyond the 22 position is important for the activity and specificity of the enzyme. Cholesterol analogs which terminate at the 22 position with either a methyl or an alcohol are not converted in appreciable quantities to pregnenolone (MORISAKI et al., 1980; SHEETS and VICKERY, 1983b; TAIT, 1972). Rather, hydroxylation occurs at the 20 or at the 21 position (TAIT, 1972). It is not difficult to envision that without the bulky side-chain, the 21-carbon rotates freely around the 17—20 bond to occupy the same position near the iron-bound activated oxygen which is normally occupied by the 22 carbon, thus allowing hydroxylation at this position.

As mentioned above, hydroxylation at the 22(R)-position causes a 20-fold enhancement in binding of the steroid to the cytochrome (HEYL et al., 1986). The presence of a 20-hydroxyl group confers the same degree of enhanced binding, and the 20,22-dihydroxycholesterol is also tightly bound. The 25-hydroxyl, in contrast, causes only a 2-fold binding enhancement. These data suggest a specific interaction of enzyme with the newly formed 22 and/or 20 hydroxyls. That this interaction is with a site other than the heme iron (i.e., probably a hydrogen bonding amino acid residue) is indicated by the fact that when oxygen is bound to the heme, the same structural specificity for the 22(R)-hydroxyl persists (HEYL et al., 1986).

A study in which the binding of the side-chain hydroxycholesterols was compared in the oxidized, reduced, and oxygen-bound forms of the enzyme has revealed an additional feature of the binding interaction (HEYL et al., 1986). Whereas the oxidized enzyme interacts strongly with analogs containing either the 22(R)- or the 20(α)-hydroxyls (or both), the reduced form of the

---

[1] The binding enhancement with 20α- and 22R-hydroxycholesterols as well as the binding differential between 22R- and 22S-hydroxysteroid was not apparent in other studies (ORME-JOHNSON et al., 1979; TSUBAKI et al., 1987b) in which the purified enzyme was used in the absence of a micellar or membrane environment. We believe that this apparent discrepancy is due to the dominating hydrophobic binding effect which drives the substrate interactions in these studies, as detailed in Section 7 and Figure 3.

enzyme shows enhanced specificity only for the 22(R)-hydroxycholesterol. Thus, the substrate binding site exhibits a degree of "conformational flexibility", depending upon the reduction state of the heme iron. Such altered specificity may occur either by a movement of the bonding residue away from the 20 position, or by a small displacement of the steroid within its binding cleft.

8.5. Proximity of side-chain positions to the heme iron

Because the normal substrate, cholesterol is hydroxylated in the 22(R)- and 20(α) positions, it is reasonable to assume that these positions are in close proximity to the heme iron to which the active oxygen is bound. Unlike with some other cytochromes P-450, the position of hydroxylation is independent of the source of the active oxygen, since 20(α)-hydroperoxycholesterol (which donates oxygen in the appropriate valence for hydroxylation and is thus independent of electron transfer and oxygen binding steps) supports a single hydroxylation in the 22(R) position (LARROQUE and VAN LIER, 1986). Some current mechanisms propose that an oxygen atom bound to the heme iron participates in the abstraction of a hydrogen atom from the substrate prior to oxygenation, thus placing these positions within 2 or 3 Å of the iron.

Direct evidence for this arrangement comes from several studies. One approach has utilized a series of amino-containing substrate analogs. Amines have high affinity as heme ligands, and produce complexes with characteristic absorbance and ESR features. VICKERY and colleagues (SHEETS and VICKERY, 1982; SHEETS and VICKERY, 1983b; KELLIS et al., 1984; SHEETS and VICKERY, 1983a) utilized a series of 3β-hydroxy-Δ5,6 analogs in which the sterol structure terminates with an amino group at the 17α-, 20, 22, 23, or 24 positions. The 22- and 23-amino analogs were very potent inhibitors of side-chain cleavage and produced characteristic optical changes in the heme spectrum which were indicative of direct ligation of the amino group to the heme iron. In some studies, the 23-amine was slightly more potent than the 22-amine. Such an effect is not unexpected, since in the absence of an extended side-chain, the 23-carbon should have considerably more rotational freedom and might be expected to move so as to allow close interaction of the amino group with the heme iron. Using the physiological substrate, the 23-position is not hydroxylated, so that the positioning of the remainder of the side-chain probably normally restricts the access to this position. Two groups (BURSTEIN et al., 1976; NAGAHISA et al., 1985) utilized amino groups at the 22-position of the intact cholesterol structure (i.e., with the complete side-chain) to study the stereochemical requirements at this position. The 22(R)-amino derivative bound 3 orders of magnitude more tightly than did the 22(S)-amino sterol, and the former produced a spectrum indicative of direct heme ligation. These studies are consistent with our observations using the 22(S)- versus 22(R)-hydroxycholesterols, and support a steric hindrance of bulky groups in the 22(S) position. Interestingly, the 22(R)-aminocholesterol (intact side-chain) bound less tightly than did the derivative which terminated in the 22-amine. Thus, the side-chain positions this group so as to restrain somewhat its access to the heme. Although indicative of close proximity to the heme iron, amino derivatives of cholesterol may not provide accurate positional distance information, since the very strong ligation of the amine probably distorts or perturbs slightly the normal binding configuration.

Other approaches have provided additional direct information as to the distances from the heme iron to various steroid positions. ORME-JOHNSON and co-workers (GROH et al., 1983) utilized 22-deuterated derivatives in spin echo studies to reveal that the 22(S)-deuterium of 22(S)-D,22(R)-hydroxy-cholesterol is located approximately 4 Å from the heme iron. Although these studies failed to detect evidence of spin modulation from the unnatural isomer [22(R)-D,22(S)-hydroxy-cholesterol], we suggest that this may have been due to the poor fit of the 22(S)-hydroxysterol in the active site, as described above.

Using another approach (HEYL et al., 1986), we compared the binding of a series of hydroxycholesterol analogs to the oxidized, reduced, oxygen-bound, and carbon monoxide-liganded forms of the enzyme. The presence of carbon monoxide, but not oxygen, interferes with the binding of analogs containing a 22(R)-hydroxyl group. Calculations reveal that the interference produces a greater than 100-fold effect on the binding constant. In addition, there is a smaller degree of interference of carbon monoxide with the binding of cholesterols with a 20(α)-hydroxyl group. This effect is on the order of 7-fold. Data can be explained by the model shown in Figure 5. Because oxygen is known to bind in a bent conformation, whereas carbon monoxide binds in a linear orientation, the 22(R)-hydroxyl group can be accommodated into the active site when oxygen is bound (upper panel of Fig. 5.), but occupies the same space as the terminal oxygen of carbon monoxide when CO is bound to the heme (lower panel). According to this model, the 22(R)-hydroxyl is located between 2 and 3 Å from the heme iron, along a line approximately perpendicular to the plane of the heme ring system. The 22(R)-hydrogen is located slightly further, approximately 2.5 to 3.5 Å, in good agreement with the spin echo studies which place the 22(S)-hydrogen slightly further from the iron. Such a location positions the 22(R)-hydrogen at an optimal distance for a stereospecific hydrogen abstraction by an active oxygen bound to the heme iron and for subsequent oxygen addition. Such a geometry thus accounts for the initial hydroxylation at this position. The model requires that the side-chain of cholesterol be rotated at the 20—22 bond as shown in Fig. 4., so as to place the 22(R)-hydrogen adjacent to the 22(S)-hydrogen. Such an arrangement allows a second hydroxylation to occur at the adjacent position with minimal conformational readjustment. Although other configurations are possible, the model is best accommodated if the sterol rings are drawn as lying more or less parallel to the heme, since such an orientation allows an optimal approach of both the 20 and 22 positions, and better accommodates

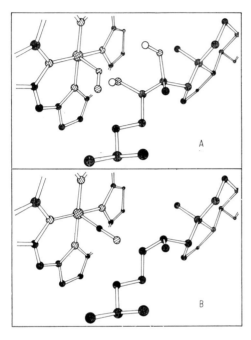

Fig. 5. A model for the orientation and proximity of the side-chain of cholesterol and analogs with respect to the heme. The orientation of the side-chain region of 20,22-dihydroxycholesterol (right side of panel) with respect to the heme of cytochrome P-450$_{\text{scc}}$ (upper left of panel) is shown in panel A. Oxygen is depicted as bound to the heme iron in its bent conformation (120° angle), and does not interfere with the 22(R) hydroxyl. In panel B, the same view is shown, except that the bound steroid is cholesterol, and carbon monoxide is depicted as bound to the heme iron in its linear conformation (180° angle). The oxygen of carbon monoxide occupies the same position as the oxygen of the 22R-hydroxyl group in panel A. Reproduce from (HEYL et. al. 1986) with permission from the publisher.

the sterol side-chain than the more perpendicular model which was suggested based on studies using amino cholesterols (SHEETS and VICKERY, 1983a; NAGAHISA et al., 1985).

Recent studies have provided additional verification of the above model (see Fig. 5.). Bound cholesterol increases the photodissociability of heme-bound carbon monoxide (MITANI et al., 1988), and it was concluded that the steroid interfered sterically with heme-bound carbon monoxide. Resonance Raman spectroscopy has been used to study the C−O stretch frequency of heme-associated carbon monoxide and the ability of various steroids to perturb the carbon monoxide binding conformation (TSUBAKI et al., 1987b). Consistent with the binding studies, the 22(R)-hydroxycholesterol failed to produce any C−O stretch frequency at all, due to the marked instability and photodissociability of the CO complex. Cholesterol itself produced some perturbation, indicating some steric interference. The data were interpreted as indicating a strong steric interaction between CO and substituents at the 22(R)-position. The authors also provided evidence for a significant con-

J. D. LAMBETH

formational change in the active site induced by the binding of 20(α)-hydroxy-steroids. The same group also used ESR to investigate the effects of cholesterol analogs on the nitroxide-liganded form of the cytochrome (TSUBAKI et al., 1987a). Similar conclusions were obtained, including the observation of significant conformational flexibility in the active site. Thus, several physical techniques provide data which are consistent with the geometrical model shown in Fig. 5., and predict the initial hydroxylation at this position.

8.6. Implications of the properties of the cholesterol binding site for the side-chain cleavage mechanism

The structural features of the steroid binding and active site described in the above sections have allowed us to propose (LAMBETH, 1986) a reasonable scenario for the sequence of oxidations catalyzed by cytochrome P-450$_{scc}$. The specific interactions of enzyme residues with the 3β- and 22-positions are suggested to have important roles in substrate positioning. Although hypothetical, the following sequence of events is consistent with the data:

1. Interaction of an amino acid residue in the substrate binding site with the oxygen of the 3β-hydroxyl provides much of the initial binding energy for the enzyme-substrate interaction. This interaction positions the steroid in a binding cleft with the 22(R)-hydrogen close to the heme iron, so that once generated, the active oxygen is expected to hydroxylate stereo-selectively at the 22(R) position.

2. The 22(R)-hydroxyl group, once formed, interacts strongly with a second amino acid residue in the active site very near the heme iron. This inter-action helps to prevent the dissociation of hydroxysteroid intermediates, and thus allows the hydroxylation/cleavage sequence to proceed to com-pletion. This strong interaction probably provides sufficient energy to shift the steroid position slightly along its binding cleft relative to the heme iron, as is consistent with the reported conformational flexibility of the active site. This shift allows a second heme-bound active oxygen to hydroxylate specifically at the 20(α) position. It should be noted that if the side-chain is rotated about the 20—22 carbon-carbon bond, as is depicted in Fig. 4., a very small binding adjustment will account for the observed change in specificity.

3. According to this model, the newly formed 20(α)- and 22(R)-hydroxyl groups are now both in close proximity to the heme iron, and more to the point, to the heme-bound active oxygen. The 20 position has no ab-stractable hydrogens and steric and other evidence argues against a further hydrogen abstraction or hydroxylation at the 22(S) position. A more likely mechanism is that the active oxygen, located between the 20 and 22 hydro-

xyls, acts directly on these positions to abstract hydrogen atom(s) and/or electrons, so as to allow cleavage of the 20—22 carbon-carbon bond. A plausible mechanism involving hydrogen abstraction (but by no means the only possible mechanism) is depicted in Figure 6.

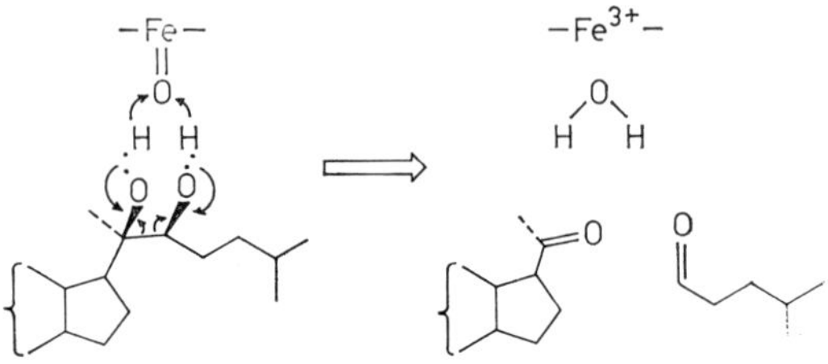

Fig. 6. Possible mechanism for oxidative cleavage of the side-chain of cholesterol. Following hydroxylation at the 22(R) and 20(α) positions to yield the dihydroxycholesterol, a third oxidative sequence again results in the generation of active oxygen bound to the heme. The orientation of the active oxygen is depicted as adjacent to one or both hydroxyl groups from which a hydrogen atom (or an electron) is proposed to be abstracted, resulting in the electronic rearrangements shown, which result in cleavage of the side-chain and generation of a water molecule at the active site. Such a mechanism would allow both hydroxylations and the cleavage reaction with minimal conformational readjustment in either the active site or in the bound steroid.

9. Oxygen binding to cytochrome P-450scc

The ferrous heme of cytochrome P-450$_{scc}$ binds ligands such as carbon monoxide and oxygen. The K_m for oxygen, using membrane-reconstituted cytochrome P-450$_{scc}$, decreases with increasing contents of cholesterol in the membrane (STEVENS et al., 1984). At saturating cholesterol, the K_m is 4 μM, but is significantly higher than this value in isolated mitochondria, consistent with the reported low content of inner membrane cholesterol in adrenal mitochondria.

The complex with oxygen can be detected spectrophotometrically. While it is labile at temperatures above 4 °C, it can be stabilized and observed at subzero temperatures (LARROQUE and VAN LIER, 1980). The effect of cholesterol and cholesterol analogs on the binding and stability of the oxy-ferrous complex has been studied (TUCKEY and KAMIN, 1982b; TUCKEY and KAMIN, 1983). Cholesterol extended the half-time for autoxidation by a factor of 15, and hydroxycholesterols further extended this half-time by factors of 4 and 14, for 22(R)- and 20(α),22(R)-dihydroxycholesterol, respectively. The increasing stabilization was also reflected in the K_d value for oxygen binding. Thus, as the cytochrome proceeds through its oxidative sequence (Fig. 1.), not only do the steroid intermediates become more tightly bound, but also the oxy-ferrous complex becomes progressively stabilized. This mechanism undoubtedly contributes to protecting the system against nonproductive autoxidation and favors commitment to complete conversion of cholesterol through its three oxidative steps to form pregnenolone.

10. Electron transfer to cytochrome P-450scc

As described above and diagrammed in Fig. 3., the electron transfer system for cytochrome P-450$_{scc}$ consists of an FAD-containing protein (NADPH-adrenodoxin reductase) and an iron-sulfur protein (adrenodoxin). Early studies of the electron transfer mechanism utilized cytochrome c and ferricyanide as model electron acceptors, and are summarized in greater detail in an earlier review (LAMBETH et al., 1982a) to which the reader is referred for additional information. The present review will summarize briefly some of the conclusions of these earlier studies, in particular as they pertain to electron transfer to cytochrome P-450$_{scc}$.

10.1. Interaction of pyridine nucleotide with adrenodoxin reductase

Adrenodoxin reductase is highly specific for NADPH compared with NADH, with respective K_m values of about 2 µM and 6 mM (CHU and KIMURA, 1973a). In early studies (LAMBETH and KAMIN, 1976), we demonstrated the formation of a stable 1:1 complex upon anaerobic titration of the flavoprotein with NADPH, but not NADH. The species, characterized by long wavelength absorbance beyond 750 nm, was most likely a charge transfer complex, probably between oxidized pyridine nucleotide and reduced flavin. Stopped flow studies revealed that formation of this complex occurs with a first order rate constant of $18 \, s^{-1}$, but is preceded by the very rapid formation of another species, undoubtedly the reduced pyridine nucleotide-oxidized flavoprotein complex. Consistent with the proposed assignments, a deuterium isotope effect of up to 7.7 was demonstrated for formation of the second species (LIGHT and WALSH, 1980), indicating that the hydride transfer (reduction) step coincides with its formation.

Studies of the oxidation-reduction potential of the flavoprotein in the presence and absence of $NADP^+$ revealed that the reduced cytochrome binds $NADP^+$ two orders of magnitude more tightly than does the oxidized (LAMBETH and KAMIN, 1976). The result is that this pyridine nucleotide remains tightly bound to the complex until the flavin has given up one or more electrons to adrenodoxin or alternative electron acceptors. It was later demonstrated (SUGIYAMA et al., 1979) that the spontaneous oxidase activity of the flavoprotein was considerably lower when NADPH rather than NADH was used as the electron donor, suggesting that one role of the complex may be to limit the air oxidation of the reduced flavoprotein.

The tight binding of pyridine nucleotide is also seen when NADPH rather than NADP⁺ is added to the reduced enzyme (LAMBETH and KAMIN, 1976), indicating that it is the redox state of the flavoprotein rather than the pyridine nucleotide which dictates the strength of binding. The NADPH can exchange slowly with bound NADP⁺ on the reduced enzyme, with a rate constant of about 1 s^{-1}, too slow to be catalytically competent. Thus, partial or complete reoxidation of the flavoprotein must precede loss of the oxidized pyridine nucleotide during catalysis. A complex of pyridine nucleotide with the semiquinone form of the flavoprotein has also been observed (NONAKA et al., 1986), and data suggest that the binding is similar to that seen in the oxidized flavoprotein. Thus, reoxidation to the semiquinone state may then allow dissociation of the NADP⁺ with subsequent binding of a second molecule of NADPH. This scenario is consistent with data (LAMBETH, 1982b; LAMBETH and KAMIN, 1979) which implicate the bound NADPH as an effector which facilitates electron transfer from the semiquinone flavin to an acceptor such as adrenodoxin.

10.2. The complex between adrenodoxin reductase and adrenodoxin

At low ionic strength, adrenodoxin forms a rather tight ($K_d \sim 10^{-8} \text{ M}$) complex with adrenodoxin reductase (CHU and KIMURA, 1973b; LAMBETH et al., 1976a; LAMBETH et al., 1979). Despite this tight binding, the **off** rate constant is rapid, about 300 s^{-1} (LAMBETH et al., 1980b), thus allowing rapid exchange of adrenodoxin molecules. NADP⁺ binding to both the oxidized and reduced flavoprotein is unaffected by the presence of bound adrenodoxin, indicating that the flavoprotein possesses separate binding sites for its electron donor and acceptor (LAMBETH et al., 1976a). The complex between the flavoprotein and the iron sulfur protein is sensitive to ionic strength, with decreased binding at increased ionic strength (CHU and KIMURA, 1973b; LAMBETH et al., 1979).

The adrenodoxin reductase-adrenodoxin complex, but not the flavoprotein alone, has significant catalytic activity for NADPH: cytochrome **c** reduction, a reaction which has been used as a model to investigate the catalytic mechanism of electron transfer within the complex. Using this acceptor, electron transfer to the cytochrome occurs from the iron-sulfur center. While the adrenodoxin appears to remain associated with the reductase (LAMBETH et al., 1976a), recent studies implicate what has been termed a partially dissociated state as the actual reductant (LAMBETH et al., 1981b).

In this model system, the rate-limiting step is the electron transfer from the flavin to the iron-sulfur center (LAMBETH and KAMIN, 1979). Upon reaction of an NADPH-reduced 2-electron-containing complex with excess cytochrome **c**, only one electron is transferred to the cytochrome at a rate which is compatible with catalysis. In the 2-electron-reduced species, both electrons reside primarily on the flavin, but this form is in rapid equilibrium with a reduced iron-sulfur/flavin semiquinone species. Electron transfer from the iron-sulfur center to cytochrome **c** results in production of the flavin semiquinone, but a second electron transfer from flavin to the iron-sulfur center does not occur rapidly. Additional electron transfers to the cytochrome require a second mole of bound NADPH, which apparently serves as an effector to promote the intra-complex electron transfer from the flavin semiquinone to the iron-sulfur group. Thus, during its normal catalytic cycle the complex never achieves its fully oxidized state.

Despite its utility for investigating the intra-complex electron transfer mechanisms, cytochrome **c** reduction is not a realistic model system with which to study the electron transfers to the physiologic electron acceptor, cytochrome P-450$_{scc}$, since important aspects of the kinetics are not mimicked. For example, while cytochrome c reduction is favored by a tightly associated 1:1 complex, cytochrome P-450 reduction (and associated P-450-dependent activities) fail to occur at the 1:1 molar ratio, and an excess of adrenodoxin is required (SEYBERT et al., 1978; LIGHT and ORME-JOHNSON, 1981; HANU-KOGLU and JEFCOATE, 1980). In addition, activity is favored by ionic conditions which promote dissociation of the reduced adrenodoxin (LAMBETH et al., 1979). However, by acetylating the lysines on cytochrome **c**, the electron transfer rate is slowed, and the kinetic mechanism becomes similar to that for electron transfer to cytochrome P-450 (LAMBETH et al., 1981b). The kinetics of reduction of acetylated cytochrome **c** can be explained by assuming that adrenodoxin must dissociate prior to interaction with and transfer of an electron to the acetylated cytochrome. This "shuttle mechanism" for electron transfer will be described in greater detail below.

The formation of the adrenodoxin-adrenodoxin reductase complex alters the midpoint oxidation-reduction potential of the iron-sulfur center by more than 100 mV (LAMBETH et al., 1976a; LAMBETH et al., 1979). While the potential of the reductase was about -295 mV both in the absence and presence of adrenodoxin, the potential of the adrenodoxin changed from -270 mV to about -360 mV or more in the complex. Despite the effect of this shift to make **intra**-complex electron transfer less favorable, this electron transfer occurs at an appreciable rate (about 5 s^{-1}) (LAMBETH and McCASLIN, 1976b), and the more rapid rate of reverse electron transfer from the iron-sulfur center to the flavin is similar to the **off** rate constant for dissociation of adrenodoxin from the complex (both about 300 s^{-1}). Thus, after roughly half of the intra-complex electron transfers, dissociation of the reduced adrenodoxin occurs. The reduced, dissociated state of adrenodoxin is then thermodynamically much more favorable than the associated state. Thus, the reduction of adrenodoxin can be thought of as providing the driving energy for dissociation of the complex. This property of the system is crucial to the physiological mechanism of electron transfer to the cytochrome, as detailed below.

10.3. The complex between adrenodoxin and cytochromes P-450

Adrenodoxin binds in a 1:1 ratio to purified cytochrome P-450$_{scc}$ (KATAGIRI et al., 1977). Like the complex with the flavoprotein, that with cytochrome P-450$_{scc}$ is relatively strong at low ionic strength (LAMBETH and KRIENGSIRI, 1985) (K$_d \sim 3 \times 10^{-8}$) but can also undergo rapid exchange (LAMBETH et al., 1982). The complex is salt-sensitive, but somewhat less so than that with the reductase (LAMBETH and KRIENGSIRI, 1985). The interaction of the two proteins also perturbs the midpoint potential of both redox centers (LAMBETH and PEMBER, 1983). For the cytochrome P-450$_{scc}$-cholesterol complex, bound adrenodoxin shifts the redox potential of the heme from -284 mV to -314 mV, thus making its reduction less favorable. Nevertheless, the potential of the iron-sulfur center is also shifted in the same direction, from -273 mV to -291 mV. Thus, the unfavorable effect of complex formation on the heme potential is nearly compensated by a parallel potential shift in the iron-sulfur center.

Formation of the adrenodoxin-cytochrome P-450 complex blocks access of both of the redox centers to aqueous-dissolved paramagnetic complexes (KAMIN et al., 1985), since dysprosium perturbation of the ESR spectrum of the heme was diminished by complex formation. The data were initially interpreted as indicating that the heme center was not deeply buried in the membrane, but an alternative and more likely explanation (see below) is that in the absence of adrenodoxin, the lanthanide-EDTA complexes had access to within 15 Å or so of the heme via an aqueous-exposed channel which becomes blocked by adrenodoxin. In either case, data are consistent with an aqueous-exposed face of the cytochrome which interacts with the soluble adrenodoxin.

10.4. Geometry of the interactions between adrenodoxin and its redox partner proteins

The geometry of the interactions between adrenodoxin and its complexing proteins (adrenodoxin reductase, cytochrome P-450$_{scc}$, and the model electron acceptor cytochrome c) was studied using fluorescence energy transfer (TULS et al., 1987). The single free sulfhydryl of adrenodoxin, Cys-95, was derivatized with a fluorescent label, and the distance from this group to the light-absorbing redox centers of the complexing partner proteins was estimated using Förster energy transfer calculations. The modification did not affect the activity or the interaction of adrenodoxin with any of its partner proteins. The location of the labelled residue was estimated by alignment of homologous sequences with the [2 Fe—2S] ferredoxin from *Spirulina platensis*, the crystal structure of which is known (FUKUYAMA et al., 1980). As in adrenodoxin, this ferredoxin has a highly acidic region which participates in electron transfer complexes (see below). Alignment of homologous regions leads to the prediction that the fluorescently labelled group is on the opposite side of the protein from the acidic binding region, and about 32 Å from it, as diagrammed in Figure 7.

As a test case, the distance was estimated in the complex between cytochrome c and adrenodoxin (ADX). Cytochrome c, the crystal structure of which is known, has one highly positively charged surface which pairs with the negative surface on a redox partner protein. In addition, its heme edge is exposed to the surface and is surrounded by the positive charges, so that the center of the heme is located 9 Å from the surface. Thus, the predicted distance from the fluorescent group to the heme is 41 Å. Using fluorescence energy transfer, the distance was calculated to be 42 Å (Fig. 7., top left). Thus, in a well-characterized system, the calculated distance was in excellent agreement with the prediction, serving as a verification of the structural model.

Using the same technique, the distances from the fluorescent group to the flavin in adrenodoxin reductase and the heme in cytochrome P-450$_{scc}$ were estimated to be 36 Å and 58 Å, respectively. If it is assumed that this same acidic interaction surface is utilized in the complexes with both of these physiologic partner (as is supported from chemical derivatization studies, see below), then the following conclusions may be drawn. Flavin (or a portion of the flavin) is predicted to lie at or near the surface of the adrenodoxin reductase (AR) (right side of Fig. 7.). In contrast, the heme of cytochrome P-450$_{scc}$ is buried about 26 Å from the adrenodoxin interaction site (bottom of Fig. 7.). This model is consistent with the crystal structure of cytochrome P-450$_{cam}$, which shows a deeply buried heme at depths from the protein surface ranging from 15 to 30 Å.

10.5. Regulation of electron transfer to cytochrome P-450scc by substrate binding

At least two mechanisms control the flux of electrons into cytochrome P-450$_{scc}$. Both are triggered by substrate binding to the cytochrome, thus permitting electron transfer only when the cholesterol binding site is occupied. Such regulation seems to be important to prevent the generation of free reactive reduced oxygen intermediates which would be formed if the heme center were to become reduced and then autoxidize in the absence of substrate.

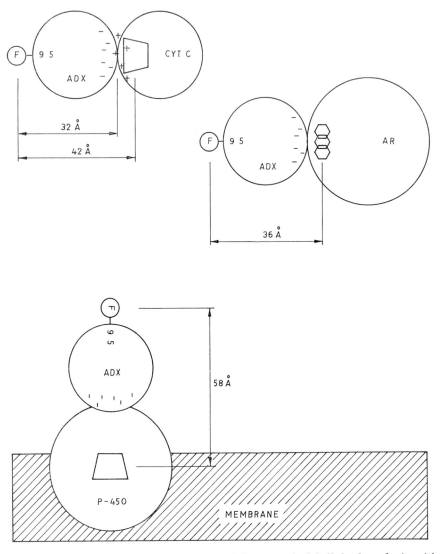

Fig. 7. Proposed geometry for interaction of fluorescently labelled adrenodoxin with cytochromes c and P-450scc and adrenodoxin reductase. The acidic binding face of adrenodoxin (ADX) is depicted with negative charges (—). Also depicted are the heme groups of the two cytochromes and the flavin of adrenodoxin reductase (AR). Calculated distances between the fluorescent group and the prosthetic groups of the proteins are shown.

J. D. LAMBETH

The first of these mechanisms involves substrate control of the protein-protein interactions between adrenodoxin and cytochrome P-450. In the absence of substrate, the binding is relatively weak, but is enhanced about 20-fold with the binding of cholesterol or its analogs (LAMBETH et al., 1980a). The presence of adrenodoxin likewise enhances the binding of cholesterol. The mutual facilitation of binding is seen both in the K_d and K_m values for cholesterol and adrenodoxin interactions with the cytochrome. A kinetic analysis (HANUKOGLU et al., 1981b) was consistent with a random order sequence with positive cooperativity (synergism) for the interaction of the cholesterol and adrenodoxin with the cytochrome during catalysis.

A second effect of substrate binding is on the rate of electron transfer from reduced adrenodoxin to the cytochrome, as investigated in stopped flow studies (LAMBETH and KRIENGSIRI, 1985). While the protein-protein interaction was sensitive to solution conditions (salt, detergent), the rate of heme reduction was relatively insensitive to such factors. Kinetic analysis indicated very rapid formation of the adrenodoxin-cytochrome P-450 complex, followed by a somewhat slower intra-complex electron transfer from the iron-sulfur to the heme[1]. The electron transfer was dependent upon the presence and type of sterol substrate.

The factors controlling this substrate-induced enhancement of electron transfer were investigated. Neither the strength of substrate binding nor the substrate-induced spin-state of the heme were found to be important factors governing electron transfer rates (a series of substrates produced both low, high, and intermediate spin states of the cytochrome[2], but the induced spin-state did not correlate with reduction rate). However, this series of substrates perturbed the midpoint potential of the cytochrome heme to different extents, producing redox potentials ranging from -265 to $-350 \, mV$ for steroid-bound forms of the cytochrome (LAMBETH and KRIENGSIRI, 1985; LAMBETH and PEMBER, 1983), and $-412 \, mV$ for the substrate-free cytochrome (LIGHT and ORME-JOHNSON, 1981). Using this series, there was a linear correlation between the rate of electron transfer from the iron-sulfur center to the heme and the substrate-induced midpoint potential. Thus, the rate of electron transfer is controlled by substrate binding via an effect of substrate on the midpoint potential of the cytochrome.

---

[1] The cytochrome reduction was biphasic. The second phase was too slow to participate catalytically, and was consistent with either a partially inactivated form of the cytochrome, or a less active conformation which slowly converted to the active form. The correlations summarized herein refer to the more rapid reductive phase.

[2] It should be noted that the strength of binding also did not correlate with the ability of a substrate to produce the high-spin state. Apparently, these parameters are independent in this system, as might be predicted from the large size (and therefore multiple binding determinants) of the substrate. Likewise, the midpoint potential in this system failed to correlate with the induced spin-state.

In summary, two factors are important in regulating electron transfer from the iron-sulfur center of adrenodoxin to the heme of cytochrome P-450$_{scc}$: (1) substrate regulation of adrenodoxin-cytochrome interactions, and (2) substrate regulation of electron transfer within the associated adrenodoxin-cytochrome P-450$_{scc}$ complex. We have suggested that these mechanisms work in concert to control electron transfer reactions and minimize the generation of toxic oxygen radicals.

10.6. Mechanism of electron transfer: adrenodoxin as a mobile electron shuttle

The structures described above and referred to in Fig. 7. imply that the same face of adrenodoxin — the acidic region — is involved in the interaction not only in the model electron transfer complex with cytochrome c, but also in the natural electron transfer complexes with adrenodoxin reductase and cytochrome P-450$_{scc}$ (or other cytochromes P-450). The protein interaction site(s) on adrenodoxin may either be essentially identical, or may overlap. If this is the case, then binding to one electron transfer partner protein precludes binding to another, and the electron transfer mechanism must therefore involve adrenodoxin functioning as a mobile electron shuttle. According to this mechanism, an electron is first accepted by the iron-sulfur center of adrenodoxin from a reduced flavin in the flavoprotein-adrenodoxin complex. The reduced adrenodoxin must then dissociate and form a new complex with cytochrome P-450, prior to an electron transfer to the heme. Such a mechanism makes teleologic sense, in view of the lower ratio of adrenodoxin reductase to the other two components; current estimates are 1:3:8 for flavoprotein:iron-sulfur protein:cytochrome P-450 in adrenal (HANUKOGLU and HANUKOGLU, 1986). Thus, a single flavoprotein must provide electrons to up to 8 cytochromes, and can do so only by sequentially interacting with many molecules of adrenodoxin. Such a mechanism is also supported by several direct lines of evidence, as follows:

1. The formation of the 1:1 complexes between adrenodoxin and its two partner proteins is mutually exclusive (SEYBERT et al., 1978; SEYBERT et al., 1979; LIGHT and ORME-JOHNSON, 1981). When a 1:1 molar ratio of adrenodoxin reductase:cytochrome P-450 is titrated with adrenodoxin, distinct optical changes can be used to detect specifically the formation of each of the complexes. The tighter complex between flavoprotein and iron sulfur-protein forms initially. The complex of adrenodoxin with cytochrome P-450 is detected only after the binding sites on the reductase are filled. Thus, instead of forming a ternary complex, adrenodoxin forms

sequential 1:1 complexes, first with the flavoprotein, and then with the cytochrome. This finding explains kinetic studies in which P-450-dependent metabolic conversions are minimal until the 1:1 ratio of adrenodoxin to its reductase is exceeded (HANUKOGLU and JEFCOATE, 1980). This is because at low adrenodoxin-to-reductase ratios, the binding of reduced adrenodoxin to reductase competes for its interaction with the cytochrome. At ratios of adrenodoxin exceeding reductase, oxidized adrenodoxin (which binds the reductase more tightly than reduced) effectively exchanges with the reduced iron-sulfur protein on the flavoprotein, thus allowing reduced adrenodoxin to interact with the cytochrome. In addition, the finding that the rate of cholesterol side-chain cleavage correlates with the % reduction of free adrenodoxin rather than its absolute concentration (HANUKOGLU and JEFCOATE, 1980) implies that free oxidized adrenodoxin competes with free reduced adrenodoxin for binding to P-450$_{scc}$.

The conclusion of mutually exclusive binding is also supported by cross-linking studies (LAMBETH et al., 1984a) using a water-soluble carbodiimide, EDC. This reagent activates carboxyl groups for covalent reaction with nearby nucleophiles such as the ε-amino groups of lysines. Since the EDC only activates and does not itself participate in the cross-link, the covalent complexes formed are likely to utilize the correct protein orientations so as to allow the geometrically correct cross-linking of residues which normally participate in the electrostatic interactions (e.g., the negatively charged carboxyls on one protein and positively charged amino groups on the other). Treatment of solutions containing either adrenodoxin plus adrenodoxin reductase or adrenodoxin plus cytochrome P-450$_{scc}$ with EDC resulted in the formation of larger molecular weight complexes between adrenodoxin and each of the partner proteins. Molecular weights indicated proteins are cross-linked in each of the 1:1 complexes. When a solution containing all three proteins was treated with EDC, only the two binary complexes were seen and no higher order complexes (1:1:1 or 1:2:1) were formed. Thus, the electron transfer protein-protein complexes are mutually exclusive, so that sequential interactions are required to move electrons.

2. The kinetics indicate that free adrenodoxin rather than the adrenodoxin reductase-adrenodoxin complex acts as the physiological reductant of cytochrome P-450. As described above, unless an excess of free adrenodoxin is present, complex formation with the reductase competes for binding and electron transfer to the cytochrome. In addition, we have studied the reduction of free adrenodoxin by adrenodoxin reductase (free referring to the excess adrenodoxin not bound in the complex with flavoprotein (LAMBETH et al., 1979)). We found that increased ionic strength first stimulated, and then inhibited the rate of reduction of free adrenodoxin. When the reduction system was limiting, parallel ionic effects on cytochrome P-450-dependent reactions such as 11β-hydroxylation and cholesterol side-chain

cleavage[1] were seen. The data were interpreted in terms of a requirement for free adrenodoxin to support the cytochrome P-450-dependent reactions.

3. Chemical modification studies indicate that the same or overlapping site(s) on adrenodoxin participate in the interaction with both partner proteins (LAMBETH et al., 1984a). In addition to promoting cross-linking, water-soluble carbodiimides can rearrange to modify carboxyl groups. By incubating adrenodoxin (in which the lysine groups had previously been methylated to prevent internal cross-linking) with EDC, the free carboxyl groups of aspartate and glutamate residues in the highly acidic region from residues 72 to 86 were modified. The modified residues were 72, 73, 74, 76, 79, and 86. These modifications reduced the ability of adrenodoxin to interact with both adrenodoxin reductase and cytochrome P-450$_{scc}$. The magnitude of the effect was similar for the interaction with both proteins (about 10 to 20-fold). Thus, as expected for complexes in which electrostatic interactions play a role, modification of residues in the highly acidic region of adrenodoxin weakens the interaction with both of the electron transfer partner proteins, supporting the hypothesis that both protein interactions involve the same acidic face of adrenodoxin.

Modification of a variety of other residues does not affect significantly the interactions of adrenodoxin with either the reductase or cytochrome P-450. Modification of all the lysines using maleic anhydride, or of all the arginine residues using p-hydroxyphenylglyoxal had no effect on the interaction with either protein (TULS et al., 1987). However, modification of histidines with diethylpyrocarbonate resulted in a small differential effect on binding: the K_d for interaction with the cytochrome was increased by a factor of 3, while the interaction with the reductase was unaffected. The data are consistent with either overlapping rather than identical interaction sites on adrenodoxin, or with a modification-induced change in the conformation of the iron-sulfur protein which affects binding differentially. These data, together with subtle differences in ion sensitivities of the two complexes (LAMBETH and KRIENGSIRI, 1985) and the differential effects of reduction of the iron-sulfur center on binding, suggest that although the major determinant is the highly acidic region, the two partner proteins have slightly different binding requirements. These differences may be important in allowing the proteins to discriminate between the oxidized vs. reduced forms of adrenodoxin in such a way as to promote the production of productive electron transfer complexes.

---

[1] Additional ion effects on the kinetics of the system have been investigated by HANUKOGLU et al., (HANUKOGLU et al., 1981a), and are exerted at the level of the adrenodoxin-cytochrome P-450 interaction and the Vmax for side-chain cleavage.

J. D. LAMBETH

10.7. Thermodynamics of protein-protein interactions: control of the adrenodoxin shuttle mechanism by reduction state and substrate binding

As described above, although the protein-protein complexes which participate in the adrenodoxin shuttle are tightly associated, their **off** rate constants are rapid, and these rates exceed by many fold the electron transfer rates for individual steps. Thus, the relative strengths of binding of adrenodoxin to the other two proteins determines the instantaneous distribution of adrenodoxin in protein complexes, and the shuttle mechanism will be regulated by thermodynamic rather than kinetic factors. The thermodynamics of the various protein-protein and substrate-protein interactions have been reported previously (LAMBETH et al., 1980a; LAMBETH and PEMBER, 1983; LAMBETH et al., 1976a), and many of these have been described above. The various binding energies for the possible complexes are summarized graphically in Figure 8. in which the interaction strengths are expressed for convenience as a function of the reduction state of adrenodoxin. The Y-axis in this figure represents the free energy change required to go from one complex or state to another (with the most stable state at the bottom of the diagram). Thus, when adrenodoxin is oxidized, the most stable complex is that with the flavoprotein, and this

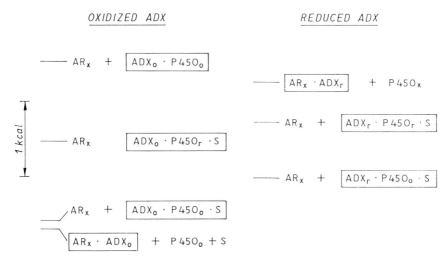

Fig. 8. Energetics of formation of the possible complexes of oxidized (o) versus reduced (r) adrenodoxin (ADX) with forms of adrenodoxin reductase (AR) and cytochrome P-450scc. "S" refers to cholesterol substrate bound to the sterol binding site on cytochrome P-450. The subscript "x" refers to any of the reduction states of the flavoprotein. The horizontal lines represent relative free energy levels, with more negative free energies (more stable complexes) towards the bottom of the diagram. Reproduced from (LAMBETH and PEMBER, 1983).

---

association is independent of the oxidation state of the flavin. Thus, reduced as well as oxidized flavoprotein associates avidly with the oxidized adrenodoxin, and this electron transfer complex is the most favored. However, upon reduction of the iron-sulfur center, the complex with the reductase is less favored, and that with the cytochrome P-450-cholesterol complex now becomes the most favorable.[1] The adrenodoxin associates poorly with the substrate-free cytochrome (LAMBETH et al., 1980a); this interaction is off scale in this diagram. Thus, substrate binding is a major determinant of adrenodoxin binding. Of the two possible complexes of reduced adrenodoxin with the cytochrome-substrate complex, the most favored is that with oxidized rather than reduced cytochrome P-450. Thus, at least two factors regulate the sequential protein-protein interactions which comprise the shuttle mechanism: 1) the reduction state of the proteins (both adrenodoxin and cytochrome P-450$_{scc}$), and 2) substrate binding to the cytochrome. These interactions conspire to favor the formation of productive electron transfer complexes in which the down-stream component is an oxidized species and the upstream component is reduced. Using this remarkable enzymatic machinery, electrons are meticulously delivered from pyridine nucleotide to cytochrome to support the oxidative reactions which culminate in the side-chain cleavage of cholesterol.

Acknowledgements

This chapter is dedicated to the memory of Dr. Henry KAMIN, whose advice, optimism, friendship and inspiration are gratefully acknowledged.

11. References

AKHREM, A. A., V. I. VASILEVSKY, T. B. ADAMOVICH, A. G. LAPKO, V. M. SHKUMATOV, and V. L. CHASHCHIN, (1980a), in: Biochemistry, Biophys. and Regulation of Cytochrome P 450 Elsevier-North, Amsterdam, pp. 57—64.

AKHREM, A. A., V. I. VASILEVSKY, V. M. SHKUMATOV, and V. L. CHASHCHIN, (1980b), in: Microsomes, Drug Oxidations, and Chemical Carcinogenesis, (COON, M. J., CONNEY, A. H., ESTABROOK, R. W., GELBOIN, H. V., GILLETTE, J. R., and O'BRIEN, P. J., eds.), Academic Press, New York, pp. 77—84.

ARINGER, L., P. ENEROTH, and L. NORDSTROM, (1979), J. Steroid Biochem. 11, 1271 to 1285.

ARTHUR, J. R., H. A. F. BLAIR, G. S. BOYD, J. I. MASON, and K. E. SUCKLING, (1976), Biochem. J. 158, 47—51.

---

[1] Comparison of the binding of oxidized **versus** reduced adrenodoxin to the oxidized P-450-substrate complex reveals very similar energetics of binding which in fact slightly favor binding of the oxidized adrenodoxin. This explains the ability of oxidized adrenodoxin to act as a competitive inhibitor with respect to reduced adrenodoxin (HANUKOGLU and JEFCOATE, 1980).

BURNSTEIN, S. and M. GUT, (1976), Steroids 28, 115—131.
BURNSTEIN, S., Y. LETOURNEUX, H. L. KIMBALL, and M. GUT, (1976), Steroids 27, 361 to 382.
CAMMER, W. and R. W. ESTABROOK, (1967), Arch. Biochem. Biophys. 122, 735—747.
CHASHCHIN, V. L., V. I. VASILEVSKY, V. M. SHKUMATOV, and A. A. AKHREM, (1984), Biochim. Biophys. Acta 787, 27—38.
CHENG, B. and T. KIMURA, (1983), Lipids 18, 577—584.
CHU, J. W. and T. KIMURA, (1973a), J. Biol. Chem. 248, 2089—2094.
CHU, J. W. and T. KIMURA, (1973b), J. Biol. Chem. 248, 5183—5187.
DEAN, W. L. and R. D. GRAY, (1982), J. Biol. Chem. 257, 14679—14685.
DEGENHART, H. J., G. J. ALSEMA, J. HOOGERBRUGGE, B. G. WOLTHERS, and R. KAPTEIN, (1984), J. Steroid Biochem. 21, 447—451.
DUBOIS, R. N., E. R. SIMPSON, J. TUCKEY, J. D. LAMBETH, and M. R. WATERMAN, (1981), Proc. of Natl. Acad. Sci. 78, 1028—1032.
FARKASH, Y., R. TIMBERG, and J. ORLY, (1986), Endocrinology 118, 1353—1365.
FOSTER, R. P. and L. D. WILSON, (1975), Biochemistry 14, 1477—1484.
FUKUYAMA, K., T. HASE, S. MATSUMOTO, T. TSUKIHARA, Y. KATSUBE, N. TANAKA, M. KAKUDO, K. WADA, and H. MATSUBARA, (1980), Nature 286, 522—524.
GASPARINI, F., A. WOLFSON, R. HOCHBERG, and S. LIEBERMAN, (1979), J. Biol. Chem. 254, 6650—6656.
GOTOH, O., Y. TAGASHIRA, K. MOROHASHI, and Y. FUJII-KURIYAMA, (1985), FEBS Lett. 188, 8—10.
GREENFIELD, N. J., R. PARSONS, M. WELSH, and B. GEROLIMATOS, (1985), J. Steroid Biochem. 23, 313—321.
GREENFIELD, N. J. and R. PARSONS, (1986), J. Steroid Biochem. 24, 909—916.
GROH, S. E., A. NAGAHISA, S. L. TAN, and W. H. ORME-JOHNSON, (1983), J. Am. Chem. Soc. 105, 7445—7446.
HALL, P. F., (1976), in: Iron and Copper Proteins, (YASUNOBU, K. T., MOWER, H. F. and HAYAISHI, O., eds.), Plenum Pub. Corp., New York, pp. 303—313.
HALL, P. F., M. WATANUKE, and B. A. HAMKALO, (1979), J. Biol. Chem. 254, 547.
HANUKOGLU, I. and C. R. JEFCOATE, (1980), J. Biol. Chem. 255, 3057—3061.
HANUKOGLU, I., C. T. PRIVALLE, and C. R. JEFCOATE, (1981a), J. Biol. Chem. 256, 4329—4335.
HANUKOGLU, I., V. SPITSBERG, J. A. BUMPUS, K. M. DUS, and C. R. JEFCOATE, (1981b), J. Biol. Chem. 256, 4321—4328.
HANUKOGLU, I. and Z. HANUKOGLU, (1986), Eur. J. Biochem. 157, 27—31.
HEYL, B. L., D. J. TYRRELL, and J. D. LAMBETH, (1986), J. Biol. Chem. 261, 2743 to 2749.
HIWATASHI, A. and Y. ICHIKAWA, (1982), J. Biochem. 92, 335—342.
HUME, R. and G. S. BOYD, (1978), Biochemical Society Transactions 6, 893—898.
HUME, R., R. W. KELLY, P. L. TAYLOR, and G. S. BOYD, (1984), Eur. J. Biochem. 140, 583—591.
ISHIMURA, Y., T. YOSHINAGA, H. FUJITA, S. SUGANO, M. OKAMOTO, and T. YAMANO, (1988), Arch. Histol. Jap. 48, 541—546.
JACOBS, R. E., J. SINGH, and L. E. VICKERY, (1987), Biochemistry 26, 4541—4545.
KAMIN, H., C. BATIE, J. D. LAMBETH, J. R. LANCASTER, Jr., L. GRAHAM, and J. C. SALERNO, (1985), Biochemical Society Trans. 13, 615—618.
KAMIN, H. and J. D. LAMBETH, (1982), in: Flavins and Flavoproteins, (MASSEY, V. and WILLIAMS, C. H., eds.), Elsevier, New York, pp. 655—666.
KATAGIRI, M., O. TAKIKAWA, H. SATO, and K. SUHARA, (1977), Biochem. Biophys. Res. Comm. 77, 804—809.

KELLIS, J. R., Jr., J. J. SHEETS, and L. E. VICKERY, (1984), J. Steroid Biochem. **20**, 671—676.
KIMURA, T., (1968), Structure and Bonding **5**, 1—40.
KOBAYASHI, S. and S. ICHII, (1969), J. Biochem. **66**, 51—56.
KRAMER, R. E., R. N. DuBOIS, E. R. SIMPSON, C. M. ANDERSON, K. KASHIWAGI, J. D. LAMBETH, C. R. JEFCOATE, and M. R. WATERMAN, (1982), Arch. Biochem. Biophys. **215**, 478—485.
LAMBETH, J. D. and H. KAMIN, (1976), J. Biol. Chem. **251**, 4299—4306.
LAMBETH, J. D., D. R. McCASLIN, and H. KAMIN, (1976a), J. Biol. Chem. **251**, 7545 to 7550.
LAMBETH, J. D. and D. R. McCASLIN, (1976b), Fed. Proc. **35**, 1599—1599. (Abstract)
LAMBETH, J. D., D. W. SEYBERT, and H. KAMIN, (1979), J. Biol. Chem. **254**, 7255—7264.
LAMBETH, J. D. and H. KAMIN, (1979), J. Biol. Chem. **254**, 2766—2774.
LAMBETH, J. D., H. KAMIN, and D. W. SEYBERT, (1980), J. Biol. Chem. **255**, 8282—8288.
LAMBETH, J. D., D. W. SEYBERT, and H. KAMIN, (1980a), J. Biol. Chem. **255**, 138—143.
LAMBETH, J. D., D. W. SEYBERT, and H. KAMIN, (1980b), J. Biol. Chem. **255**, 4667 to 4672.
LAMBETH, J. D., (1981a), J. Biol. Chem. **256**, 4757—4762.
LAMBETH, J. D., J. R. LANCASTER, Jr., and H. KAMIN, (1981b), J. Biol. Chem. **256**, 3674—3678.
LAMBETH, J. D., S. E. KITCHEN, A. A. FAROOQUI, R. TUCKEY, and H. KAMIN, (1982), J. Biol. Chem. **257**, 1876—1884.
LAMBETH, J. D., D. W. SEYBERT, J. R. LANCASTER, Jr., J. C. SALERNO, and H. KAMIN, (1982a), Mol. Cell. Biochem. **45**, 13—31.
LAMBETH, J. D., (1982b), in: Flavins and Flavoproteins (MASSEY, V. and WILLIAMS, C. H., eds.), Elsevier, New York, pp. 689—694.
LAMBETH, J. D. and S. O. PEMBER, (1983), J. Biol. Chem. **258**, 5596—5602.
LAMBETH, J. D., L. M. GEREN, and F. MILLETT, (1984a), J. Biol. Chem. **259**, 10025 to 10029.
LAMBETH, J. D., (1984b), in: Phospholipids and Cellular Regulation (KUO, J. F., ed.), CRC Press, New York, pp. 189—228.
LAMBETH, J. D. and V. L. STEVENS, (1985), Endocrine Res. **10**, 283—309.
LAMBETH, J. D. and S. KRIENGSIRI, (1985), J. Biol. Chem. **260**, 8810—8816.
LAMBETH, J. D., (1986), Endocrine Res. **12**, 371—392.
LANGE, R., L. MAURIN, C. LARROQUE, and A. BIENVENUE, (1988), Eur. J. Biochem. **172**, 189—195.
LARROQUE, C. and J. E. VAN LIER, (1980), FEBS Lett. **115**, 175—177.
LARROQUE, C. and J. E. VAN LIER, (1986), J. Biol. Chem. **261**, 1083—1087.
LIGHT, D. R. and C. WALSH, (1980), J. Biol. Chem. **255**, 4264—4277.
LIGHT, D. R. and N. R. ORME-JOHNSON, (1981), J. Biol. Chem. **256**, 343—350.
MILLER, W. L., (1987), J. Steroid Biochem. **27**, 759—766.
MITANI, F., T. IIZUKA, H. SHIMADA, R. UENO, and Y. ISHIMURA, (1988), J. Biol. Chem. **260**, 12042—12048.
MORISAKI, M., S. SATO, and N. IKEKAWA, (1976), FEBS Lett. **72**, 337—340.
MORISAKI, M., C. DUQUE, N. IKEKAWA, and M. SHIKITA, (1980), J. Steroid Biochem. **13**, 545—550.
MORISAKI, M., C. DUQUE, K. RAKANE, N. IKEKAWA, and M. SHIKITA, (1982), J. Steroid Biochem. **16**, 101—105.
MOROHASHI, K., Y. FUJII-KURIYAMA, Y. OKADA, K. SOGAWA, T. HIROSE, S. INAYAMA, and T. OMURA, (1984), Proc. of Natl. Acad. Sci. **81**, 4647—4651.
MOROHASHI, K., H. YOSHIOKA, O. GOTOH, Y. OKADA, K. YAMAMOTO, T. MIYATA, K. SOGAWA, Y. FUJII-KURIYAMA, and T. OMURA, (1987), J. Biochem. **102**, 559—568.

NAGAHISA, A., T. FOO, M. GUT, and W. H. ORME-JOHNSON, (1985), J. Biol. Chem. **260**, 846–851.

NAKAJIN, S., Y. ISHII, M. SHINODA, and M. SHIKITA, (1979), Biochem. Biophys. Res. Comm. **87**, 524–531.

NONAKA, Y., S. FUJII, and T. YAMANO, (1986), J. Biochem. **99**, 803–814.

NONAKA, Y., H. MURAKAMI, Y. YABUSAKE, S. KURAMITSU, H. KAGAMIYAMA, T. YAMANO, and M. OKAMOTO, (1987), Biochem. Biophys. Res. Comm. **145**, 1239–1247.

OMURA, T., E. SANDERS, and R. W. ESTABROOK, (1966), Arch. Biochem. Biophys. **117**, 660–673.

ORME-JOHNSON, N. R., D. R. LIGHT, R. W. WHITE-STEVENS, and W. H. ORME-JOHNSON, (1979), J. Biol. Chem. **254**, 2103–2111.

PEDERSEN, R. C., (1985), Endocrine Res. **10**, 533–562.

PEMBER, S. O., G. L. POWELL, and J. D. LAMBETH, (1983), J. Biol. Chem. **258**, 3198 to 3206.

POULOS, T. L., B. C. FINZEL, I. C. GUNSALUS, G. C. WAGNER, and J. KRAUT, (1985), J. Biol. Chem. **260**, 16122–16130.

PRIVALLE, C. T., B. C. McNAMARA, M. S. DHARIWAL, and C. R. JEFCOATE, (1987), Mol. Cell Endocrinol. **53**, 87–101.

RAGGATT, P. R. and M. W. WHITEHOUSE, (1966), Biochem. J. **101**, 819–839.

ROBERTS, K. D., L. BANDY, and S. LIEBERMAN, (1967), Biochem. Biophys. Res. Commun. **29**, 741–746.

SATO, H., M. ASHIDA, K. SUHARA, F. ITAGAKI, S. TAKEMORI, and M. KATAGIRI, (1978), Arch. Biochem. Biophys. **190**, 307–314.

SEYBERT, D. W., J. D. LAMBETH, and H. KAMIN, (1978), J. Biol. Chem. **253**, 8355 to 8358.

SEYBERT, D. W., J. R. LANCASTER, Jr., J. D. LAMBETH, and H. KAMIN, (1979), J. Biol. Chem. **254**, 12088–12098.

SHEETS, J. J. and L. E. VICKERY, (1982), Proc. of Natl. Acad. Sci. **79**, 5773–5777.

SHEETS, J. J. and L. E. VICKERY, (1983a), J. Biol. Chem. **258**, 11446–11452.

SHEETS, J. J. and L. E. VICKERY, (1983b), J. Biol. Chem. **258**, 1720–1725.

SHIKITA, M. and P. F. HALL, (1973), J. Biol. Chem. **248**, 5598–5604.

STEVENS, V. L., T. Y. AW, D. P. JONES, and J. D. LAMBETH, (1984), J. Biol. Chem. **259**, 1174–1179.

SUGIYAMA, T., R. MIURA, and T. YAMANO, (1976), in: Iron and Copper Proteins, (YASUNOBU, K. T., MOWER, H. F. and HAYAISHI, O., eds.), Plenum Publishing Corporation, New York, pp. 290–302.

SUGIYAMA, T., R. MIURA, and T. YAMANO, (1979), J. Biochem. **86**, 213–223.

SUHARA, K., T. GOMI, H. SATO, E. ITAGAKI, S. TAKEMORI, and M. KATAGIRI, (1978), Arch. Biochem. Biophys. **190**, 290–299.

SUZUKI, K. and T. KIMURA, (1965), Biochem. Biophys. Res. Comm. **19**, 340–345.

TAIT, A. D., (1972), Biochem. J. **128**, 467–470.

TAKEMORI, S., H. SATO, T. GOMI, K. SUHARA, and M. KATAGIRI, (1975a), Biochem. Biophys. Res. Comm. **67**, 1151–1157.

TAKEMORI, S., K. SUHARA, S. HASHIMOTO, M. HASHIMOTO, H. SATO, T. GOMI, and M. KATAGIRI, (1975b), Biochem. Biophys. Res. Comm. **63**, 588–593.

TAKIKAWA, O., T. GOMI, K. SUHARA, E. ITAGAKI, S. TAKEMORI, and M. KATAGIRI, (1978), Arch. Biochem. Biophys. **190**, 300–306.

TANAKA, M., M. HANIU, and K. T. YASUNOBU, (1970), Biochem. Biophys. Res. Comm. **39**, 1182–1188.

TANFORD, C., (1980), in: The Hydrophobic Effect: Formation of micelles and biological membranes. John Wiley and sons, New York, pp. 106–127.

TILLEY, B. E., M. WATANUKI, and P. F. HALL, (1977), Biochim. Biophys. Acta **488** 330—339.

TROUT, E. C. and W. ARNETT, (1971), Proc. Soc. Exp. Biol. Med. **136**, 469—472.

TSUBAKI, M., A. HIWATASHI, Y. ICHIKAWA, and H. HORI, (1987a), Biochemistry **26**, 4527—4534.

TSUBAKI, M., A. HIWATASHI, and Y. ICHIKAWA, (1987b), Biochemistry **26**, 4535—4540.

TSUBAKI, M., H. OHKUBO, Y. TSUNEOKA, S. TOMITA, A. HIWATASHI, and Y. ICHIKAWA, (1987c), Biochim. Biophys. Acta **914**, 246—258.

TUCKEY, R. C. and H. KAMIN, (1982a), J. Biol. Chem. **257**, 2887—2893.

TUCKEY, R. C. and H. KAMIN, (1982b), J. Biol. Chem. **257**, 9309—9314.

TUCKEY, R. C. and H. KAMIN, (1983), J. Biol. Chem. **258**, 4232—4237.

TUCKEY, R. C. and P. M. STEVENSON, (1985), Eur. J. Biochem. **148**, 379—384.

TULS, J., L. GEREN, J. D. LAMBETH, and F. MILLETT, (1987), J. Biol. Chem. **262**, 10020—10025.

UZGIRIS, V. I., P. E. GRAVES, and H. A. SALHANICK, (1977), Biochemistry **16**, 593—600.

VICKERY, L. E. and J. T. KELLIS, (1983), J. Biol. Chem. **258**, 1720—1725.

VIK, S. B. and R. A. CAPALDI, (1977), Biochemistry **16**, 5755—5755.

WARBURTON, R. J. and D. W. SEYBERT, (1988), Biochem. Biophys. Res. Commun. **152**, 177—183.

WOLFSON, A. J. and S. LIEBERMAN, (1979), J. Biol. Chem. **254**, 4096—4100.

YOUNG, D. G. and P. F. HALL, (1969), Biochemistry 8, 2987—2996.

YU, C., L. YU, and T. E. KING, (1975), J. Biol. Chem. **250**, 1383—1392.

Chapter 3
Mechanisms of Regulation
of Steroid Hydroxylase Gene Expression

M. R. WATERMAN and E. R. SIMPSON

1. Introduction

In the broadest sense, the cytochrome P-450 supergene family (NEBERT et al., 1987) consists of two general classes of mixed-function oxidases, those which metabolize xenobiotic or exogenous substrates and those which metabolize endogenous substrates. This latter group of enzymes is involved in key biological pathways including cholesterol biosynthesis, bile acid formation, vitamin D metabolism, prostaglandin biosynthesis and steroid hormone production. The cytochromes P-450 which metabolize endogenous substrates are generally not inducible, unlike those forms which metabolize xenobiotic compounds, and are frequently found to be present in relatively low abundance. Exceptions to this are the steroid hydroxylases involved in steroid hormone biosynthesis which are found in relatively high levels, particularly in the adrenal cortex (HANUKOGLU and HANUKOGLU, 1986). Consequently these forms of P-450 have been purified and studied in considerable detail, both biophysically and biochemically as indicated in other chapters in this volume. Furthermore, the steroid hydroxylases were the first of the forms of P-450 which metabolize endogenous substrates whose regulation of gene expression was examined.

While our understanding of steroid hydroxylase gene expression is not as detailed as that of TCDD regulation of P_1-450 gene expression (JONES et al., 1986), it will be seen in this chapter that such detailed understanding is on the horizon. Furthermore, it will become apparent that this is a very complex process involving at least three distinct mechanisms of regulation of steroid hydroxylase gene expression; one leading to optimal steroidogenic capacity, one associated with tissue specific expression and one required for developmental regulation.

2. Peptide hormones and steroid hydroxylases

The steroidogenic pathways in the adrenal cortex and the gonads are outlined in Figure 1. These pathways illustrate several interesting features associated with steroid hydroxylase gene expression and biosynthesis. For example, it will be noted that tissue-specific gene expression leads to the appearance of $P-450_{11\beta}$ and $P-450_{C21}$ only in the adrenal cortex while $P-450_{scc}$ and $P-450_{17\alpha}$ are expressed in all steroidogenic tissues. Also $P-450_{11\beta}$ and $P-450_{scc}$ are localized in mitochondria while $P-450_{17\alpha}$ and $P-450_{C21}$ are found in the endoplasmic reticulum. In the adrenal cortex, steroidogenesis leads to production of glucocorticoids, mineralocorticoids, and C_{19}-precursors of sex hormones (androgens), while in the gonads the steroidogenic pathways lead only to production of C_{19} steroids which are then converted to the sex hormones estrogen or testosterone. However, it can be seen from Fig. 1., that the first

M. R. WATERMAN; E. R. SIMPSON

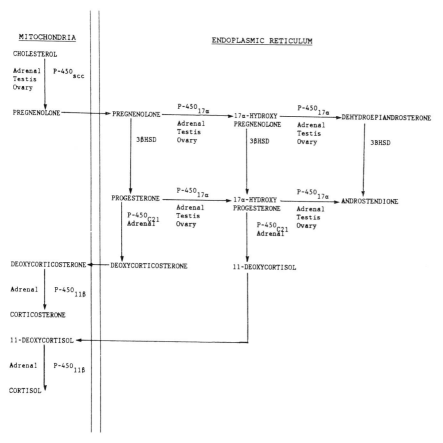

MITOCHONDRIA

ENDOPLASMIC RETICULUM

CHOLESTEROL

Adrenal	P-450$_{scc}$
Testis	
Ovary	

PREGNENOLONE ⟶ PREGNENOLONE ⟶ 17α-HYDROXY ⟶ DEHYDROEPIANDROSTERONE
PREGNENOLONE

P-450$_{17α}$
Adrenal
Testis
Ovary

3βHSD   3βHSD   3βHSD

PROGESTERONE ⟶ 17α-HYDROXY ⟶ ANDROSTENDIONE
PROGESTERONE

P-450$_{17α}$
Adrenal
Testis
Ovary

P-450$_{C21}$
Adrenal

DEOXYCORTICOSTERONE ⟵ DEOXYCORTICOSTERONE   11-DEOXYCORTISOL

| Adrenal | P-450$_{11β}$ |

CORTICOSTERONE

11-DEOXYCORTISOL ⟵

| Adrenal | P-450$_{11β}$ |

CORTISOL

Fig. 1. Steroidogenic pathways in adrenal cortex, testis and ovary. Included is the subcellular localization of the enzymatic reactions. 3βHSD = 3β-hydroxysteroid dehydrogenase.

step in all these pathways is the conversion of cholesterol to pregnenolone in the mitochondrion, catalyzed by P-450$_{scc}$. The biochemistry and regulation of this side chain cleavage reaction is described elsewhere in the volume. In all steroidogenic pathways, pregnenolone leaves the mitochondrion and travels to the endoplasmic reticulum where it undergoes 17α-hydroxylation. In the gonads, no additional steroid hydroxylation takes place until after the 17,20-lyase reaction cleaves the side chain from 17α-hydroxypregnenolone or 17α-hydroxyprogesterone. The resultant respective C$_{19}$ steroids, dehydroepiandrosterone or androstendione, are then committed to production of sex hormones by other reactions including those catalyzed by aromatase cytochrome P-450. In the adrenal cortex, a portion of the 17α-hydroxypregnenolone is converted to dehydroepiandrosterone by the same 17,20-lyase reaction while the remainder undergoes an additional steroid hydroxylation reaction in the endoplasmic reticulum, steroid 21-hydroxylation catalyzed by P-450$_{C21}$, to yield 11-deoxycortisol. The final step in glucocorticoid biosynthesis in the adrenal cortex takes place in the mitochondrion to yield cortisol, the potent glucocorticoid in man. This reaction, 11β-hydroxylation, is catalyzed by P-450$_{11\beta}$. A number of species variations on the general pathways shown in Fig. 1. are known to exist. For example, in rodents, P-450$_{17\alpha}$ is not found in the adrenal cortex (although it is present in the gonads) and the major glucocorticoid produced by these species is corticosterone rather than cortisol. Also in man and cows, the primary product of androgen biosynthesis is dehydroepiandrosterone while in rodents it is androstendione, reflecting a differential preference of P-450$_{17\alpha}$ in these species for either pregnenolone or progesterone. Evidence obtained from both in vitro and in situ experiments prove that a single polypeptide chain, P-450$_{17\alpha}$, catalyzes both the 17α-hydroxylase and the 17,20-lyase reactions (NAKAJIN and HALL, 1981a; NAKAJIN et al., 1981b; NAKAJIN et al., 1984; SUHARA et al., 1984; ZUBER et al., 1986b).

The steroid hydroxylases localized in the endoplasmic reticulum (P-450$_{17\alpha}$ and P-450$_{C21}$) are reduced by NADPH via the ubiquitous microsomal flavoprotein, NADPH-cytochrome P-450 reductase (HIWATASHI and ICHIKAWA, 1979). The mitochondrial steroid hydroxylases are reduced by NADPH via a mini-electron transport chain localized in the mitochondrial matrix. Reducing equivalents are transferred from NADPH to a flavoprotein, adrenodoxin reductase (OMURA et al., 1966), which in turn reduces an iron-sulfur protein, adrenodoxin (OMURA et al., 1966; SUZUKI and KIMURA, 1965). Adrenodoxin then transfers electrons to either P-450$_{scc}$ or P-450$_{11\beta}$, both of which are localized in the inner mitochondrial membrane.

The amino acid sequences of P-450$_{scc}$, P-450$_{11\beta}$, P-450$_{17\alpha}$, P-450$_{C21}$, NADPH-cytochrome P-450 reductase, adrenodoxin and adrenodoxin reductase have been deduced, and it is evident that the steroid hydroxylases are members of distinct P-450 gene families (NEBERT et al., 1987). P-450$_{scc}$ is tentatively classified as the product of the P450XIA gene and P-450$_{11\beta}$ as the product of the P450XIB gene, P-450$_{17\alpha}$ is the product of the P450XVII gene, P-450$_{C21}$ is the product of the P450XXI gene. Whether each of these gene families consist of a single gene product is not yet firmly established, but it will not be surprising if this is the case.

Steroidogenesis is regulated in the adrenal cortex, testis and ovary by peptide hormones derived from the anterior pituitary (Table 1). In each tissue, the hormone binds to its cell surface receptor and activates adenylate cyclase leading to elevated levels of intracellular cyclic AMP. This leads to mobilization of cholesterol to the mitochondrion and the acute regulation of steroidogenesis (JEFCOATE et al., 1986) as described elsewhere in this volume. In addition, cAMP also exerts a chronic action on the steroidogenic pathway which involves

M. R. WATERMAN; E. R. SIMPSON

regulation of expression of genes encoding the steroid hydroxylases and related enzymes. It is our belief that this chronic action of peptide hormones, mediated via cAMP, is required for maintenance of optimal steroidogenic capacity in the adrenal cortex and testes. Thus, rather than leading to induction of P-450 gene expression as is observed with xenobiotic compounds such as TCDD or

Table 1. Steroidogenic tissues and their specific peptide hormones

Tissue	Peptide hormones

Adrenal cortex
  Fasiculata-reticularis
  Glomerulosa
	ACTH

Testis Leydig cells	LH, hCG

Ovary
  Granulosa cells
  Theca interna cells
  Corpus luteum
	LH, FSH

phenobarbital, we imagine that in these tissues cAMP serves to maintain optimal levels of steroid hydroxylases. On the other hand in the ovary, steroidogenesis is episodic and occurs in a precisely coordinated cyclical fashion throughout the ovarian cycle. Clearly, cAMP-mediated processes, controlled by the gonadotropins LH and FSH, are important in regulation of this pattern of events; but as will be discussed later in this chapter, other factors which do not act via cAMP may also involved.

The initial evidence that peptide hormones were involved in maintenance of normal levels of steroid hydroxylases was obtained using hypophysectomized rats. Kimura (KIMURA, 1968) and Purvis (PURVIS et al., 1973a) both demonstrated that following removal of the pituitary, levels of steroid hydroxylase activities and P-450 levels as measured spectrophotometrically declined. Upon administration of exogenous ACTH, P-450 levels and steroid hydroxylase activities were restored. PURVIS and colleagues demonstrated similar phenomena in rat testis, using hCG as the exogenous hormone (PURVIS et al., 1973b). These results led to the conclusion that peptide hormones regulated steroid hydroxylase levels. Utilizing the adrenal tumor cell line, Y-1, ASANO and HARDING (1976) demonstrated that ACTH enhanced adrenodoxin biosynthesis and KOWAL and colleagues (KOWAL, 1969; KOWAL et al., 1970) showed a similar effect of ACTH on 11β-hydroxylase activity in the same cell line.

These results set the stage for more detailed analysis of regulation of steroid hydroxylase levels. However, before such studies could be initiated, it was necessary for two experimental developments to occur. First, the steroid hydroxylases and related enzymes had to be purified such that antibodies could be raised against them. Second, development of a cell culture system which contained all the steroidogenic enzymes, namely bovine adrenocortical cells in primary monolayer culture, was necessary. As a result of these developments, it has been possible to undertake detailed investigation of steroid hydroxylase gene expression.

3. Immunological studies of the action of peptide hormones on steroid hydroxylase levels

Utilizing primary cultures of bovine adrenocortical cells (GOSPODAROWICZ et al., 1977) it has been established that ACTH treatment leads to increased rates of synthesis of P-450$_{scc}$ (DuBois et al., 1981a), P-450$_{11\beta}$ (KRAMER et al., 1983), P-450$_{17\alpha}$ (ZUBER et al., 1985), P-450$_{C21}$ (FUNKENSTEIN et al., 1983a), adrenodoxin reductase (KRAMER et al., 1982a), adrenodoxin (KRAMER et al., 1982b) and NADPH-cytochrome P-450 reductase (DEE et al., 1985). These experiments were executed by treating the primary cell cultures with ACTH for varying periods of time, pulse radiolabeling total cell protein with ^{35}S-methionine and immunoisolating the newly synthesized steroid hydroxylase protein using monospecific antibodies (KRAMER et al., 1982c). The rates of synthesis of the mitochondrial steroid hydroxylases optimized about 36 hours after initiation of ACTH treatment while those for the microsomal steroid hydroxylases optimized at about 24 hours after initiation of ACTH treatment. In most instances there was a corresponding increase in enzymatic activity; an exception to this being P-450$_{C21}$ where an increase in the rate of synthesis was observed which was not associated with an increase in activity (FUNKENSTEIN et al., 1983a). Why no increase in steroid 21-hydroxylase activity is associated with the increase in synthesis remains an unanswered question. It was originally thought that time differences of rate of synthesis observed between mitochondrial and microsomal enzymes might represent significant differences between enzymes destined for these two subcellular compartments, however this difference has not been observed using more sensitive techniques for studying gene expression, as described later.

In the case of each of the steroid hydroxylases and related enzymes noted above, the increase in the rate of synthesis was associated with an increase in the level of translatable mRNA. This was determined by treatment of primary cell cultures with ACTH for varying periods of time followed by isolation of total RNA from the cell cultures, in vitro translation of the RNA

M. R. WATERMAN; E. R. SIMPSON

in a rabbit reticulocyte lysate translation system and immunoisolation of newly synthesized steroid hydroxylases using monospecific antibodies. The temporal appearance of optimal levels of in vitro translation products was the same as noted for cell labeling experiments, 36 hours for mitochondrial steroid hydroxylases and 24 hours for microsomal steroid hydroxylases.

During the course of the in vitro translation experiments it was found that the mitochondrial components of the steroidogenic pathways were synthesized as higher molecular weight precursors, a common feature of proteins encoded by the nuclear genome and destined for mitochondrial localization. Thus bovine P-450$_{scc}$ (DuBois et al., 1981b), P-450$_{11\beta}$ (KRAMER et al., 1982c), adrenodoxin and adrenodoxin reductase (NABI and OMURA, 1980; KRAMER et al., 1982c) are all synthesized as higher molecular weight precursors, while the microsomal components of this pathway are not. Likewise, human P-450$_{scc}$ is also synthesized as a higher molecular weight precursor form (OHASHI et al., 1983). A detailed description of these precursor proteins and their uptake by mitochondria is found elsewhere in this volume.

In some instances, increased levels of steroid hydroxylases were also detected by immunoblot analysis and immunofluorescence. Thus it can be concluded from these initial studies that ACTH treatment of primary adrenocortical cells leads to increased levels of translatable RNA for the steroid hydroxylases and related enzymes. This increase in translatable level of RNA is accompanied by an increase in the rates of synthesis of these enzymes which in turn leads to increased levels of these proteins and, in most cases, increased levels of enzymatic activity. Each of these actions of ACTH were found to be mimicked by addition of cyclic AMP analogs to the cell cultures in place of ACTH (KRAMER et al., 1984). Thus it is apparent that the chronic action of ACTH to maintain optimal levels of steroid hydroxylases is mediated by cAMP whose synthesis by adenylate cyclase is regulated by binding of ACTH to its adrenal cell receptor. Based on this data, reasonable speculation was that cAMP played a role in the transcriptional activation of the genes encoding the steroid hydroxylases and related enzymes. However, proof of such a hypothesis required identification and characterization of complementary DNA (cDNA) probes specific for each of these enzymes.

Regulation of synthesis of steroid hydroxylases in other steroidogenic cells has been examined utilizing methodologies similar to those described above. Treatment with human chorionic gonadotropin (hCG) or cyclic AMP analogs enhances the synthesis of P-450$_{scc}$ and adrenodoxin in rat Leydig cells (ANDERSON and MENDELSON, 1985) and in Leydig cells from immature pig testis (MASON et al., 1984). FSH or cyclic AMP analogs have a similar action on the synthesis of P-450$_{scc}$, adrenodoxin and NADPH-cytochrome P-450 reductase in cultured bovine granulosa cells (FUNKENSTEIN et al., 1983b; FUNKENSTEIN et al., 1984) or rat granulosa cells (TRZECIAK et al., 1986). Presumably then, the same general cAMP-mediated mechanism is at work to regulate the synthesis of steroid hydroxylases in stereoidogenic cells in adrenal cortex, ovary and testis, and the specificity associated with this process is derived from the peptide hormones and the localization of their specific receptors.

4. Regulation of levels of steroid hydroxylase mRNA by peptide hormones

In order to obtain a more detailed understanding of the mechanism(s) by which cAMP regulates steroid hydroxylase levels it was necessary to identify and characterize cDNA probes specific for the steroid hydroxylases and related enzymes. This has been accomplished in several laboratories, leading to probes specific for bovine (MOROHASHI et al., 1984; JOHN et al., 1984), and human (CHUNG et al., 1986a) and rat (MCMASTERS et al., 1987) P-450$_{scc}$, bovine (JOHN et al., 1985; MOROHASHI et al., 1987a; CHUA et al., 1987) and human (CHUA et al., 1987) P-450$_{11\beta}$, bovine (ZUBER et al., 1986a), human (CHUNG et al., 1987; BRADSHAW et al., 1987) P-450$_{17\alpha}$ and chicken (ONO et al., 1988) P-450$_{17\alpha}$ and bovine (WHITE et al., 1984a; JOHN et al., 1986a, YOSHIAKA et al., 1986) P-450$_{C21}$. In addition, cDNA probes have been characterized for bovine (OKA-MURA et al., 1985), human (PICADO-LEONARD et al., 1988; MITTAL et al., 1988) and chicken (KAGIMOTO et al., 1988a) adrenodoxin, bovine adrenodoxin reductase (SAGARA et al., 1987; HANUKOGLU et al., 1987) and rat liver NADPH-cytochrome P-450 reductase (GONZALEZ and KASPER, 1982). These probes have provided the means with which to investigate the effect of cAMP on steroid hydroxylase mRNA levels, and as will be described subsequently, have led to identification and characterization of genes encoding the steroid hydroxylases. As can be imagined, an additional benefit derived from the cDNA clones was the elucidation of the primary sequence of the steroid hydroxylases. The sequence deduced for P-450$_{scc}$ (MOROHASHI et al., 1984) played an important role in establishing that HR2, the conserved cysteine-containing region nearest the C-terminal end of all cytochromes P-450, contains the heme binding cysteine residue. The sequence homology amongst these four forms of P-450 steroid hydroxylases is 35% or less indicating that each is a member of a distinct gene family within the P-450 supergene family.

The specific cDNA probes have been used in conjunction with primary cultures of bovine adrenocortical cells to investigate regulation of mRNA levels for these various enzymes. As in the immunological experiments described above, the cell cultures were treated with ACTH or cAMP analogs for various periods of time, RNA isolated from the cells, and specific mRNA levels determined by Northern analysis. The mRNA sizes for most P-450s are about 2 kb, a reasonable size considering that 1.5 kb of coding information is required for a protein containing 500 amino acids (Fig. 2). The mRNA for both bovine

P-450$_{\text{scc}}$ (JOHN et al., 1984) and P-450$_{17\alpha}$ (ZUBER et al., 1986a) is about 1.9 kb
and only a single mRNA band is apparent upon Northern analysis (Fig. 2).
Bovine P-450$_{C21}$ mRNA consists of 2 sizes, 2.3 kb and 2.0 kb, the size differences perhaps being due to utilization of two different poly A addition sites
(JOHN et al., 1986a). The most unusual of the steroid hydroxylase mRNA's is

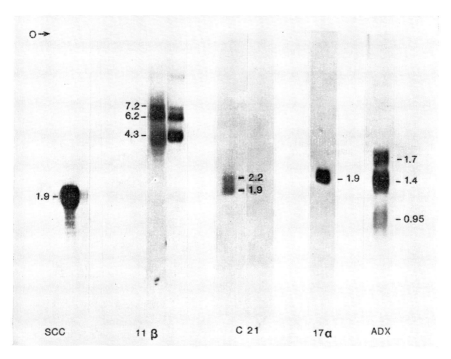

Fig. 2. Sizes of bovine mRNA species encoding steroid hydroxylases and adrenodoxin
as determined by Northern analysis. SCC = P-450$_{\text{scc}}$ and Adx = adrenodoxin.

that for P-450$_{11\beta}$. The smallest P-450$_{11\beta}$ mRNA is 4.3 kb in length, twice as
long as necessary for a 50,000 molecular weight P-450 (JOHN et al., 1985). Presumably this mRNA contains a very long 3'-untranslated region. Even more
curious is the observation of two larger forms of P-450$_{11\beta}$ mRNA, one of 6.2 kb
and one of 7.3 kb, both of which produce immunodetectable P-450$_{11\beta}$ when
translated in an in vitro system. Perhaps these forms have even longer 3'-un-
translated regions. Adrenodoxin is encoded by 3 or 4 mRNAs utilizing distinct
poly A addition sites (OKAMURA et al., 1985; OKAMURA et al., 1987). In fact
the profile of adrenodoxin mRNA is even more complex because of the presence of two types of mRNA each encoding distinct precursor sequences, and
each type utilizing 3 or 4 poly A addition sites (OKAMURA et al., 1987). Fol-

---

lowing ACTH treatment of primary bovine adrenocortical cell cultures, increased mRNA levels for P-450$_{scc}$ (JOHN et al., 1984), P-450$_{11\beta}$ (JOHN et al., 1985), P-450$_{17\alpha}$ (ZUBER et al., 1986a), P-450$_{C21}$ (JOHN et al., 1986a) and adrenodoxin (OKAMURA et al., 1985) are observed. This action is mimicked by cyclic AMP analogs. The increases in RNA levels range from about 4-fold to 20-fold, the larger increase being observed for P-450$_{17\alpha}$ which has virtually undetectable levels of mRNA in unstimulated cell cultures (Fig. 3). Thus it can be concluded that the increase in steroid hydroxylase levels observed immunochemically (see Section 3.) results from increased levels of mRNA. There are two general mechanisms by which ACTH treatment could lead to enhanced levels of

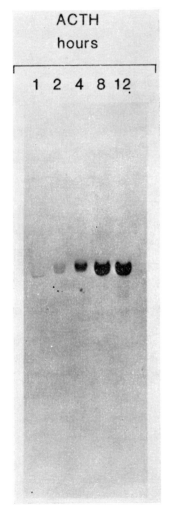

Fig. 3. Temporal pattern of ACTH mediated accumulation of P-450$_{17\alpha}$ mRNA in primary cultures of bovine adrenocortical cells (ZUBER et al., 1986a).

M. R. WATERMAN; E. R. SIMPSON

steroid hydroxylase mRNA. On the one hand, cAMP could enhance transcription of steroid hydroxylase genes while on the other hand it could stabilize existing steroid hydroxylase mRNA species.

The levels of mRNA specific for steroidogenic enzymes in other steroidogenic cells can also be increased by peptide hormones via cAMP. For example, P-450$_{scc}$ mRNA in human fetal adrenal cells (JOHN et al., 1986b; DiBLASIO et al., 1987), in human granulosa cells (VOUTILAINEN et al., 1986) and in rat granulosa cells (GOLDRING et al., 1987; TRZECIAK et al., 1987a) and P-450$_{17\alpha}$ mRNA in human fetal adrenal and Leydig cells (DiBLASIO et al., 1987) all show increases similar to those noted above for bovine adrenocortical cells in response to the appropriate stimuli. Thus it can be imagined that the same general mechanisms function in all steroidogenic cells to regulate the expression of genes encoding steroid hydroxylases thereby maintaining optimal steroidogenic capacity.

5. Studies on the mechanism of cAMP-mediated enhancement of steroid hydroxylase mRNA levels

In order to evaluate in detail the mechanism by which cAMP regulates steroid hydroxylases mRNA levels, two types of experiments have been performed; one to investigate the effect of cAMP on RNA stability and one to investigate the effect of cAMP on the transcription of steroid hydroxylase genes. The half-lives of steroid hydroxylase mRNAs were determined by using a pulse-chase technique. Primary bovine adrenocortical cells were maintained in the presence or absence of ACTH for 24 hours and then incubated with ^{3}H-uridine, to generate total radiolabeled RNA. The ^{3}H-uridine was chased with unlabeled uridine and at selected time intervals, RNA was isolated from the cells and hybridized with unlabeled cDNA probes. By determining the amount of radiolabeled, specific RNA remaining at various times, the half-lives of the RNAs were estimated (BOGGARAM et al., 1989). As seen in Table 2, ACTH had no effect on the half-lives of RNA encoding P-450$_{11\beta}$, P-450$_{C21}$, P-450$_{17\alpha}$ and adrenodoxin. However, ACTH treatment led to a 5-fold increase in the half life of P-450$_{scc}$ RNA. Thus ACTH has a post-transcriptional effect on P-450$_{scc}$ RNA, but not on the other RNAs examined. The effect of ACTH on transcription of the steroid hydroxylase genes was determined from nuclear run-on experiments. In this type of experiment, nuclei were isolated from ACTH-stimulated and unstimulated primary adrenocortical cell cultures. The isolated nuclei were then incubated with a mixture of nucleotides including ^{32}P-labeled UTP, RNA molecules whose transcription had been initiated prior to nuclei isolation were elongated and consequently radiolabeled. The RNA was then isolated from the nuclei and hybridized to unlabeled cDNA probes. By isolating nuclei from cells at various times following initiation of ACTH treatment it was found that within approximately 8 hours increased levels of radiolabeled RNA encoding P-450$_{scc}$, P-450$_{11\beta}$, P-450$_{17\alpha}$, P-450$_{C21}$ and adrenodoxin were observed (JOHN et al., 1986c). These results indicate that a major action of cAMP in bovine adrenocortical cells is to increase the transcription of the

Table 2. Half-lives of steroid hydroxylase RNA, molecules in bovine adrenocortical cells

RNA	$t_{1/2}$ − ACTH	$t_{1/2}$ + ACTH
P-450$_{11\beta}$	6 hr	8 hr
P-450$_{17\alpha}$	10 hr	13 hr
P-450$_{C21}$	13 hr	17 hr
Adrenodoxin	7 hr	12 hr
P-450$_{scc}$	5 hr	25 hr

genes encoding the steroid hydroxylases. An additional role of the second messenger is to stabilize P-450$_{scc}$ mRNA. It should be emphasized that we do not consider these actions of ACTH to lead to induction of steroid hydroxylases in normal adrenal cortex, but rather to maintain optimal steroidogenic capacity in this tissue. However, it would not be surprising if under prolonged periods of stress associated with elevated levels of ACTH or other conditions of abnormally high production of ACTH, that induced levels of steroid hydroxylases might be observed. However, this would not be a normal role of ACTH. Similar experiments to investigate the transcriptional and post-transcriptional actions of cAMP in ovary and testis have not been carried out, but we imagine that cAMP has similar actions on steroid hydroxylase genes in these tissues.

The detailed mechanism by which cAMP enhances the transcription of steroid hydroxylase genes has not yet been elucidated. However, it is known that protein synthesis is required for this event to occur in the adrenal cortex, as established by use of the protein synthesis inhibitor, cycloheximide. Upon treatment of bovine adrenocortical cells with ACTH or cAMP in the presence of cycloheximide, accumulation of steroid hydroxylase mRNA is inhibited (Fig. 4). This is found not to be a general effect on adrenocortical RNA, but rather is limited to a small number of genes including those which encode P-450$_{scc}$, P-450$_{11\beta}$, P-450$_{17\alpha}$, P-450$_{C21}$ and adrenodoxin (JOHN et al., 1986c).

M. R. WATERMAN; E. R. SIMPSON

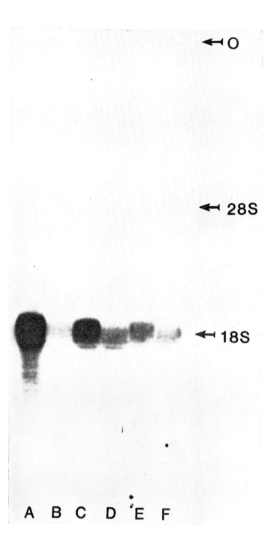

Fig. 4. Inhibition of accumulation of P-450$_{scc}$ mRNA in primary cultures of bovine adrenocortical cells by cycloheximide as determined by Northern analysis. A — ACTH treatment; B — control (untreated); C — dibutyryl cAMP treatment; D — ACTH + cycloheximide; E — dibutyryl cAMP + cycloheximide; F — control + cycloheximide.

The ability of cycloheximide to inhibit steroid hydroxylase mRNA accumulation is observed for both nuclear and cytoplasmic RNA accumulation. Also cycloheximide has no effect on the transport of steroid hydroxylase mRNA from the nucleus to the cytosol. Therefore, we have developed a working hypothesis for the transcriptional activation of steroid hydroxylase genes which is illustrated schematically in Figure 5. In this hypothesis, ACTH binds to its receptor on the surface of adrenocortical cells and activates adenylate cyclase leading to elevated levels of intracellular cAMP. Either directly or mediated by as yet unidentified factors, elevated levels of cAMP lead to increased levels

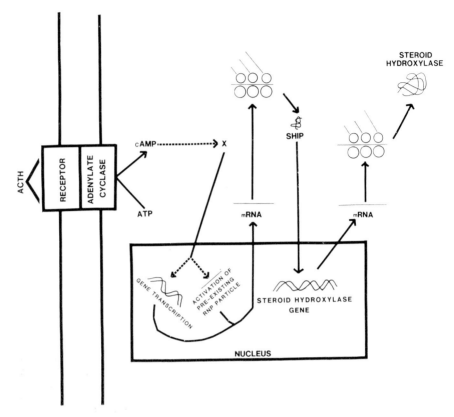

Fig. 5. Working hypothesis of ACTH-mediated transcription of steroid hydroxylase gene expression (WATERMAN and SIMPSON, 1985).

of mRNA which are translated into a protein or proteins designated as steroid hydroxylase inducing proteins (SHIP) (WATERMAN and SIMPSOM, 1985). SHIP then activates the transcription of steroid hydroxylase genes leading to the production of steroid hydroxylase mRNA and subsequently the steroid hydroxylases themselves. It is the translational synthesis of SHIP which is presumed to be inhibited by cycloheximide. It has also been shown that cycloheximide has a similar effect on P-450$_{scc}$ mRNA levels in human fetal adrenal cells in culture (JOHN et al., 1986b). A number of unanswered questions are raised by this hypothesis. For example, does one SHIP regulate transcription of all steroid hydroxylases or is there a specific SHIP for each gene? Also does cAMP directly mediate SHIP synthesis or are there several steps occurring in between elevation of cAMP levels and SHIP synthesis? Experiments designed to provide answers to these questions are presently underway in our laboratories.

The hypothesis outlined in Figure 5. indicates that SHIP is a trans-acting factor which regulates steroid hydroxylase gene expression by interacting with a cis-regulatory element associated with the gene, perhaps even an enhancer. The first step toward testing this aspect of the hypothesis is to identify and characterize steroid hydroxylase genes. The structure of the human P-450$_{scc}$ gene has recently been reported (MOROHASHI et al., 1987) and the P-450$_{C21}$ gene as well as a closely linked pseudogene has been characterized in several species. In fact the genes encoding mouse (WHITE et al., 1984b; CARROLL et al., 1985; AMOR et al., 1985), bovine (CHUNG et al., 1986b) and human (WHITE et al., 1986; HIGASHI et al., 1986) P-450$_{C21}$ have all been characterized structurally. Most recently the human P-450$_{17\alpha}$ gene (PICADO-LEONARD and MILLER, 1987; KAGIMOTO et al., 1988b) and the bovine P-450$_{17\alpha}$ gene (BHASKER et al., 1989) have been structurally characterized. Also the 5'-region of the bovine adrenodoxin gene has been characterized (KAGIMOTO et al., 1988c). The genes encoding P-450$_{C21}$ are located on human chromosome 6 within the HLA gene complex (WHITE et al., 1984b). One P-450$_{C21}$ pseudogene and the expressed P-450$_{C21}$ gene are closely linked within this locus. The human P-450$_{scc}$ gene is localized on chromosome 15 (CHUNG et al., 1986a) while that for P-450$_{17\alpha}$ is found on chromosome 10 (MATTESON et al., 1986). Most recently, the human P-450$_{11\beta}$ gene has been localized to chromosome 8 (CHUA et al., 1987).

Studies on the regulatory region of the mouse P-450$_{C21}$ genes demonstrated that only the complete gene was functional and that transcription of this gene could be increased by ACTH (PARKER et al., 1983). However when the promoter region of the pseudogene was used in place of the 5'-flanking region of the functional gene, this chimeric gene was functional indicating that the pseudogene contains a functional promoter (CHAPLIN et al., 1986). Further evaluation of the mouse 21-hydroxylase gene has shown that deletion of sequences from -1100 to -230 bp in the 5'-flanking region of this gene had no significant effect on its transcription. However the region between -230 and -180 bp contains an important regulatory sequence. Within this region is located a sequence which is highly conserved between the mouse, bovine and human P-450$_{C21}$ genes (PAARKER et al., 1986). These investigators have defined a region between -330 and -150 bp which serves as a tissue specific enhancer in Y1 cells. By DNAase I footprinting analysis, binding sites for several nuclear proteins have been found in this region, including one specific for steroidogenic cells (RICE et al., 1988). FUJII-KURIYAMA and colleagues have linked 5.4 kb of upstream untranscribed sequence of the human P-450$_{scc}$ gene to the chloramphenicol acetyltransferase (CAT) gene and have transfected Y1 cells as well as mouse fibroblast (L929), rat hepatoma (HTC) and mouse hepatoma (Hepa-1) cells with this chimeric gene. Expression was observed only in Y1 cells and enhanced CAT activity resulting from increased transcription was observed in these cells upon treatment with cAMP (INOUE et al., 1988). Recently CHUNG and colleagues have reported that a 230 bp segment of the 5'-flanking region of this gene is sufficient to direct CAT expression in steroidogenic cells (CHUNG et al., 1988). Utilizing the 5'-flanking region of the bovine P-450$_{17\alpha}$ gene coupled to the CAT gene we have observed that the sequence between -288 bp and -88 bp is required for cAMP-dependent expression of CAT activity in transfected bovine adrenocortical cells (D. WU, unpublished). This 200 bp nucleotide sequence does not contain the consensus cAMP-regulator sequence (LIN and GREEN, 1988) suggesting a new mechanism for cAMP-dependent regulation of transcription of genes encoding steroid hydroxylases. While no trans-acting factors associated with expression of these genes have yet been identified (SHIP or otherwise), it can be imagined that such experiments may be completed by appearance of this volume.

As noted above, sequence analysis of the four adrenocortical steroid hydroxylases has led to the conclusion that each protein is the product of a distinct gene family. The schematic diagrams of the genes encoding P-450$_{C21}$, P-450$_{17\alpha}$ and P-450$_{scc}$ clearly illustrate that these belong to different gene families (Fig. 6). It can be estimated that these genes diverged from one another more than 500 million years ago (NEBERT et al., 1987). It is interesting to consider that while these various forms of P-450 have diverged from one another to meet specific metabolic needs leading to specific steroid hydroxylase activities, a common mechanism for their regulation (cAMP via SHIP) has been conserved. Viewed in this light it will be extremely interesting to decipher

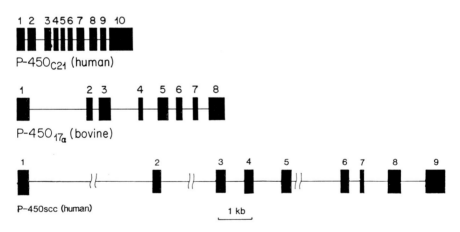

Fig. 6. Schematic diagram of steroid hydroxylase gene structure. Human P-450$_{C21}$ gene (WHITE et al., 1986); bovine P-450$_{17\alpha}$ gene (BHASKER et al., 1989); human P-450$_{scc}$ gene (MOROHASHI et al., 1987).

M. R. WATERMAN; E. R. SIMPSON

and compare the SHIP proteins and regulatory elements required for transcription of each of these steroid hydroxylase genes.

It appears that SHIP proteins may not be required for transcriptional activation of steroid hydroxylase genes in all steroidogenic tissues. In both bovine (JOHN et al., 1986c) and human (JOHN et al., 1986b) adrenal cortex, cycloheximide sensitivity of cAMP-mediated mRNA accumulation is observed. However, in human granulosa cells it has been reported that the cAMP-mediated accumulation of P-450$_{scc}$ and adrenodoxin mRNA is not inhibited by cycloheximide (GOLOS et al., 1987). Furthermore, in the transformed human trophoblast cell line JEG-3, cycloheximide treatment inhibits cAMP-mediated accumulation of P-450$_{scc}$ mRNA but enhances accumulation of adrenodoxin mRNA (PICADO-LEONARD et al., 1988).

6. Cell-specific expression of steroid hydroxylase genes

One of the most intriguing aspects of steroid hydroxylase gene expression is the cell-specific regulation noted in Figure 1. In the fasciculata-reticularis of the adrenal cortex all four steroid hydroxylases are expressed. In the glomerulosa cells of the adrenal cortex only P-450$_{scc}$, P-450$_{11\beta}$ and P-450$_{C21}$ are expressed, P-450$_{17\alpha}$ being absent. In Leydig cells of the testis only P-450$_{scc}$ and P-450$_{17\alpha}$ are expressed. In the ovarian follicle, granulosa cells contain only P-450$_{scc}$ while cells of the theca interna contain both P-450$_{scc}$ and P-450$_{17\alpha}$ (RODGERS et al., 1986a). Of course, in non-steroidogenic tissues such as liver and kidney, there is no expression of these genes. Thus there is a very intriguing all-or-none type of expression of steroid hydroxylase genes in various cell types. Studies are presently underway by modifying the 5'-flanking region of the mouse P-450$_{C21}$ gene to identify cis-acting elements involved in cell-specific gene expression (PARKER et al., 1985; PARKER et al., 1986).

In addition to the all-or-none type of cell-specific regulation of gene expression noted above, there is one very intriguing example of physiological regulation of cell-specific expression. In the ovarian follicle, following ovulation the cells of the granulosa and theca interna differentiate into luteal cells making up the mass of the corpus luteum, a very active progesterone producing steroidogenic tissue. During this differentiation process, there is a pronounced increase in the level of P-450$_{scc}$ in both cell types accounting in part for the dramatic increase in progesterone production (RODGERS et al., 1986b). This is associated with disappearance of P-450$_{17\alpha}$ from the cells of the bovine theca interna (RODGERS et al., 1986c; RODGERS et al., 1987). Thus as the cells of the bovine theca interna differentiate into luteal cells, P-450$_{scc}$ gene expression is enhanced while P-450$_{17\alpha}$ gene expression is repressed. This physiologically-mediated transition in steroid hydroxylase gene expression within a single cell type occurs monthly within cycling women and would seem to be quite distinct from the all-or-none types of cell-specific steroid hydroxylase gene expression noted above. The mechanisms of both all-or-none and physiologically-mediated cell specific regulation will be extremely interesting to resolve.

It should be noted that the potential exists for expression of some or all of these enzymes in tissues not normally considered to be steroidogenic. This is emphasized by the appearance of P-450$_{C21}$ RNA in mouse liver (AMOR et al., 1985) and by the detection by immunochemical methods of P-450$_{scc}$ and adrenodoxin in white matter throughout the rat brain as well as in cells of the entorhinal and cingulate cortex and in the olfactory bulb (DE COSCOGNE et al., 1987). The role of steroid hydroxylases in the brain is unclear, but their presence in the brain adds a new dimension to studies aimed at understanding the molecular basis of cell-specific steroid hydroxylase gene expression.

7.   Ontogeny of steroid hydroxylase gene expression

Dating back to the pioneering studies of Kimura and Purvis, it has been apparent that the chronic action of peptide hormones mediated by cAMP to regulate levels of steroid hydroxylase activity is the major mechanism throughout adult life by which optimal steroidogenic capacity in various tissues is maintained. However, the results described so far in this Chapter do not address the issue of developmental regulation of steroid hydroxylase gene expression. We have had an opportunity to examine steroid hydroxylase protein and mRNA levels in adrenals from anencephalic human fetuses. Anencephalic fetuses have no functional hypothalamic-pituitary axis, little or no detectable ACTH in fetal blood, and extremely low levels of adenylate cyclase in their adrenal membranes (CARR, 1986). We reasoned, based on these data and results of our studies on cAMP regulated gene expression, that low or undetectable levels of steroid hydroxylases would be found in anencephalic adrenals compared to those from normal human fetal adrenal. Much to our surprise, this hypothesis did not stand the test of experiment. We found similar, relatively high levels of steroid hydroxylase RNA and protein in anencephalic fetal adrenals and normal fetal adrenals (JOHN et al., 1987). This was also true for adrenodoxin. We conclude that there is a third, important mechanism of steroid hydroxylase gene expression which is manifest early in fetal development leading to the original expression of steroid hydroxylase gene or their fetal imprinting. This mechanism might be cAMP-independent and thus precede, temporally, the cAMP-dependent or adult mechanism of maintenance of optimal steroidogenic capacity.

Recently we have initiated studies on the ontogeny of steroid hydroxylase gene expression in bovine fetuses. The results of these studies indicate that both cAMP-dependent and cAMP-independent regulatory mechanisms are important during fetal development in this species (LUND et al., 1988). Western

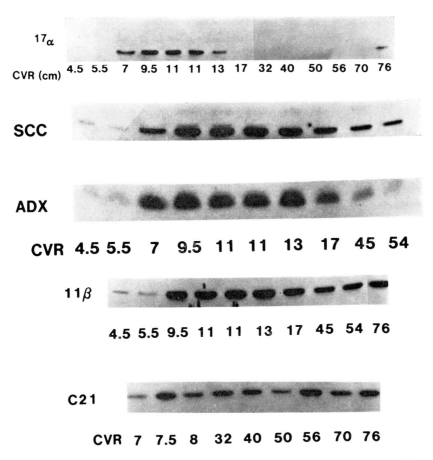

Fig. 7. Immunoblot analysis of steroid hydroxylases in fetal bovine adrenal homogenates. CVR (cm) is the length of each fetus measured from the crown along the vertebrae to the rump and is related to gestational age (20 cm $\simeq$ 100 days and 80 cm $\simeq$ 280 days).

blot analysis of bovine fetal adrenals ranging in gestational age from about 40 days to 280 days (term $\simeq$ 280 days) shows that P-450$_{scc}$, P-450$_{C21}$, P-450$_{11\beta}$ and adrenodoxin are present throughout gestation (Fig. 7.). However, P-450$_{17\alpha}$ is detectable only during early and late gestation and is absent between about 120 days and 250 days during midgestation. The physiological consequence of this episodic pattern is the absence of fetal cortisol production during midgestation. The levels of P-450$_{17\alpha}$ in the fetal adrenal follow the pattern of immunodetectable ACTH in fetal plasma. During the midgestation period ACTH is undetectable. Presumably then, all steroid hydroxylases in bovine

---

fetal adrenal except $P-450_{17\alpha}$ are regulated by both cAMP-dependent and cAMP-independent mechanisms. On the other hand, $P-450_{17\alpha}$ is strictly dependent on cAMP levels and is not expressed during the period at which ACTH is not present or at a very low level. It is interesting to note, however, that $P-450_{17\alpha}$ is readily detectable in fetal bovine testis at times that is undetectable in fetal adrenal, indicating a tissue specific element associated with the ontogeny of bovine steroid hydroxylases. Thus, the fetal bovine model system has several interesting features which make it excellent for investigation of steroid hydroxylase ontogeny.

8. Summary of mechanisms of steroid hydroxylase gene expression

From the discussion presented above it is apparent that there are at least three major mechanisms of regulation of steroid hydroxylase gene expression, summarized in Table 3. Each of these mechanisms probably involves specific

Table 3. Major mechanisms of steroid hydroxylase gene expression

1. cAMP Dependent — SHIP

2. Cell-specific
    a) All-or-none regulation
    b) Physiological regulation

3. Developmental — fetal imprinting

trans-acting protein factors which interact with specific cis-regulatory elements associated with steroid hydroxylase genes. The mechanisms by which normal steroidogenic capacity is maintained throughout adult life in the adrenal cortex requires cAMP regulation of SHIP factors, which are hypothesized to be trans-acting factors. Whether a single SHIP protein regulates all steroid hydroxylase genes or each gene is regulated by its own SHIP remains to be determined, but it is certainly likely that more than one factor is involved. Secondly, there are two classes of cell-specific steroid hydroxylase gene expression, all-or-none and physiologically regulated. Given the complexity of the cell types involved, it seems likely that there will be a large number of trans-acting factors involved in this type of regulation. Finally the developmental aspects of steroid hydroxylase gene expression will lead to discovery of another group of trans-acting factors whose regulation may be cAMP-

M. R. WATERMAN; E. R. SIMPSON

independent. It is our thought that a complex array of protein factors, some being enhancers of gene expression and some repressors of gene expression, are required for control of steroid hydroxylase activity.

8.1. Are there other types of regulation of steroid hydroxylase gene expression?

A variety of observations made within the last few years indicate that additional mechanisms beyond those described above may play roles in regulation of steroid hydroxylase gene expression, particularly in the ovary. While the expression of P-450$_{17\alpha}$ in the theca interna noted above may be regulated by cAMP, under the control of LH, the mechanism underlying the changes occuring during luteinization are poorly understood. In a similar fashion, the increase in estrogen formation by the granulosa cells of the preovulatory follicle is clearly the result of the expression of aromatase cytochrome P-450 under the control of FSH, whose actions are mediated by cAMP. However, once again, the mechanism whereby estrogen formation declines at the time of ovulation in not understood.

It is apparent, however, that in vitro, a number of factors have been identified that regulate the activity of and the expression of the genes encoding these steroidogenic P-450s by mechanisms not involving cAMP. Thus in rat ovarian granulosa cells (TRZECIAK et al., 1987 b) and human fetal adrenal cells (MASON et al., 1986; McALLISTER and HORNSBY, 1987) phorbol esters are markedly inhibitory of the induction of P-450$_{scc}$ by FSH on the one hand and ACTH on the other, as well by cAMP. By contrast, phorbol esters potentiate the action of cAMP analogues to induce the expression of aromatase P-450 in human adipose stromal cells (EVANS et al., 1987). Since phorbol esters presumably act via protein kinase C activation, it appears that the signal transduction mechanism initiated by this enzyme has profound differential effects on the regulation of expression of these steroidogenic forms of P-450.

Another group of substances which regulate the expression of these enzymes in ovarian granulosa cells in vitro are the growth factors. Thus IGF-I appears to stimulate the synthesis of both P-450$_{scc}$ (VELDHUIS et al., 1986) and aromatase P-450 (STEINKAMP et al., 1986) in granulosa cells. On the other hand, EGF potentiates the action of FSH to induce the synthesis of P-450$_{scc}$ (TRZECIAK et al., 1987 c), whereas it inhibits the actions of both FSH and dibutyryl cAMP to induce the expression of aromatase P-450 in both granulosa cells (MERRILL et al., 1987) and adipose stromal cells (EVANS et al., 1987). Thus we observe that within the same cell, a regulatory factor has opposite effects on the expression of two steroidogenic P-450 enzymes.

In addition, several other factors have been shown to alter steroidogenic enzyme activities in granulosa cells although it has not yet been demonstrated that these effects are via induction of synthesis. Thus, estrogens and androgens increase aromatase activity, whereas GnRH is inhibitory of both aromatase activity and progesterone formation (HSUEH et al., 1984).

We envision that in the ovary locally-produced agents, such as growth factors which may be produced at the time of ovulation in response to the gonadotropin surge, act via mechanisms other than that involving cAMP to bring about the differential expression of steroidogenic cytochromes P-450, which is a characteristic feature of the luteinization process.

9. Concluding statement

The multifactorial regulation of steroid hydroxylase gene expression serves as a wonderful model system for understanding mechanisms of regulation of P-450 gene expression, particularly at the cell-specific and developmental levels. In addition, however, because this system is so complex, elucidation of the various mechanisms outlined above will also inevitably play an important role in our understanding of the overall mechanism of gene expression in eukaryotes.

Acknowledgements

The authors wish to acknowledge the intellectual and experimental insights of their colleagues who have contributed so heavily to this research program. Also the support of USPHS Grants DK 28350, and HD 13234 and Grant I-624 from The Robert A. Welch Foundation is appreciated.

10. References

AMOR, M., M. TOSI, G. CUPONCHEL, M. STEINMETZ, and T. MEO, (1985), Proc. Natl. Acad. Sci. USA **82**, 4453—4457.

ANDERSON, C. M. and C. R. MENDELSON, (1985), Arch. Biochem. Biophys. **238**, 379 to 387.

ASANO, K. and B. W. HARDING, (1976), Endocrinol. **99**, 977—987.

BHASKER, C. R., B. S. ADLER, A. DEE, M. E. JOHN, M. KAGIMOTO, M. X. ZUBER, R. AHLGREN, X. WANG, E. R. SIMPSON, and M. R. WATERMAN, (1989), Arch. Biochem. Biophys., **271**, 479—487.

BOGGARAM, V., M. E. JOHN, E. R. SIMPSON, and M. R. WATERMAN, (1989), Biochem. Biophys. Res. Commun. **160**, 1227—1232.

BRADSHAW, K. D., M. R. WATERMAN, R. T. COUCH, E. R. SIMPSON, and M. X. ZUBER, (1987), Molec. Endocrinol. **1**, 348—354.

CARR, B. R., (1986), J. Clin. Endocrinol. Metab. **63**, 51—55.

CARROLL, M. C., D. CAMPBELL, and R. R. PORTER, (1985), Proc. Natl. Acad. Sci. USA **82**, 521—525.

CHAPLIN, D. D., J. GAILRAITH, J. G. SEIDMAN, P. C. WHITE, and K. L. PORTER, (1986), Proc. Natl. Acad. Sci. USA **83**, 9601—9605.

CHUA, S. C., P. SZABO, A. VITEK, K. H. GRZESCHIK, M. E. JOHN, and P. C. WHITE, (1987), Proc. Natl. Acad. Sci. USA **84**, 7193—7197.

CHUNG, B.-C., K. J. MATTESON, R. VOUTILAINEN, T. K. MOHANDAS, and W. L. MILLER, (1986a), Proc. Natl. Acad. Sci. USA **83**, 8962–8966.

CHUNG, B.-C., K. J. MATTESON, and W. L. MILLER, (1986b), Proc. Natl. Acad. Sci. USA **83**, 4243–4247.

CHUNG, B.-C., J. PICADO-LEONARD, M. HANIU, M. BIENKOWSKI, P. F. HALL, J. E. SHIVELY, and W. L. MILLER, (1987), Proc. Natl. Acad. Sci. USA **84**, 407–411.

CHUNG, B.-C., C.-C. LAI, M.-C. HU, and C.-H. LINN, (1988), J. Cell Biochem., Supplement 12D, p. 176 (Abstract).

DEE, A., G. CARLSON, C. SMITH, B. S. MASTERS, and M. R. WATERMAN, (1985), Biochem. Biophys. Res. Commun. **128**, 650–656.

DIBLASIO, A. M., R. VOUTILAINEN, R. B. JAFFE, and W. L. MILLER, (1987), J. Clin. Endocrinol. Metab. **65**, 170–175.

DUBOIS, R. N., E. R. SIMPSON, R. E. KRAMER, and M. R. WATERMAN, (1981a), J. Biol. Chem. **256**, 7000–7005.

DUBOIS, R. N., E. R. SIMPSON, J. TUCKEY, J. D. LAMBETH, and M. R. WATERMAN, (1981b), Proc. Natl. Acad. Sci. USA **78**, 1028–1032.

EVANS, C. T., C. J. CORBIN, C. T. SAUNDERS, E. R. SIMPSON, and C. R. MENDELSON, (1987), J. Biol. Chem. **262**, 6914–6920.

FUNKENSTEIN, B., J. L. MCCARTHY, K. M. DUS, E. R. SIMPSON, and M. R. WATERMAN, (1983a), J. Biol. Chem. **258**, 9398–9405.

FUNKENSTEIN, B., M. R. WATERMAN, B. S. S. MASTERS, and E. R. SIMPSON, (1983b), J. Biol. Chem. **258**, 10187–10191.

FUNKENSTEIN, B., M. R. WATERMAN, and E. R. SIMPSON, (1984), J. Biol. Chem. **259**, 8572–8577.

GOLDRING, N. B., J. M. DURICA, J. LIFKA, L. HEDIN, S. L. RATOOSH, W. L. MILLER, J. ORLY, and J. S. RICHARDS, (1987), Endocrinol. **120**, 1942–1950.

GOLOS, T. G., W. L. MILLER, and J. F. STRAUSS, (1987), J. Clin. Invest. **80**, 896–899.

GONZALEZ, F. J. and C. B. KASPER, (1982), J. Biol. Chem. **257**, 5962–5968.

GOSPODAROWICZ, D., C. R. ILL, P. J. HORNSBY, and G. N. GILL, (1977), Endocrinol. **100**, 1080–1089.

HANUKOGLU, I. and Z. HANUKOGLU, (1986), Eur. J. Biochem. **157**, 27–31.

HANUKOGLU, I., T. GUTFINGER, M. HANIU, and J. E. SHIVELY, (1987), Eur. J. Biochem. **169**, 449–485.

HIGASHI, Y., H. YOSHIOKA, M. YAMANE, O. GOTOH, and Y. FUJII-KURIYAMA, (1986), Proc. Natl. Acad. Sci. USA **83**, 2841–2845.

HIWATASHI, A. and Y. ICHIKAWA, (1979), Biochim. Biophys. Acta **580**, 44–63.

HSUEH, A. J., E. Y. ADESHI, P. B. JONES, and T. H. WELSH, (1984), Endocrine Rev., 76–127.

INOUE, H., Y. HIGASHI, K. MOROHASHI, and Y. FUJII-KURIYAMA, (1988), Eur. J. Biochem. **171**, 435–440.

JEFCOATE, C. R., B. C. MCNAMARA, and M. J. DIBARTOLOMEIS, (1986), Endo. Res. **12**, 315–350.

JOHN, M. E., M. C. JOHN, P. ASHLEY, R. J. MACDONALD, E. R. SIMPSON, and M. R. WATERMAN, (1984), Proc. Natl. Acad. Sci. USA **81**, 5628–5632.

JOHN, M. E., M. C. JOHN, E. R. SIMPSON, and M. R. WATERMAN, (1985), J. Biol. Chem. **260**, 5760–5767.

JOHN, M. E., T. OKAMURA, A. DEE, B. ADLER, M. C. JOHN, P. C. WHITE, E. R. SIMPSON, and M. R. WATERMAN, (1986a), Biochemistry **25**, 2846–2853.

JOHN, M. E., E. R. SIMPSON, M. R. WATERMAN, and J. I. MASON, (1986b), Mol. Cell. Endocrinol. **45**, 197–204.

JOHN, M. E., M. C. JOHN, V. BOGGARAM, E. R. SIMPSON, and M. R. WATERMAN, (1986c), Proc. Natl. Acad. Sci. USA **83**, 4715–4719.

JOHN, M. E., E. R. SIMPSON, B. C. CARR, R. R. MAGNESS, C. R. ROSENFELD, M. R. WATERMAN, and J. I. MASON, (1987), Mol. Cell. Endocrinol. **50**, 263—268.

JONES, D. B. C., L. K. DURRIN, D. R. GALEAZZI, and J. P. WHITLOCK, Jr., (1986), Proc. Natl. Acad. Sci. USA **83**, 2802—2806.

KAGIMOTO, M., K. KAGIMOTO, E. R. SIMPSON, and M. R. WATERMAN, (1988), J. Biol. Chem. **263**, 8925—8928.

KAGIMOTO, K., J. L. McCARTHY, M. R. WATERMAN, and M. KAGIMOTO, (1988a), Biochem. Biophys. Res. Commun. **155**, 379—383.

KAGIMOTO, M., J. S. D. WINTER, K. KAGIMOTO, E. R. SIMPSON, and M. R. WATERMAN, (1988b), Mol. Endocrinol. **2**, 564—570.

KIMURA, T., (1969), Endocrinol. **85**, 492—499.

KOWAL, J., (1969), Biochemistry **8**, 1821—1831.

KOWAL, J., E. R. SIMPSON, and R. W. ESTABROOK, (1970), J. Biol. Chem. **245**, 2438 to 2443.

KRAMER, R. E., C. M. ANDERSON, J. L. McCARTHY, E. R. SIMPSON, and M. R. WATERMAN, (1982a), Fed. Proc. **41**, 1298.

KRAMER, R. E., C. M. ANDERSON, J. A. PETERSON, E. R. SIMPSON, and M. R. WATERMAN, (1982b), J. Biol. Chem. **257**, 14921—14925.

KRAMER, R. E., R. N. DuBOIS, E. R. SIMPSON, C. M. ANDERSON, K. KASHIWAGI, J. D. LAMBETH, C. R. JEFCOATE, and M. R. WATERMAN, (1982c), Arch. Biochem. Biophys. **215**, 478—485.

KRAMER, R. E., E. R. SIMPSON, and M. R. WATERMAN, (1983), J. Biol. Chem. **258**, 3000—3005.

KRAMER, R. E., W. E. RAINEY, B. FUNKENSTEIN, A. DEE, E. R. SIMPSON, and M. R. WATERMAN, (1984), J. Biol. Chem. **259**, 707—713.

LE COSCOGNE, C., P. ROBEL, M. MOUEZOU, N. SANANES, E.-F. BAULIEU, and M. R. WATERMAN, (1987), Science **237**, 1212—1215.

LIN, Y.-S. and M. R. GREEN, (1988), Proc. Natl. Acad. Sci. USA **85**, 3396—3400.

LUND, J., D. J. FAUCHER, S. P. FORD, J. C. PORTER, M. R. WATERMAN, and J. I. MASON, (1988), J. Biol. Chem. **263**, 16195—16201.

MASON, J. I., A. A. MacDONALD, and A. LAPTOOK, (1984), Biochim. Biophys. Acta **795**, 504—512.

MASON, J. I., B. R. CARR, and W. E. RAINEY, (1986), Endocrine Res. **12**, 447—467.

MATTESON, K. J., J. PICADO-LEONARD, B.-C. CHUNG, T. K. MOHANDAS, and W. L. MILLER, (1986), J. Clin. Endocrinol. Metab. **63**, 789—791.

McALLISTER, J. M. and P. J. HORNSBY, Endocrinology **121**, 1908—1910.

McMASTERS, K. M., L. A. DICKSON, R. V. SHAMY, K. ROBINSON, G. J. MacDONALD, and W. R. MOYLE, (1987), Gene **57**, 1—9.

MERRILL, J. C., M. P. STEINKAMPF, C. R. MENDELSON, and E. R. SIMPSON, Absts. 69th Ann. Meeting Endocrine Soc., Indianapolis, (1987), p. 267.

MITTAL, S., Y.-Z. ZHU, and L. E. VICKERY, (1988), Arch. Biochem. Biophys. **264**, 383 to 391.

MOROHASHI, K., Y. FUJII-KURIYAMA, Y. OKADA, Y. SOGAWA, T. HIROSE, S. INAYAMA, and T. OMURA, (1984), Proc. Natl. Acad. Sci. USA **81**, 4647—4651.

MOROHASHI, K., H. YOSHIOKA, O. GOTOH, Y. OKADA, K. VAMAMOTO, T. MIYATA, K. SOGAWA, Y. FUJII-KURIYAMA, and T. OMURA, (1987a), J. Biochem. **102**, 559—568.

MOROHASHI, K., K. SOGAWA, T. OMURA, and Y. FUJII-KURIYAMA, (1987b), J. Biochem. **101**, 879—887.

NABI, H. and T. OMURA, (1980), Biochem. Biophys. Res. Commun. **97**, 680—686.

NAKAJIN, S. and P. F. HALL, (1981a), J. Biol. Chem. **256**, 3871—3876.

NAKAJIN, S., P. F. HALL, and M. ONODA, (1981b), J. Biol. Chem. **256**, 6134—6139.

Nakajin, S., M. Shinoda, M. Haniu, J. Shively, and P. F. Hall, (1984), J. Biol. Chem. **259**, 3971—3976.

Nebert, D. W., M. Adesnik, M. J. Coon, R. W. Estabrook, F. J. Gonzalez, F. P. Guengerich, I. C. Gunsalus, E. F. Johnson, B. Kemper, W. Levin, I. R. Phillips, R. Sato, and M. R. Waterman, (1987), DNA **6**, 1—11.

Nonaka, Y., H. Murakami, Y. Yabusaki, S. Kuramitsu, H. Kagamiyama, T. Yamano, and M. Okamoto, (1987), Biochem. Biophys. Res. Commun. **145**, 1239—1247.

Ohashi, M., E. R. Simpson, J. I. Mason, and M. R. Waterman, (1983), Endocrinol. **112**, 2039—2045.

Okamura, T., M. E. John, M. X. Zuber, E. R. Simpson, and M. R. Waterman, (1985), Proc. Natl. Acad. Sci. USA **82**, 5705—5709.

Okamura, T., M. Kagimoto, E. R. Simpson, and M. R. Waterman, (1987), J. Biol. Chem. **262**, 10335—10338.

Omura, T., E. Sanders, R. W. Estabrook, D. Y. Cooper, and O. Rosenthal, (1966), Arch. Biochem. Biophys. **117**, 660—673.

Ono, H., M. Iwasaki, N. Sakanoto, and S. Mizuno, (1988), Gene **66**, 77—85.

Parker, K. L., D. D. Chaplin, M. Wong, J. G. Seidman, J. A. Smith, and B. P. Schimmer, (1985), Proc. Natl. Acad. Sci. USA **82**, 7860—7864.

Parker, K. L., B. P. Schimmer, D. D. Chaplin, and J. G. Seidman, (1986), J. Biol. Chem. **261**, 5353—5355.

Picado-Leonard, J. and W. L. Miller, (1987), DNA **6**, 439—448.

Picado-Leonard, J., R. Voutilainen, L.-C. Kao, B.-C. Chung, J. F. Strauss, and W. L. Miller, (1988), J. Biol. Chem. **263**, 3240—3244.

Purvis, J. L., J. A. Canick, J. I. Mason, R. W. Estabrook, and J. L. McCarthy, (1973a), Ann. N.Y. Acad. Sci. **212**, 319—342.

Purvis, J. L., J. A. Canick, S. A. Latie, J. H. Rosenbaum, J. Hologgitas, and R. H. Menard, (1973b), Arch. Biochem. Biophys. **159**, 39—49.

Rice, D. A., M. S. Kronenberg, B. P. Schimmer, and K. L. Parker, (1988), J. Cell Biochem., Supplement 12D, pg. 126 (Abstract).

Rodgers, R. J., H. F. Rodgers, P. F. Hall, M. R. Waterman, and E. R. Simpson, (1986a), J. Reprod. Fert. **78**, 627—638.

Rodgers, R. J., H. F. Rodgers, M. R. Waterman, and E. R. Simpson, (1986b), J. Reprod. Fert. **78**, 639—652.

Rodgers, R. J., M. R. Waterman, and E. R. Simpson, (1986c), Endocrinol. **118**, 1366 to 1374.

Rodgers, R. J., M. R. Waterman, and E. R. Simpson, (1987), Molec. Endocrinol. **1**, 274—279.

Sagara, Y., Y. Takata, T. Miyata, T. Hara, and T. Horiuchi, (1987), J. Biochem. **102**, 1333—1336.

Suhara, K., Y. Fujmara, M. Shiroo, and M. Katagiri, (1984), J. Biol. Chem. **259**, 8729—8736.

Suzuki, K. and T. Kimura, (1965), Biochem. Biophys. Res. Commun. **19**, 340—345.

Trzeciak, W. H., M. R. Waterman, and E. R. Simpson, (1986), Endocrinol. **119**, 323 to 330.

Trzeciak, W. H., M. R. Waterman, E. R. Simpson, and S. R. Ojeda, (1987a), Mol. Endocrinol. **1**, 500—504.

Trzeciak, W. H., T. Duda, M. R. Waterman, and E. R. Simpson, (1987b), J. Biol. Chem., In Press.

Trzeciak, W. H., T. Duda, M. R. Waterman, and E. R. Simpson, (1987c), Mol. Cell. Endocrinol. **52**, 43—50.

VOUTILAINEN, R., J. TAPANAINEN, B.-C. CHUNG, K. J. MATTESON, and W. L. MILLER, (1986), J. Clin. Endocrinol. Metab. **63**, 202—207.

WATERMAN, M. R. and E. R. SIMPSON, (1985), in: Microsomes and Drug Oxidations, (A. R. BOOBIS, J. CALDWELL, F. DE MATTEIS, and C. R. ELCOMBE, eds.), Taylor and Francis, Ltd., London, 136—144.

WHITE, P. C., M. I. NEW, and B. DUPONT, (1984a), Proc. Natl. Acad. Sci. USA **81**, 1986—1990.

WHITE, P. C., D. D. CHAPLIN, J. H. WEIS, B. DUPONT, M. I. NEW, and J. G. SEIDMAN, (1984b), Nature **312**, 465—467.

WHITE, P. C., M. I. NEW, and B. DUPONT, (1984), Proc. Natl. Acad. Sci. USA **81**, 7505—7509.

WHITE, P. C., M. I. NEW, and B. DUPONT, (1986), Proc. Natl. Acad. Sci. USA **83**, 5111—5115.

YOZHIOKA, H., K. MOROHASHI, K. SOGAWA, M. YAMANE, S. KOMINAMI, S. TAKENORI, Y. OKADA, T. OMURA, and Y. FUJII-KURIYAMA, (1986), J. Biol. Chem. **261**, 4106 to 4109.

ZUBER, M. X., E. R. SIMPSON, P. F. HALL, and M. R. WATERMAN, (1985), J. Biol. Chem. **260**, 1842—1848.

ZUBER, M. X., M. E. JOHN, T. OKAMURA, E. R. SIMPSON, and M. R. WATERMAN, (1986a), J. Biol. Chem. **261**, 2475—2482.

ZUBER, M. X., E. R. SIMPSON, and M. R. WATERMAN, (1986b), Science **234**, 1258—1261.

Chapter 4

Structure and Function
of Adrenal Mitochondrial Cytochrome P-450$_{11\beta}$

M. Okamoto and Y. Nonaka

1. Introduction

The hydroxylation of a steroid nucleus at the 11β-position is an essential step for the biosynthesis of glucocorticoids and the most potent mineralocorticoid, aldosterone. The reaction occurs in the mitochondria of the adrenal cortex and requires the presence of NADPH and molecular oxygen (SWEAT and LIPSCOMB, 1955; GRANT and BROWNIE, 1955). On the basis of the observation that ^{18}O from $^{18}O_2$ is incorporated at the 11β-position of 11-deoxycorticosterone, it was suggested that steroid 11β-hydroxylase belongs to a class of enzymes known as mixed-function oxidases (HAYANO et al., 1955; SWEAT et al., 1956). OMURA and coworkers undertook an intensive study aimed at characterizing the 11β-hydroxylase and discovered that it could be separated into at least three components — a flavoprotein, a non-heme iron protein and cytochrome P-450 (OMURA et al., 1965). The flavoprotein, now called NADPH: adrenocortical ferredoxin oxido-reductase (adrenodoxin reductase, EC 1.18.1.2), containing 1 FAD per mole, and the iron-sulfur (Fe_2-S_2) protein, now called adrenodoxin, form an electron transport chain from NADPH to the cytochrome P-450-dependent steroid transforming enzyme. That the enzyme responsible for the 11β-hydroxylation is of a cytochrome P-450 nature was clearly demonstrated by the results of photochemical action spectral studies in which it was found that CO inhibition of the 11β-hydroxylase enzyme system was relieved by light, the maximum relief being achieved at the wavelength of 450 nm (COOPER et al., 1967).

In this article we will discuss recently acquired knowledge of P-450$_{11\beta}$. Special emphasis will be placed upon studies conducted since its purification by TAKEMORI et al. in 1975 (TAKEMORI et al., 1975b). Those readers who are interested in the general aspects of steroidogenic P-450s should refer to the recent reviews by KATAGIRI et al. (1978a), MITANI (1979), KATAGIRI (1982), LAMBETH et al. (1982), HALL (1984), LIEBERMAN et al. (1984), WATERMAN and SIMPSON (1985), HALL (1986), and MILLER (1987).

2. Molecular properties of P-450$_{11\beta}$

2.1. Purification

It has been known for many years that the cytochrome P-450 of adrenocortical mitochondria exhibits enzymatic activity with regard to steroid 11β-hydroxylation and the side-chain cleavage of cholesterol. Various attempts have been made to extract, purify and characterize the P-450 (MITANI and HORIE, 1969; JEFCOATE et al., 1970; ANDO and HORIE, 1972; SCHLEYER et al., 1972a; BOYD et al., 1972; RAMSEYER and HARDING, 1973; SHIKITA and HALL, 1973; WANG and KIMURA, 1976). It was soon found that the P-450 could be extracted from sonicated mitochondria of bovine adrenal cortex with organic solvents or detergents. Based on the results of both enzymatic and spectral studies on partially purified preparations obtained from bovine adrenocortical mitochondria, it was suggested that two distinct cytochromes P-450 are involved in the steroid 11β-hydroxylation (P-450$_{11\beta}$) and the cholesterol side-chain cleavage (P-450$_{SCC}$) (MITANI and HORIE, 1970; MITANI et al., 1973; JEFCOATE et al., 1974; MITANI et al., 1975). Further attempts to isolate the respective P-450s in pure forms were hampered by the difficulty encountered in the solubilization of the P-450s as fully active preparations and in their separation from each other. In particular, P-450$_{11\beta}$ was found to be extremely unstable when released from the mitochondrial membrane.

In 1975, TAKEMORI and coworkers found that the presence of the substrate, deoxycorticosterone, protects P-450$_{11\beta}$ (but not P-450$_{SCC}$) almost completely from inactivation (TAKEMORI et al., 1975b). Testosterone is equally effective but cholesterol has no effect. They also found that P-450$_{11\beta}$ is much more unstable than P-450$_{SCC}$, and that P-450$_{11\beta}$ more readily aggregates than P-450$_{SCC}$ when the detergent is removed from the solution. Thus, dialysis against detergent-free buffer containing deoxycorticosterone selectively precipitates P-450$_{11\beta}$ without a serious loss of activity, while P-450$_{SCC}$ remains in the supernatant. Taking advantage of these unique properties of P-450$_{11\beta}$ and using an aniline-substituted Sepharose column chromatography (TAKEMORI et al., 1975a), they were able to successfully purify P-450$_{11\beta}$ in a fully active form (KATAGIRI et al., 1976; KATAGIRI et al., 1978b; SUHARA et al., 1978).

The specific heme content of the purified P-450$_{11\beta}$ was 12 nmoles per mg of protein (KATAGIRI et al., 1978b). The preparation was free from contaminants, as judged with SDS-polyacrylamide gel electrophoresis. The molecular weight, estimated on electrophoresis under reducing conditions, was about 50,000. On the basis of the assumption that this molecular weight corresponds to that of a protomeric form of P-450$_{11\beta}$, about 40% of the purified preparation should be in an apo-form. The specific heme content of the purified cytochrome varies with the preparation, that most recently purified having a specific content of $16-18$ nmoles per mg, but not the 20 nmoles expected for a fully holo-form. A measure for the approximate purity of the cytochrome (the substrate-complexed form) is the ratio of the absorbance at 393 nm to that at 280 nm ($A_{393}:A_{280}$), which is around 0.95 for the pure cytochrome. The turnover number for 11β-hydroxylation of deoxycorticosterone was about 110 moles of product (corticosterone) formed per mole heme per min (110 min^{-1}).

The hydrophobic nature of P-450$_{11\beta}$ requires the presence of detergents in the preparation. (In the usual purification procedure, the final preparation contains both Tween 20 and cholate.) Since all efforts to remove the detergents from a preparation resulted equally in denaturation of the protein, all the assay mixtures for P-450$_{11\beta}$ inevitably contained nonionic detergents at concentrations above their critical micelle concentration. Further addition of the same nonionic detergent apparently did not affect the rate of the P-450-catalyzed hydroxylase reaction. However, a variety of anionic detergents, such as laurate, myristate and laurylsulfate, when added in the presence of 0.3% Emulgen 913, increased, by about 2-fold, the rate of the 11β-hydroxylation of deoxycorticosterone (KATAGIRI et al., 1982). The maximum activity was obtained with a laurate : Emulgen 913 ratio of 1:4 (M/M), regardless of the absolute concentration of Emulgen 913 in the range of 0.1% to 0.3%. The anionic detergent appeared to be co-micelled with the nonionic detergent and to exert its stimulatory action by increasing the V_{max} and decreasing the K_m for adrenodoxin. Further details of the mechanisms of this action remain to be elucidated.

The purified P-450$_{11\beta}$ can be stored at a concentration of less than 5 μM in 50 mM potassium phosphate buffer, pH 7.4, containing 100 μM EDTA, 100 μM DTT, 10 μM deoxycorticosterone, 0.3% cholate and 0.3% Tween 20 at 4 °C for a few weeks; it can also be stored at $-80\,°C$ for several months without loss of activity.

The introduction of the above-mentioned purification method for the first time allowed P-450$_{11\beta}$ to be the subject of enzyme-chemical and protein-chemical research, and almost all the investigators presently involved in research on the purified P-450$_{11\beta}$ are adopting this method with a variety of modifications. Thus, instead of aniline-substituted Sepharose, phenyl-Sepharose 4B (SHIMIZU et al., 1981), aminooctyl-Sepharose (YANAGIBASHI et al., 1986) or octyl-Sepharose (WADA et al., 1985) has been used for the hydrophobic chromatography. Deoxycorticosterone, as a substrate-stabilizer, seems to be an essential ingredient in the buffers used throughout all the purification steps.

2.2. Spectral properties

The ferrous CO-bound form showes absorption maxima at 448 nm (millimolar extinction value of 102) and 550 nm.

Since deoxycorticosterone had been added as a stabilizing agent during the preparation and storage, the purified P-450$_{11\beta}$ in its ferric state showed high-spin type absorption maxima at 394 nm (millimolar extinction value of 92), 510 nm and 645 nm. A substrate-free preparation was obtained by incubation of the high-spin type preparation with an NADPH-adrenodoxin reductase-adrenodoxin electron transfer system for a couple of minutes, followed by passage through a short column of Sephadex G-25 at 4 °C. P-450$_{11\beta}$, in the absence of a substrate, showes an absorption spectrum characteristic of a low-spin hemoprotein, having maxima at 418 nm (millimolar extinction value of 105), 539 nm and 570 nm. This low-spin P-450$_{11\beta}$ is very labile and cannot be stored for longer than 30 min at 4 °C.

When an 11β-hydroxylatable substrate, such as deoxycorticosterone or 11-deoxycortisol, was added to the substrate-free form of P-450$_{11\beta}$, the absorption peaks of low-spin type rapidly and completely changed to those of a typical high-spin or substrate-complexed type spectrum. Since the magnitude of such spectral changes was dependent on the concentration of the steroid, one can estimate the affinity of the steroid to P-450$_{11\beta}$ from the results of titration experiments. From the results of such experiments, KATAGIRI and SUHARA (1982) concluded that there is a positive correlation between the calculated affinity constant for the steroid and the ability of the steroid to undergo hydroxylation. Thus, deoxycorticosterone, 11-deoxycortisol, 4-androstene-3,17-dione, testosterone and progesterone are highly 11β-hydroxylatable substrates as well as efficient high-spin inducers, while 11β-hydroxycorticosteroids, such as cortisol, corticosterone, 11β-hydroxy-4-androstene-3,17-dione and 11β-hydroxytestosterone, are hydroxylated at a position other than C-11 at much slower rates and are inefficient high-spin inducers (see section 3 for further discussion). In contrast, nonsubstrates for P-450$_{11\beta}$, such as cholesterol, pregnenolone, aldosterone, estrogens and D-camphor, all have no significant effect regarding the induction of the high-spin type absorption spectrum of ferric P-450$_{11\beta}$.

The results of electron paramagnetic resonance (EPR) studies conducted at 4.2 °K and 77 °K clearly demonstrated that the conversion from a low to high spin state corresponds to the binding of P-450$_{11\beta}$ with the steroid substrate (KOMINAMI et al., 1979). They also compared the EPR spectra of P-450$_{11\beta}$ with those of P-450$_{SCC}$, and suggested that both the high and low spin forms of P-450$_{11\beta}$ show lower rhombicity than those of P-450$_{SCC}$.

The heme-environment of P-450$_{11\beta}$ was studied by circular dichroism (CD) spectrometry (SHIMIZU et al., 1981). The magnitude of the molecular ellipticity ($[\Theta]$) at 400 nm of the ferric high-spin, deoxycorticosterone-bound form was $-100,000$, which was very similar to that of P-450$_{CAM}$ but much larger than those of high-spin type P-450s (P-448$_1$) from phenobarbital- and 3-methylcholanthrene-induced rabbit liver microsomes. The conversion of high-spin ferric P-450$_{11\beta}$ to the high-spin ferrous form resulted in an increase in the minus Soret CD band and a shift of the trough position to the lower wavelength of 384 nm. On the basis of theoretical calculations it has been shown that a coupled oscillator interaction of the π-π^* transition between porphyrin and nearby aromatic amino acid groups is responsible for most of the Soret CD band of hemoproteins. Therefore, SHIMIZU and coworkers suggested that a structural change in the heme-environment induced by the reduction may strengthen the π-π^* type interaction between the heme plane and nearby aromatic amino acid groups. An attempt to obtain the CD spectrum of the ferric low-spin, substrate-free P-450$_{11\beta}$ failed, because the sample became opaque in the heme concentration range of $5-20$ µM.

A substrate-induced change in the heme-environment of P-450$_{11\beta}$ was demonstrated by Mitani et al., (1985) using another approach, a flash photolysis technique. It was assumed that the binding of a steroid to the substrate site of P-450$_{11\beta}$ alters the bonding mode of CO with the heme-iron, thereby affecting the kinetic properties of CO-P-450$_{11\beta}$ complex formation. Based on this assumption, they measured the kinetic constants for the recombination reaction of CO with the substrate-free and substrate-bound forms of the cytochrome. The results showed that the CO-recombination rate constant (k_{on}) of substrate-free ferrous P-450$_{11\beta}$ was 7.0×10^5 M^{-1} s^{-1}, which was comparable to those of CO-myoglobin and CO-hemoglobin. In the presence of deoxycorticosterone, however, the k_{on} value was reduced 100 fold (5.4×10^3 M^{-1} s^{-1}). Corticosterone lowered the value more remarkably than deoxycorticosterone (5.3×10^2 M^{-1} s^{-1}).

The rate constants for the dissociation of CO (k_{off}) from the CO-P-450$_{11\beta}$ complex in the presence and absence of a substrate were calculated from the recombination rate constants (k_{on}) and equilibrium dissociation constants (K_d). The results indicated that k_{off} became smaller upon association of the substrate with the cytochrome. Thus, the k_{off} values were 0.20 s^{-1} for the substrate-free P-450$_{11\beta}$, 0.015 s^{-1} for the deoxycorticosterone-bound form and 0.002 s^{-1} for the corticosterone-bound form, respectively. These results suggested that, by binding in close proximity to the heme-iron, the substrate affects the configuration of Fe—C—O in the complex through steric pressure.

Flash photolysis studies on the CO-P-450$_{11\beta}$ complex further revealed a curious property of P-450$_{11\beta}$ (Mitani et al., 1985). The magnitude of the observed absorption change in the Soret region after irradiation with a flash of light was unexpectedly small. Only about 20% of the CO-P-450$_{11\beta}$ complex was photodissociable. These results were reproducible with at least 5 purified P-450$_{11\beta}$ preparations, which were judged to be homogeneous on SDS-polyacrylamide gel electrophoresis. The cytochrome was in a reasonably pure state, because it contained about 12.7 nmoles heme per mg of protein and the turnover number was 87 min^{-1} for the 11β-hydroxylation of deoxycorticosterone under their assay conditions. Changes in the substrate concentration, pH, between 6.5 and 8.5 with several buffer systems, temperature, from 5 to 30 °C, ionic strength of the medium and concentrations of detergents did not improve this low dissociation percentage. Therefore, these investigators concluded that the CO complex of the otherwise homogeneous P-450$_{11\beta}$ preparation is heterogeneous in terms of photodissociability; it contains two species or states which are photodissociable and photoindissociable under experimental conditions. The heterogeneity of P-450$_{11\beta}$ preparations in terms of the photodissociability might be related to the multifunctions of P-450$_{11\beta}$ described below.

2.3. Primary structure

On screening 1,200 cDNA clones prepared from bovine adrenocortical poly $(A)^+$ RNA, JOHN and coworkers (1985) identified two cDNA clones corresponding to P-450$_{11\beta}$. The insert DNAs of these clones were 0.35 kb and 1.3 kb in length and therefore too short to code for the entire P-450$_{11\beta}$ peptide. However, using one of these clones as a probe for Northern blot analysis they were able to demonstrate that at least three molecular weight RNA species from adrenal cortex (7.2, 6.2 and 4.4 kb) were specific to P-450$_{11\beta}$. They also confirmed that these RNA species could direct P-450$_{11\beta}$ synthesis in the in vitro translation system.

Based on this knowledge of the size of the mRNAs for P-450$_{11\beta}$, two laboratories recently succeeded in isolating cDNAs encoding the entire P-450$_{11\beta}$ peptide (MOROHASHI et al., 1987; CHUA et al., 1987). The insert DNA of the clone isolated by MOROHASHI et al. (1987) consists of 4,105 nucleotides (without poly (A) or artificial G/C and A/T tails) and codes for 503 amino acid residues (Mr: 57,924). On comparison of the predicted amino acid sequence of this peptide with the previously reported NH$_2$-terminal sequence of P-450$_{11\beta}$ (OGISHIMA et al., 1983), MOROHASHI and coworkers assumed the existence of an extension peptide consisting of 24 amino acids (Mr: 2,760) at the NH$_2$-terminus of the mature P-450$_{11\beta}$ molecule, which in turn is a peptide of 479 amino acid residues.

Comparison of the primary structures of P-450$_{11\beta}$ and other forms of cytochrome P-450, including P-450$_{SCC}$, indicated that the two mitochondrial P-450s, P-450$_{11\beta}$ and P-450$_{SCC}$, were significantly different from microsomal forms of cytochrome P-450. The homology between P-450$_{11\beta}$ and P-450$_{SCC}$ was 36%, which is higher than the homology between P-450$_{11\beta}$ and various microsomal P-450s. Alignment of P-450$_{11\beta}$ and P-450$_{SCC}$ to give maximum matching revealed four highly conserved regions, designated C-1 (Val$_{203}$-Leu$_{212}$), C-2 (Trp$_{253}$-Phe$_{264}$), C-3 (Pro$_{363}$-Leu$_{429}$) and C-4 (Phe$_{438}$-Glu$_{474}$) (the amino acid numbering is that of the precursor peptide). The homology values of these regions were 58—70%, i.e., considerably higher than the overall homology between these two mitochondrial P-450s.

A putative heme binding site was located in the C-4 region, in which Cys$_{450}$ may be the 5th ligand of the heme. On comparison of the amino acid sequences of P-450$_{SCC}$ and prostatic steroid binding protein C2, a putative steroid binding site in P-450$_{SCC}$ was found by GOTOH et al. (1985). A sequence homologous to the putative steroid binding site is also present in the C-1 region of P-450$_{11\beta}$ (MOROHASHI et al., 1987). The amino acid sequences of the steroid binding sites of P-450$_{11\beta}$ and P-450$_{SCC}$ are highly homologous, although two other steroidogenic P-450s in microsomes, 21-hydroxylase P-450 and 17α-hydroxylase P-450, seem to contain no sequence highly homologous to the putative steroid binding site.

The hydropathy profiles of P-450$_{11\beta}$ and P-450$_{SCC}$ were very similar. When these profiles of the mitochondrial P-450 precursors were compared with those of microsomal P-450s, however, a definite difference was noted in the NH$_2$-terminal portion. The microsomal type cytochrome P-450s had a hydrophobic sequence consisting of about 20 amino acids, whereas the mitochondrial type had an extension peptide containing many positively charged amino acids.

The amino acid sequence of P-450$_{11\beta}$ deduced on nucleotide sequence analysis of a cDNA clone isolated by CHUA et al. (1987) was essentially the same as that deduced by MOROHASHI and coworkers. Moreover, CHUA et al. used a bovine clone to isolate a homologous clone with a 3.5 kb insert from a human adrenal cDNA library. The human clone had a region of 1,100 bp in length which was 81% homologous to 767 bp of the coding sequence of the bovine cDNA, but it contained an unspliced intron of a 400 bp segment.

3. Versatile catalytic activities of P-450$_{11\beta}$

3.1. 11β-, 18- and 19-Hydroxylation and related reactions

Since the 11β-hydroxylation reaction occurring in adrenocortical mitochondria is an essential and final step for the production of glucocorticoids, such as corticosterone and cortisol, it was characterized extensively in terms of substrate-specificity before the pure P-450$_{11\beta}$ became available for study (HAYANO and DORFMAN, 1962). Both deoxycorticosterone and 11-deoxycortisol can be converted efficiently into corticosterone and cortisol respectively. The 4-en-3-oxo configuration of a C$_{21}$ steroid should be intact if 11β-hydroxylation is to proceed efficiently. Reduction of the 4,5-double bond to the 5α or 5β configuration reduced the rate by some 50%. A substituent at C-6β almost totally prevented 11β-hydroxylation, while one at C-6α had little effect.

Reduction of the C-20-oxo group of a C$_{21}$ steroid substrate resulted in a very low rate of 11β-hydroxylation. A 21-deoxy steroid was also poorly 11β-hydroxylated. Thus, progesterone was much less readily hydroxylated at C-11 than was its 21-hydroxy derivative, deoxycorticosterone. This substrate-specificity undoubtedly explains the usual order of hydroxylations in corticosteroid biosynthesis, in which hydroxylation at C-21 is more likely to occur before that at C-11.

Studies using the purified P-450$_{11\beta}$ in the reconstituted system (KATAGIRI and SUHARA, 1980) confirmed earlier observations on the substrate-specificity of 11β-hydroxylation. P-450$_{11\beta}$ is strictly specific for the 4-en-3-oxo configuration of a steroid. The cytochrome is essentially nonspecific in terms of the functional group at either C-17, C-18 or C-21 of the steroid; 11-deoxycortisol, 18-hydroxy-11-deoxycorticosterone, progesterone, 4-androstene-3,17-dione and testosterone were also shown to be 11β-hydroxylatable substrates (see Figs.

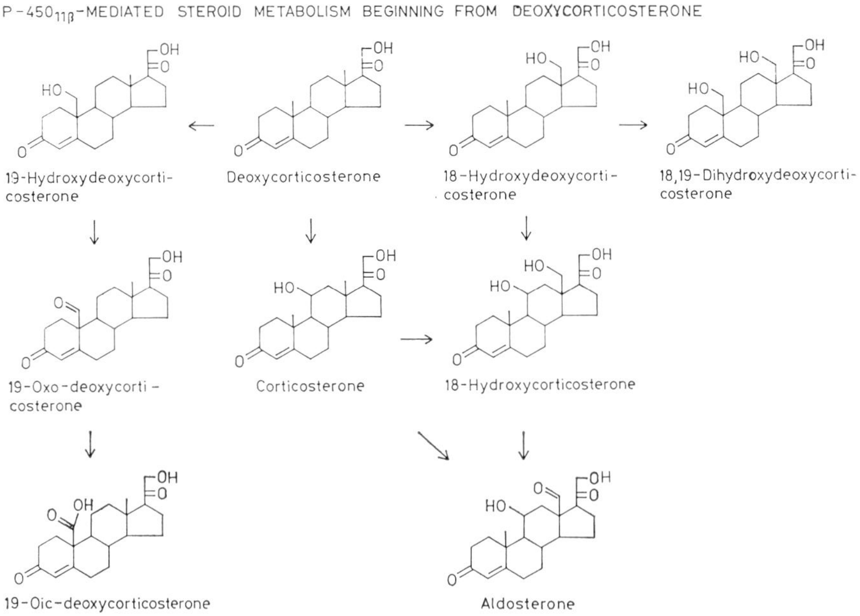

Fig. 1. P-450$_{11\beta}$-mediated steroid metabolism beginning from deoxycorticosterone.

1. and 2.). In this respect, P-450$_{11\beta}$ appears to be active toward the general structure of 3-keto-Δ^4-steroids, in which the presence of additional hydroxyl groups at C-17, C-21 and either C-11β or C-18 may not be critical. The group at C-17 of a C$_{19}$ steroid may be either a keto or hydroxyl group.

It has long been known that 18-hydroxylase occurs in the adrenal cortex. BJÖRKHEM and KARLMER (1975; 1977) studied the 18-hydroxylation as well as the 11β-hydroxylation of deoxycorticosterone using partially purified P-450-reconstituted systems from rat and bovine adrenals. Under all experimental conditions, the ratio of the 18- and 11β-hydroxylation activities was about the same, provided that the source of cytochrome P-450 was the same. Inhibition by CO, metyrapone, cytochrome **c** and different steroids did not significantly change the ratio of the two activities. For example, progesterone inhibited the 18- and 11β-hydroxylation of deoxycorticosterone to about the same extent. Conversely, when progesterone was used as the substrate for hydroxylation, it was almost exclusively hydroxylated at C-11 in the reconstituted system. The addition of deoxycorticosterone was found to inhibit the 11β-hydroxylation of progesterone. From these results, BJÖRKHEM and KARLMER concluded that the cytochrome P-450s involved in the 18- and 11β-hydroxylation of deoxycorticosterone and in the 11β-hydroxylation of progesterone must be very similar or even identical.

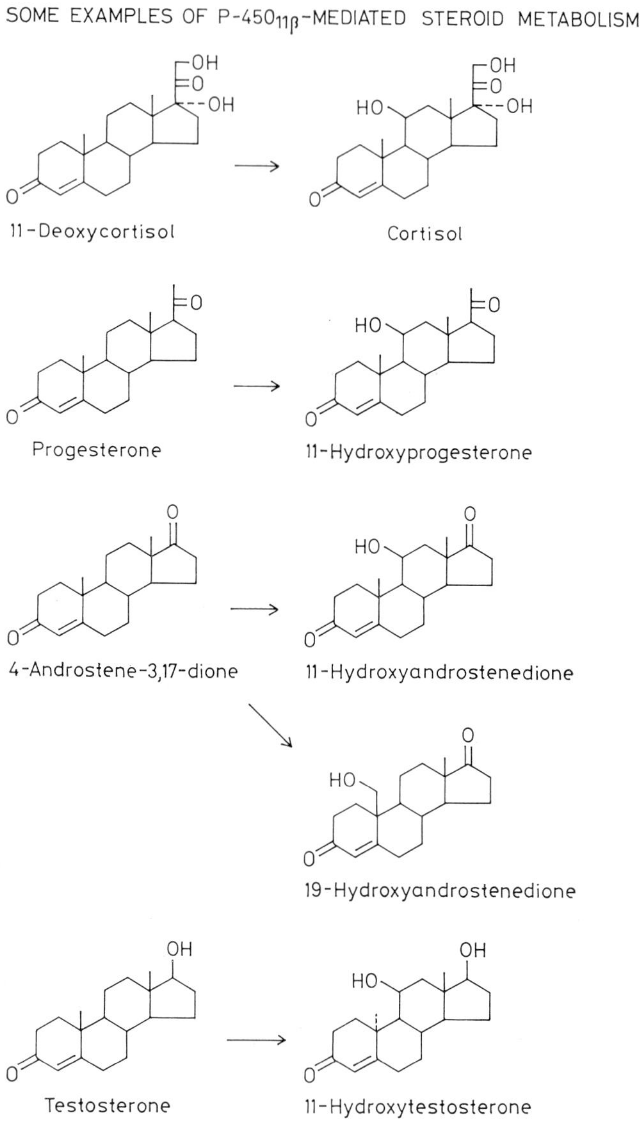
SOME EXAMPLES OF P-450₁₁β-MEDIATED STEROID METABOLISM
OH
O
OH
HO
11-Deoxycortisol
Cortisol
Progesterone
11-Hydroxyprogesterone
4-Androstene-3,17-dione
11-Hydroxyandrostenedione
19-Hydroxyandrostenedione
Testosterone
11-Hydroxytestosterone

Fig. 2. Some examples of P-$450_{11\beta}$-mediated steroid metabolism.

Further positive evidence for the proposition that a single cytochrome P-450 catalyzes the 18- as well as 11β-hydroxylation of deoxycorticosterone was provided by RAPP and DAHL. Rats were selectively bred for their blood pressure response to a high salt (NaCl) diet. Two strains were obtained; a susceptible (S) strain and a resistant (R) strain (DAHL et al., 1962). S rats responded to high salt intake with a marked increase in blood pressure, whereas R rats on the same diet showed little or no blood pressure change. RAPP and DAHL (1972) demonstrated that a single genetic locus controlling the production of 18-hydroxydeoxycorticosterone and corticosterone is mutated in the two strains. They further characterized the CO inhibition of the 18- and 11β-hydroxylation of deoxycorticosterone by using mitochondrial preparations obtained from the two rat strains (RAPP and DAHL, 1976). The CO/O_2 ratio causing 50% inhibition (Warburg's partition constant, K) was identical for both 18- and 11β-hydroxylation within a strain, but different for both 18- and 11β-hydroxylation between strains. The K values reported were; S rats, 18-hydroxylation $= 11.4 \pm 1.4$; S rats, 11β-hydroxylation $= 11.0 \pm 1.2$; R rats, 18-hydroxylation $= 56.4 \pm 13.7$; R rats, 11β-hydroxylation $= 46.7 \pm 11.7$. The strain-specific K values for 18- and 11β-hydroxylase segregated together in an F_2 population. From these findings, they suggested that the same cytochrome P-450 is involved in both 18- and 11β-hydroxylation and that this cytochrome is mutated differently in S and R rats, the P-450 of S rats being more susceptible to CO inhibition.

In an attempt to purify $P-450_{11\beta}$ from bovine adrenocortical mitochondria, WATANUKI et al. (1978) noticed that the 18-/11β-hydroxylase activity ratio does not change significantly during purification and that it is constant throughout the protein peak on the glycerol density gradient established during ultracentrifugation. Therefore, they also suggested the likely possibility that the two hydroxylase activities are catalyzed by one protein.

After the above-mentioned preludes, the dual hydroxylation at the 11β- and 18-positions catalyzed by a single cytochrome was finally established by SATO and coworkers after the purification of $P-450_{11\beta}$ in a highly purified and functionally active form from bovine adrenocortical mitochondria (SATO et al., 1978). The $P-450_{11\beta}$ purified by them exhibited steroid hydroxylase activity which at that time was the highest in the reconstituted system. The turnover numbers were $110 \min^{-1}$ and $18 \min^{-1}$ for the 11β- and 18-hydroxylase activities respectively, when deoxycorticosterone was the substrate. The apparent K_m value was 6 μM for both reactions. The ratio, about 6:1, of the two activities was constant under various experimental conditions, including those in the presence of competitive inhibitors such as deoxycortisol and testosterone.

SATO et al. (1978) further demonstrated that when 4-androstene-3,17-dione, a C_{19}-steroid, was the substrate, the sites that could be hydroxylated were its 11β- (the turnover number being $41 \min^{-1}$) and 19- (the turnover number being 12 min^{-1}) positions. The ratio of the two hydroxylation activities was about 4:1. Thus, it seems that a single cytochrome P-450 can bind a steroid in three, somewhat loose, stereospecific ways, and that binding in one way results in 11β-hydroxylation and binding in the second or third way in 18- or 19-hydroxylation. This explanation may be further rationalized by the fact that the interatomic distances between the C-11 and C-18 (C-19) are very small in the steroid structure.

The results of characterization of the catalytic activities of the purified P-450$_{11\beta}$ suggest two logically possible metabolic pathways beginning from deoxycorticosterone in the adrenal cortex. The first is one in which deoxy-corticosterone is hydroxylated at C-11β and the corticosterone so generated is then again hydroxylated at C-18 to yield 18-hydroxycorticosterone. The second possible pathway is one in which deoxycorticosterone is first hydroxy-lated at C-18 and the resultant 18-hydroxy-11-deoxycorticosterone is then converted to 18-hydroxycorticosterone through the 11β-hydroxylation reaction. It has been shown that the purified P-450$_{11\beta}$ can catalyze all four reactions in the reconstituted system (Fig. 1). Using first bovine adrenocortical mitochondria (KIM et al., 1983) and then the purified enzymes in a recon-stituted system (OKAMOTO et al., 1982; MOMOI et al., 1983), OKAMOTO and colleagues conducted kinetic studies on the 11β-hydroxylation of 18-hydroxy-11-deoxycorticosterone. During these investigations they noticed that an un-identified product appeared in addition to 18-hydroxycorticosterone. Struc-tural determination of the unidentified product was conducted by mass spec-trometry and ^{1}H NMR spectrometry. The results of the analyses indicated that the substance was 18,19-dihydroxy-11-deoxycorticosterone. According to them, the specific activity for the 19-hydroxylation of 18-hydroxy-11-deoxy-corticosterone was 71.9 nmoles per min per mg of P-450$_{11\beta}$, while that for the 11β-hydroxylation of the same steroid was 90.6 nmoles per min per mg of P-450$_{11\beta}$ (FUJII et al., 1984).

Using adrenal mitochondria prepared from various animal species, MOMOI et al. (1985) confirmed that the 19-hydroxylation of 18-hydroxy-11-deoxy-corticosterone occurs in the mitochondria of the ox and pig at rather high rates, and in those of the rabbit, guinea pig, rat and man at very low rates.

The P-450$_{11\beta}$-mediated hydroxylation of deoxycorticosterone at C-19 was studied by OHTA et al. (1987). The turnover number for 19-hydroxy-11-deoxy-corticosterone formation was 7.0 min^{-1}. Moreover, they found that 19-hydroxy-11-deoxycorticosterone was further metabolized to 19-oxo-11-deoxycortico-sterone during prolonged incubation with P-450$_{11\beta}$. A recent investigation by the same group revealed that the conversion of 19-oxo-11-deoxycorticosterone to 19-oic-11-deoxycorticosterone occurred again through the P-450$_{11\beta}$-catalyzed reaction, and that the 19-oic-11-deoxycorticosterone thus formed was non-enzymatically converted to 19-nor-11-deoxycorticosterone (OHTA et al., 1988). It appears that P-450$_{11\beta}$ is involved in the oxidative metabolism of 11-deoxycorticosterone at C-19, which finally produces one of the potent mineralo-corticoids, 19-nor-11-deoxycorticosterone (GOMEZ-SANCHEZ et al., 1979).

A similar investigation was carried out by SUHARA et al. (1986a, b), who showed that the purified P-450$_{11\beta}$ in the reconstituted system converts 19-hydroxyandrostenedione to 19-oxoandrostenedione and further converts the 19-oxoandrostenedione to estrone. Thus, they claimed that P-450$_{11\beta}$ significant-ly acts as an "aromatase", which appears to be quite different from the placental microsomal aromatase reaction.

When cortisol, one of the 11β-hydroxylated corticoids, was incubated with the purified P-450$_{11\beta}$ in the reconstituted system with a supply of O$_2$ and NADPH, cortisone, an 11-oxo-corticoid, was found to accumulate in the incubation mixture (SUHARA et al., 1986a). Thus, P-450$_{11\beta}$ seems to be respon-sible for the oxidative conversion of the 11-alcohol- to the 11-keto-group on the steroid nucleus.

3.2. Aldosterone production

Aldosterone is thought to be biosynthesized from corticosterone via 18-hydroxycorticosterone as an intermediate (see MÜLLER, 1988, for an extensive review). The conversion of 18-hydroxycorticosterone to aldosterone was generally assumed to be a dehydrogenation reaction. However, RAMAN et al. (1966) showed that NADPH or an NADPH-generating system, rather than NAD or NADP, was necessary for the conversion of added 18-hydroxycortico-sterone to aldosterone by sheep adrenal mitochondria. Also shown by MARUSIC et al. (1973) was that bovine adrenal mitochondria converted 18-hydroxy-corticosterone as well as corticosterone to aldosterone in the presence of NADPH but not in the presence of FAD, NAD, NADP or NADH. From the results of a CO-inhibition study, GREENGARD et al. (1967) presented evidence for the participation of cytochrome P-450 in the conversion of corticosterone to aldosterone by bullfrog adrenal mitochondria. The observation by MARUSIC et al. (1973) and KOJIMA et al. (1982, 1984) that both NADPH and O$_2$ were required for the conversion of 18-hydroxycorticosterone to aldosterone sug-gested that not only the 18-hydroxylation of corticosterone but also the final step of aldosterone biosynthesis involves a mixed function oxidase utilizing a cytochrome P-450 rather than a dehydrogenase. Elucidation of the enzymatic characteristics of "the aldosterone-producing cytochrome P-450" has been the focus of research for the past few years.

It was WADA and his coworkers who first found that P-450$_{11\beta}$, purified to electrophoretic homogeneity from bovine adrenocortical mitochondria con-verted radiolabeled corticosterone to 18-hydroxycorticosterone and aldo-sterone. The cytochrome also converted radiolabeled 18-hydroxycorticosterone to aldosterone (WADA et al., 1984). When a lipid extract of adrenocortical mitochondria was added to the reconstituted enzyme system the rate of production of aldosterone increased 28-fold. In a further study by the same investigators (WADA et al., 1985) the reconstitution conditions were improved, the turnover numbers and apparent K_m values being as follows: the turnover numbers for aldosterone and 18-hydroxycorticosterone formation from corticosterone were $0.23 \ \mathrm{min}^{-1}$ and $1.1 \ \mathrm{min}^{-1}$, respectively. The addition of phospholipids enhanced the turnover number to $2.0 \ \mathrm{min}^{-1}$ for aldosterone formation, with an apparent K_m of $6.9 \ \mu M$. When 18-hydroxycorticosterone was tested as a substrate, the turnover number for aldosterone production was $5.3 \ \mathrm{min}^{-1}$, with an apparent K_m of $325 \ \mu M$. CO and metyrapone inhibited the production of aldosterone from corticosterone and that from 18-hydroxy-corticosterone. These investigators concluded that the conversion of cortico-sterone and 18-hydroxycorticosterone to aldosterone occurs through a P-$450_{11\beta}$-catalyzed reaction.

The effect of phospholipids on aldosterone synthesis catalyzed by the P-$450_{11\beta}$-reconstituted system was examined by OHNISHI et al. (1984). These investigators found that in the presence of a phospholipid mixture containing commercially available lipids in the same molar ratio as found in adreno-cortical mitochondria, the rate of production of aldosterone increased 10-fold over the rate without the lipids, and that of 18-hydroxycorticosterone in-creased 3-fold. The maximal synthetic rates for aldosterone and 18-hydroxy-corticosterone were 2 and 5 nmoles per nmole P-450 per min, respectively.

YANAGIBASHI and coworkers, who purified bovine and porcine adreno-cortical P-$450_{11\beta}$, reported that three activities, those of 11β- and 18-hydroxy-lase and aldosterone synthetase, were co-purified in constant ratios, and came to the conclusion that one single enzyme catalyzes the three reactions (YANA-GIBASHI et al., 1986).

The homogeneity of the P-$450_{11\beta}$ used by these investigators was verified by the fact that the cytochrome preparation gave a single protein band on SDS-polyacrylamide gel electrophoresis, and that the NH_2-terminal amino acid sequence of the protein was consistent with the results reported by OGISHIMA et al. (1983). Although the possibility of the presence in the pre-paration of two proteins differing in only a few amino acids cannot be ex-cluded, all of the above results strongly suggest that bovine and porcine P-$450_{11\beta}$ exhibit catalytic activity with reference to the production of 18-hydroxycorticosterone and aldosterone as well as 11β-hydroxylation and 18-hydroxylation of deoxycorticosterone in the reconstituted system.

What then is the mechanism underlying the aldosterone formation from corticosterone catalyzed by a single cytochrome P-450? It has been suggested that the reaction consisted of two consecutive cytochrome P-450-catalyzed steps: first, the conversion of corticosterone to 18-hydroxycorticosterone and, second, the conversion of the 18-hydroxycorticosterone to aldosterone. Al-though the mechanism underlying the second step has not yet been clarified, it could be a second hydroxylation at C-18, yielding 18,18-dihydroxycortico-sterone, which would be unstable and decompose into aldosterone and water. The mechanism underlying a similar reaction has been discussed in the case of the conversion of testosterone to 4-androstene-3,17-dione catalyzed by liver microsomal P-450 (GUENGERICH et al., 1982) and adrenocortical micro-somal P-450 (SUHARA et al., 1984).

3.3. Inhibitors

Metyrapone is a well-known inhibitor of P-450$_{11\beta}$. This drug inhibited the 11β- as well as 18-hydroxylation of deoxycorticosterone, with a K$_i$ 0.1—0.2 μM (SATO et al., 1978). The inhibition was competitive with respect to the sub-strate. Metyrapone at less than 10 μM was also found to inhibit the produc-tion of aldosterone from 18-hydroxycorticosterone in the P-450$_{11\beta}$-reconstituted system (WADA et al., 1985). The addition of a low concentration of metyrapone (24 μM) to the substrate-bound ferric P-450$_{11\beta}$ changed its high-spin type spectrum to the typical low-spin type spectrum of a cytochrome P-450, with a Soret band at 422 nm (SATO et al., 1978). The drug is thought to interfere with the interaction of the hemoprotein and its substrate. The binding may occur at, or near, the heme group of the cytochrome, thus producing an altered ligand field. SCHLEYER and coworkers showed that there is an absorp-tion maximum at 442 nm when cytochrome P-450 (Fe^{2+}) reacts with various pyridine bases (SCHLEYER et al., 1972b) and since metyrapone is itself a pyridine derivative this may indicate that it is the nitrogenous part of the molecule that is involved in cytochrome binding.

Recently, NAGAI et al. (1986) and WADA et al. (1988) showed that imidazole antimycotic agents, such as ketoconazole, clotrimazole and miconazole, and an imidazole anesthetic, etomidate, are potent inhibitors of the P-450$_{11\beta}$-catalyzed 11β-hydroxylation of deoxycorticosterone. These drugs caused 50% inhibition of the 11β-hydroxylase activity at concentrations between 0.01 and 0.1 μM. One of the drugs, etomidate, was further tested for its potency with regard to the inhibition of aldosterone and 18-hydroxycorticosterone produc-tion under assay conditions similar to those for the 11β-hydroxylation (WADA et al., 1988). The results showed that etomidate was equally potent as an inhibitor of all the reactions mediated by P-450$_{11\beta}$. As suggested by VANDEN BOSSCHE (1985), the action of these imidazole compounds seems to involve the interaction of their imidazole ring and the heme iron of cytochrome P-450.

Spironolactone, a synthetic mineralocorticoid analog, was found to inhibit the P-450$_{11\beta}$-mediated reactions. The concentration of the drug required for 50% inhibition of the 11β-hydroxylase activity was 250 μM, that is to say almost equal to those in the cases of the aldosterone and 18-hydroxycortico-sterone biosynthetic activities (WADA et al., 1988).

3.4. Factors affecting the P-450$_{11\beta}$-mediated activities in the reconstitution system

The fact that P-450$_{11\beta}$ exhibits versatile catalytic activities in the reconstitution system raises the intriguing question of how these activities are regulated. It is possible that each catalytic activity of P-450$_{11\beta}$ is modulated in a specific manner, and that each enzymatic activity itself might be a target of the regulation. To examine this possibility, OHNISHI and coworkers searched for a factor or factors that influences a particular activity catalyzed by the P-450$_{11\beta}$-reconstituted system. As mentioned above, phospholipid enhanced the rates of production of 18-hydroxycorticosterone and aldosterone from corticosterone to different extents (OHNISHI et al., 1984). Thus, the turnover number for aldosterone formation in the presence of phospholipid was 2 min^{-1}, i.e., about 10-fold of that in the absence of the lipid. In contrast to the case of the aldosterone formation, the extent of stimulation of 18-hydroxycorticosterone formation observed was only 3-fold, the turnover number in the presence of the lipid being 5 min^{-1}. The phospholipid also affected the 11β-hydroxylation of deoxycorticosterone, but interestingly the effect was not stimulation but inhibition. Thus the turnover number for the production of corticosterone from deoxycorticosterone in the absence of phospholipid was 60 min^{-1}, while it was 10 min^{-1} in the presence of a lipid extract (100 µM lipid phosphorus) of bovine adrenocortical mitochondria under their assay conditions (OKAMOTO et al., 1985; OHNISHI et al., 1988).

How phospholipid affects the respective activities catalyzed by the single enzyme differently is difficult to determine. An explanation may be possible on further investigation of the conditions for reconstitution of the P-450$_{11\beta}$-electron-transport system. It is reasonable to entertain the idea that phospholipid profoundly affects the efficiency of the electron transfer from NADPH to P-450$_{11\beta}$, thus inducing a change in the efficiency of turnover in the P-450$_{11\beta}$ catalysis. However, the fact that the rates of the 11β- and 18-hydroxylation and aldosterone production are differently affected on the addition of phospholipid could not be explained by the effect of phospholipid on the efficiency of electron transfer alone. Its effect on the molecular conformation of P-450$_{11\beta}$ itself should also be considered. It is conceivable that phospholipid surrounding the cytochrome may influence the configuration of a steroid binding site, affecting the rates of hydroxylation at different sites of the steroid skeleton.

OHNISHI et al. (1986) found a protein factor which increases the rate of 18-hydroxycorticosterone production and decreases that of aldosterone production from corticosterone and purified the protein from bovine adrenal cortex. Its electrophoretic properties, amino acid composition and stimulatory effect on phosphodiesterase activity indicated that it was adrenocortical calmodulin. The effect of calmodulin on the P-450$_{11\beta}$-catalyzed reactions was studied in detail. When added to the $P-450_{11\beta}$-reconstituted system in the presence of dilauroyl phosphatidylcholine, calmodulin decreased the rate of aldosterone production from corticosterone from 0.8 to 0.1 nmole per nmole P-450 per min, while it increased the rate of 18-hydroxycorticosterone production from 1.8 to 4.6 nmole per nmole P-450 per min. The effect was dependent upon the presence of Ca^{2+}, and the maximal response was observed at less than 1 μM Ca^{2+}. As for 18-hydroxycorticosterone production, calmodulin increased both the maximal activity and the apparent K_m for corticosterone but it decreased the apparent K_m for adrenodoxin. Calmodulin little affected the hydroxylation of deoxycorticosterone at C-11. When calmodulin was added to the deoxycorticosterone-bound ferric $P-450_{11\beta}$, a small type I difference spectrum with a peak at 390 nm and a trough at 420 nm was observed. This difference spectrum increased significantly in the presence of both Ca^{2+} and adrenodoxin. From these results, OHNISHI et al. (1986) suggested that calmodulin interacts with $P-450_{11\beta}$ in the presence of adrenodoxin and then modulates the activity of aldosterone synthesis catalyzed by $P-450_{11\beta}$. The physiological significance of the calmodulin-$P-450_{11\beta}$ interaction remains to be elucidated. A further search for another factor, if any, in adrenocortical mitochondria that influences the catalytic activities of $P-450_{11\beta}$ should be performed in the future.

DEFAYE et al. (1982) reported that $P-450_{11\beta}$ is a good substrate for cyclic AMP (cAMP)-dependent protein kinase. When the purified $P-450_{11\beta}$ was incubated with the catalytic subunit of the cAMP-dependent protein kinase in the presence of $[\gamma\text{-}^{32}P]ATP$, a maximum of 0.84 mole of ^{32}P was incorporated into 1 mole of the cytochrome. Phosphorylation of $P-450_{11\beta}$ did not change the 11β-hydroxylase activity in the reconstituted system with an excess amount of adrenodoxin. However, kinetic studies showed that $P-450_{11\beta}$ phosphorylation remarkably increases the affinity of $P-450_{11\beta}$ for adrenodoxin in a phosphorylation-dependent manner. These investigators discussed the possibility that the cAMP-dependent phosphorylation of $P-450_{11\beta}$ may result in a net increase in 11β-hydroxylase activity under in vivo conditions where adrenodoxin availability is limited.

4. Studies on $P-450_{11\beta}$ from biological aspects

4.1. Zonal distribution and intracellular localization of $P-450_{11\beta}$ and its orientation in the membrane

Steroid 11β-hydroxylase occurs in the inner membrane of the mitochondria of all zones of the adrenal cortex (YAGO and ICHII, 1969; DODGE et al., 1970). The most recent studies (HAMAMOTO et al., 1986; SUHARA et al., 1987) showed that bovine adrenal cortex mitochondria (mostly from the zonae fasciculata and reticularis) contain 50—60 pmoles of $P\text{-}450_{11\beta}$ per mg protein and 130 pmoles of $P\text{-}450_{SCC}$ per mg protein, and that the two cytochromes are the P-450 enzymes comprising all of the CO-reactive P-450.

To investigate the zonal distribution as well as intracellular localization of $P\text{-}450_{11\beta}$, an immunocytochemical approach was taken by MITANI and coworkers (MITANI et al., 1982a, b). When the distribution was examined by means of an immunostaining technique involving horseradish peroxidase-labeled anti-$P\text{-}450_{11\beta}$ (Fab'), intense staining was observed in parenchymal cells of the zonae fasciculata and reticularis. The staining of the parenchymal cells of the zona glomerulosa was only faint. No staining was observed in the capsule or medulla of the adrenal glands. Electron microscopical observation revealed the intracellular localization of the cytochrome. The inner mitochondrial membrane of the parenchymal cells was intensely stained by the antibody, suggesting that $P\text{-}450_{11\beta}$ is exclusively localized in the mitochondria. These investigators noted that the mitochondria in a single cell differed markedly from one to the other in staining intensity. The question of whether the difference in staining among mitochondria suggests heterogeneity in steroidogenesis among mitochondria or uneven penetration of either the antibody or chemical reagents into the organelles remains to be answered. The immunocytochemical study conducted by SUGANO et al. (1985) involving a monoclonal antibody against $P\text{-}450_{11\beta}$ or that by GENZE et al. (1987) involving an immunogold technique provided essentially the same results as those of MITANI and colleagues.

The topography of $P\text{-}450_{11\beta}$ in the inner mitochondrial membrane was examined by LOMBARDO et al. (1986). Mitoplasts were prepared from bovine adrenocortical mitochondria by the hypo-osmotic shock method. The mitoplast preparations were subjected to sonication to yield inside-out inner membrane mitochondrial vesicles. The vesicles were then subjected to trypsin treatment and then $P\text{-}450_{11\beta}$-derived peptides were separated and detected by means of polyacrylamide gel electrophoresis and immunoblotting respectively. The trypsin treatment resulted in the rapid disappearance of the native protein moiety (Mr 47,000), while a major peptide component of Mr 34,000 appeared. A similar study conducted on the liposome-integrated purified $P\text{-}450_{11\beta}$ enabled the investigators to demonstrate that the Mr 34,000 peptide retained the heme moiety and part of the cytochrome's steroid 11β-hydroxylase activity. From these results, they concluded that $P\text{-}450_{11\beta}$ is embedded in the inner mitochondrial membrane through a major hydrophobic domain associated with the heme moiety, while a limited domain remains accessible on the matrix side of the membrane surface.

4.2. Biosynthesis of P-450$_{11\beta}$ and regulation of its biosynthesis

Most of the aspects of this subject are discussed in detail by WATERMAN in another chapter of this book. We will summarize only the important points regarding the biosynthesis of P-450$_{11\beta}$.

P-450$_{11\beta}$ is biosynthesized as a precursor molecule of higher molecular weight on free ribosomes in the cytoplasm of the adrenal cortex (NABI et al., 1983). The precursor molecule has an apparent half-life of 3.5 min, and is rapidly imported into and proteolytically processed to the corresponding mature form by the mitochondria (KRAMER et al., 1982; MATOCHA and WATERMAN, 1985). A chelator of divalent cations (o-phenanthroline) or a mitochondrial respiratory uncoupler (dinitrophenol) inhibits the maturation of the precursor molecule. Interestingly, MATOCHA and WATERMAN (1986) found that the precursor of P-450$_{11\beta}$, whose synthesis was directed by bovine adrenocortical RNA, could be processed by bovine corpus luteal mitochondria, even though P-450$_{11\beta}$ is not expressed in this tissue. The efficiency of the processing of pre-P-450$_{11\beta}$ by corpus luteal mitochondria was similar to that of pre-P-450$_{SCC}$, an endogenous enzyme in these mitochondria. The precursors of P-450$_{11\beta}$ and P-450$_{SCC}$ were not processed by mitochondria from a nonsteroidogenic tissue such as heart. They therefore suggested that the precursors of P-450$_{11\beta}$ and P-450$_{SCC}$ are processed via a common pathway in steroidogenic mitochondria, and that this pathway is absent in nonsteroidogenic mitochondria. More recently, however, FURUYA et al. (1987) reported that the precursor of bovine P-450$_{11\beta}$ could be imported into and processed by rat liver mitochondria as well.

The regulation of P-450$_{11\beta}$ gene expression by adrenocorticotropin (ACTH) or analogs of cAMP has been studied by WATERMAN and colleagues. In cultured bovine adrenocortical cells, ACTH (1 μM) or dibutyryl cAMP (1 mM) increased the concentration of P-450$_{11\beta}$ transcripts (KRAMER et al., 1983; KRAMER et al., 1984). Actinomycin D effectively blocked the ACTH-induced increase in the P-450$_{11\beta}$ mRNA level. Furthermore, cycloheximide administered prior to or along with ACTH resulted in the blockade of any new transcription of the P-450$_{11\beta}$ gene (JOHN et al., 1985; JOHN et al., 1986). These investigators concluded that the primary action of ACTH in the adrenal cortex is to activate, via cAMP, the synthesis of rapidly turning over protein factors that in turn mediate increased transcription of steroid hydroxylase genes. Exploration of the nature of these protein factors must be a focus of future research.

4.3. 11β-Hydroxylase deficiency and experimental animal model of aberrant P-450$_{11\beta}$

Inherited enzymatic defects of steroid synthesis interfere with the biosynthesis of normal amounts of steroid hormones and therefore cause a hyperplastic reaction in the adrenal gland consequent to increased ACTH secretion from the pituitary gland. These diseases are clinically called "congenital adrenal hyperplasia". 11β-Hydroxylase deficiency is the second most frequent cause of congenital adrenal hyperplasia (the most common being the 21-hydroxylase defect) and probably accounts for $5-7\%$ of all cases diagnosed (NEW and LEVINE, 1973; BONGIOVANNI, 1978). The defect results in the accumulation of 11-deoxycortisol and deoxycorticosterone. Cortisol, corticosterone and aldosterone are not formed. The decreased level of plasma cortisol induces increased ACTH secretion, with the resultant overproduction of precursors. The accumulated precursors are shunted into the androgen biosynthetic pathway, causing virilization of the patients. The external genitalia of the female fetus become masculinized due to the excess fetal adrenal androgen and female pseudohermaphroditism results. The internal female genitalia remain normal. An additional clinical symptom in many, but not all, patients with the 11β-hydroxylase defect is hypertension. The hypertension is thought to result from the excessive production of deoxycorticosterone, a weak mineralocorticoid.

A cDNA clone homologous to the bovine P-$450_{11\beta}$ cDNA was isolated from a human adrenal cDNA library (CHUA et al., 1987). Although the clone does not include the open reading frame coding for entire human P-$450_{11\beta}$ peptide, all comparisons of it with the bovine clone suggested that it indeed encodes P-$450_{11\beta}$ and not a related protein. Hybridization of the human cDNA to DNA from a panel of human-rodent somatic cell hybrid lines and in situ hybridization to metaphase spreads of human chromosomes allowed the localization of the P-$450_{11\beta}$ gene at the middle of the long arm of chromosome 8. The human cDNA clone can be used as a reagent for the prenatal diagnosis of 11β-hydroxylase deficiency. The diagnosis is achieved by analysis of fetal DNA obtained through a chorionic villus biopsy and permits prenatal therapy. Such therapy consists of the administration of dexamethasone to the mothers of affected female fetuses, which suppresses the fetal adrenal glands and prevents masculinization of the external genitalia.

A strain of rat having a mutant form of P-$450_{11\beta}$ was developed by RAPP and DAHL (1976), as was discussed above. An etiological study on the cause of the hypertension that develops in the S rat should be carried out in the future, based on the molecular knowledge of P-$450_{11\beta}$.

Experimental hypertension due to the decreased activity of 11β-hydroxylase has been known to develop in rats treated with the synthetic anabolic androgen 17α-methylandrostenediol (SKELTON, 1953; HYDE and DAIGNEAULT, 1968 COLBY et al., 1970; McCALL et al., 1978; FINK et al., 1980). Using specific antibodies against various steroidogenic enzymes, BROWNIE et al. (1988 recently showed that the treatment of rats with the drug caused a preferential decrease (65%) in P-$450_{11\beta}$. The decreases in P-450_{SCC} (15%), adrenodoxin (20%), microsomal 21-hydroxylase P-450 (35%) and NADPH:P-450 reductase (35%) were less than that observed in P-$450_{11\beta}$. The mechanism underlying

M. OKAMOTO; Y. NONAK.

the selective decrease in $P\text{-}450_{11\beta}$ is at present unknown, although, as suggested by HORNSBY and CRIVELLO (1983a, b), the androgen may damage $P\text{-}450_{11\beta}$ by acting as a pseudosubstrate for the cytochrome, thus causing lipid peroxidation.

4.4. Problems concerning the aldosterone synthesizing enzyme

The recent findings that highly purified preparations of bovine and porcine adrenocortical $P\text{-}450_{11\beta}$ catalyzed not only 11β-hydroxylation but also the two-step conversion of corticosterone to aldosterone (see section 3.2.), together with the immunohistochemical evidence that $P\text{-}450_{11\beta}$ is present in mitochondria of both the zonae fasciculata and reticularis, and the zona glomerulosa (see section 4.1.) apparently contradict many previous observations that aldosterone is exclusively produced by zona glomerulosa cells. A possible explanation for this contradiction is that in the intact zonae fasciculata and reticularis the aldosterone synthetic activity of $P\text{-}450_{11\beta}$ is inhibited by some specific inhibitors, whereas the activity of the enzyme in the zona glomerulosa is fully expressed. Quite recent studies by OHNISHI et al. (1988) indicated that this is indeed the case. They carefully prepared mitochondria from the zonae fasciculata and reticularis and the zona glomerulosa of bovine and porcine adrenal cortex respectively. When the mitochondria were incubated with corticosterone in the presence of malate as a generator of reducing equivalents, the corticosterone was converted to aldosterone by those prepared from the zona glomerulosa at a significant rate (35 pmoles per min per mg of protein) but not by those from the zonae fasciculata and reticularis. The conversion of corticosterone to 18-hydroxycorticosterone was found to occur with the incubation of mitochondria from both zones. When the mitochondria were solubilized by the addition of 1% cholate and then tested for steroidogenic activity in the presence of NADPH and the electron transfer components for P-450, aldosterone production occurred at the rates of 300 pmoles per min per mg with the mitochondria from the zona glomerulosa and 430 pmoles per min per mg with those from the zonae fasciculata and reticularis, respectively. These findings suggest that the aldosterone synthesizing activity of $P\text{-}450_{11\beta}$ is inhibited in mitochondria of the intact zonae fasciculata and reticularis and that the inhibition is abolished by detergent treatment. These investigators speculated that the active site of the cytochrome in the intact zonae fasciculata and reticularis mitochondria may be forced by phospholipids or a protein factor(s) like calmodulin (see section 3.4) to take on a configuration that prevents further oxygenation of 18-hydroxycorticosterone at C-18.

OHNISHI et al. (1988) also noted that the mitochondria of the rat zonae fasciculata and reticularis could hardly catalyze aldosterone synthesis under detergent-solubilized conditions, whereas those of the zona glomerulosa could.

Immunoblot analysis revealed that the mitochondria of the zonae fasciculata and reticularis contained a protein of Mr 51,000, which was immunocrossreactive with a monoclonal antibody directed against P-450$_{11\beta}$, whereas those of the zona glomerulosa contained another faintly-stained immunocrossreactive protein of Mr 49,000 in addition to the protein of Mr 51,000. These findings regarding the aldosterone synthesis by rat mitochondria are consistent with those previously made by the same investigators (LAUBER et al., 1987). They examined the aldosterone producing activity in mitochondria from capsular (zona glomerulosa) adrenals of sodium-deficient, potassium-replete rats. The mitochondria converted corticosterone to aldosterone at a markedly higher rate (161 pmoles per min per mg) than those from capsular adrenals of sodium-replete, potassium-deficient rats (less than 15 pmoles per min per mg). Only mitochondria of the zona glomerulosa from rats exhibiting stimulated aldosterone biosynthesis contained the Mr 49,000 protein, which showed strong immunocrossreactivity with the monoclonal antibody directed against P-450$_{11\beta}$. In contrast, the immunocrossreactive protein of Mr 51,000 was found in both mitochondria of the zonae fasciculata and reticularis, and the zona glomerulosa from rats exhibiting suppressed aldosterone biosynthesis. These investigators concluded that two different forms of P-450$_{11\beta}$ exist in the rat adrenal cortex, only one of which, the Mr 49,000 form, is capable of catalyzing the aldosterone production. It is conceivable that the Mr 49,000 form is the result of post-translational processing of the Mr 51,000 protein, P-450$_{11\beta}$ in rat adrenal cortex.

Acknowledgements

The authors thank Dr. M. KATAGIRI of Kanazawa University for encouraging them to write this article. Preparation of this article was supported by a research grant from the Ministry of Education, Science and Culture, Japan.

5.   References

ANDO, N. and S. HORIE, (1972), J. Biochem. (Tokyo) 72, 583—597.
BJÖRKHEM, I. and K.-E. KARLMAR, (1975), Eur. J. Biochem. 51, 145—154.
BJÖRKHEM, I. and K.-E. KARLMAR, (1977), J. Lipid Res. 18, 592—603.
BONGIOVANNI, A. M., (1978), in "The Metabolic Basis of Inherited Disease, 4th ed." (J. B. STANBURY, J. B. WYNGAARDEN, and D. S. FREDRICKSON, eds.), McGraw-Hill, New York, p. 868—893.
BOSSCHE, VANDEN, H., (1985), in "Current Topics in Medical Mycology, vol. 1.", (M. R. McGINNIS, ed.), Springer-Verlag, New York, p. 313—351.

Boyd, G. S., A. C. Brownie, C. R. Jefcoate, and E. R. Simpson, (1972), in "Biological Hydroxylation Mechanisms", (G. S. Boyd and R. M. S. Smellie, eds.), Academic Press, London, p. 207—226.

Brownie, A. C., C. R. Bhasker, and M. R. Waterman, (1988), Mol. Cell. Endocr. **55**, 15—20.

Chua, S. C., P. Szabo, A. Vitek, K.-H. Grzeschik, M. John, and P. C. White, (1987), Proc. Natl. Acad. Sci. USA **84**, 7193—7197.

Colby, H. D., F. R. Skelton, and A. C. Brownie, (1970), Endocrinology **86**, 1093 to 1101.

Cooper, D. Y., B. Novack, O. Foroff, A. Slade, E. Saunders, S. Narasimhulu, and O. Rosenthal, (1967), Fed. Proc. **26**, 341.

Dahl, L. K., M. Heine, and L. Tassinari, (1962), J. Exp. Med. **115**, 1173—1190.

Defaye, G., N. Monnier, G. Guidicelli, and E. M. Chambaz, (1982), Mol. Cell. Endocr. **27**, 157—168.

Dodge, A. H., A. K. Christensen, and R. B. Clayton, (1970), Endocrinology **87**, 254—261.

Fink, C. S., S. Gallant, and A. C. Brownie, (1980), Hypertension **2**, 617—622.

Fujii, S., K. Momoi, M. Okamoto, T. Yamano, T. Okada, and T. Terasawa, (1984), Biochemistry **23**, 2558—2564.

Furuya, S., M. Okada, A. Ito, H. Aoyagi, T. Kanmera, T. Kato, Y. Sagara, T. Horiuchi, and T. Omura, (1987), J. Biochem. **102**, 821—832.

Genze, H. J., W. Slot, K. Yanagibashi, J. A. McCracken, A. L. Schwartz, and P. F. Hall, (1987), Histochemistry **86**, 551—557.

Gomez-Sanchez, C. E., O. B. Holland, B. A. Murry, H. A. Lloyd, and L. Milewich, (1979), Endocrinology **105**, 708—711.

Gotoh, O., Y. Tagashira, K. Morohashi, and Y. Fujii-Kuriyama, (1985), FEBS Lett. **188**, 8—10.

Grant, J. K. and A. C. Brownie, (1955), Biochim. Biophys. Acta **18**, 433—434.

Greengard, P., S. Psychoyos, H. H. Tallan, D. Y. Cooper, O. Rosenthal, and R. W. Estabrook, (1967), Arch. Biochem. Biophys. **121**, 298—303.

Guengerich, F. P., G. A. Dannan, S. T. Wright, M. V. Martin, and L. S. Kaminsky, (1982), Biochemistry **21**, 6019—6030.

Hall, P. F., (1984), Int. Rev. Cytol. **86**, 53—95.

Hall, P. F., (1986), Steroids **48**, 131—196.

Hamamoto, I., A. Hiwatashi, and Y. Ichikawa, (1986), J. Biochem. **99**, 1743—1748.

Hayano, M. and R. I. Dorfman, (1962), in "Methods in Enzymology, vol. 5", (S. P. Colowick and N. O. Kaplan, eds.), Academic Press, New York, p. 503—512.

Hayano, M., M. C. Lindberg, R. I. Dorfman, J. E. H. Hancock, and W. von E. Doering, (1955), Arch. Biochem. Biophys. **59**, 529—532.

Hornsby, P. J. and J. F. Crivello, (1983a), Mol. Cell. Endocrinol. **30**, 1—20.

Hornsby, P. J. and J. F. Crivello, (1983b), Mol. Cell. Endocrinol. **30**, 123—147.

Hyde, P. M. and E. A. Daigneault, (1968), Steroids **11**, 721—731.

Jefcoate, C. R., R. Hume, and G. S. Boyd, (1970), FEBS Lett. **9**, 41—44.

Jefcoate, C. R., E. R. Simpson, and G. S. Boyd, (1974), Eur. J. Biochem. **42**, 539—551.

John, M. E., M. C. John, E. R. Simpson, and M. R. Waterman, (1985), J. Biol. Chem. **260**, 5760—5767.

John, M. E., M. C. John, V. Boggaram, E. R. Simpson, and M. R. Waterman, (1986), Proc. Natl. Acad. Sci. USA **83**, 4715—4719.

Katagiri, M., S. Takemori, E. Itagaki, K. Suhara, T. Gomi, and H. Sato, (1976), in "Iron and Copper Proteins", (K. T. Yasunobu, H. F. Mower, and O. Hayaishi, eds.). Plenum Publishing Corporation, New York, p. 281—289.

KATAGIRI, M., S. TAKEMORI, and K. SUHARA, (1978a), in "Cytochrome P-450", (R. SATO, ed.), Kodansha Scientific, Tokyo and Academic Press, New York, p. 164—183.

KATAGIRI, M., S. TAKEMORI, E. ITAGAKI, and K. SUHARA, (1978b), in "Methods in Enzymology, vol. 52", (S. FLEISCHER and L. PACKER, eds.), Academic Press, New York, p. 124—132.

KATAGIRI, M. and K. SUHARA, (1980), in "Biochemistry, Biophysics and Regulation of Cytochrome P-450", (J.-A. GUSTAFSSON et al., eds.), Elsevier/North-Holland Biomedical Press, Amsterdam, p. 97—100.

KATAGIRI, M., (1982), in "Transport and Bioenergetics in Biomembranes", (R. SATO and Y. KAGAWA, eds.), Japan Scientific Societies Press, Tokyo, p. 225—245.

KATAGIRI, M. and K. SUHARA, (1982), in "Experiences in Biochemical Perception". Academic Press, New York, p. 267—278.

KATAGIRI, M., K. NAKAYAMA, K. SUHARA, and K. SUZUKI, (1982), in "Oxygenase and Oxygen Metabolism", (M. NOZAKI, S. YAMAMOTO, Y. ISHIMURA, M. J. COON, L. ERNSTER, and R. W. ESTABROOK, eds.), Academic Press, New York, p. 391—401.

KIM, C. Y., T. SUGIYAMA, M. OKAMOTO, and T. YAMANO, (1983), J. Steroid Biochem. 18, 593—599.

KOJIMA, I., H. INANO, and B. TAMAOKI, (1982), Biochem. Biophys. Res. Commun. 106, 617—624.

KOJIMA, I., E. OGATA, H. INANO, and B. TAMAOKI, (1984), Acta Endocrinol. 107, 395 to 400.

KOMINAMI, S., H. OCHI, and S. TAKEMORI, (1979), Biochim. Biophys. Acta 577, 170 to 176.

KRAMER, R. E., R. N. DuBOIS, E. R. SIMPSON, C. M. ANDERSON, K. KASHIWAGI, J. D. LAMBETH, C. R. JEFCOATE, and M. R. WATERMAN, (1982), Arch. Biochem. Biophys. 215, 478—485.

KRAMER, R. E., E. R. SIMPSON, and M. R. WATERMAN, (1983), J. Biol. Chem. 258, 3000—3005.

KRAMER, R. E., W. E. RAINEY, B. FUNKENSTEIN, A. DEE, E. R. SIMPSON, and M. R. WATERMAN, (1984), J. Biol. Chem. 259, 707—713.

LAMBETH, J. D., D. W. SEYBERT, J. R. LANCASTER Jr., J. C. SALERNO, and H. KAMIN, (1982), Mol. Cell. Biochem. 45, 13—31.

LAUBER, M., S. SUGANO, T. OHNISHI, M. OKAMOTO, and J. MÜLLER, (1987), J. Steroid Biochem. 26, 693—698.

LIEBERMAN, S., N. J. GREENFIELD, and A. WOLFSON, (1984), Endocr. Rev. 5, 128—148.

LOMBARDO, A., M. LAINE, G. DEFAYE, N. MONNIER, C. GUIDICELLI, and E. M. CHAMBAZ, (1986), Biochim. Biophys. Acta 863, 71—81.

MARUSIC, E., T. A. WHITE, and A. R. AEDO, (1973), Arch. Biochem. Biophys. 157, 320 to 321.

MATOCHA, M. F. and M. R. WATERMAN, (1985), J. Biol. Chem. 260, 12259—12265.

MATOCHA, M. F. and M. R. WATERMAN, (1986), Arch. Biochem. Biophys. 250, 456—460.

McCALL, A. L., J. STERN, S. L. DALE, and J. C. MELBY, (1978), Endocrinology 103, 1—5.

MILLER, W. L., (1987), J. Steroid Biochem. 27, 759—766.

MITANI, F., (1979), Mol. Cell. Biochem. 24, 21—43.

MITANI, F. and S. HORIE, (1969), J. Biochem. (Tokyo) 65, 269—280.

MITANI, F., and S. HORIE, (1970), J. Biochem. (Tokyo) 68, 529—542.

MITANI, F., N. ANDO, and S. HORIE, (1973), Ann. N.Y. Acad. Sci. 212, 208—226.

MITANI, F., A. ICHIYAMA, A. MASUDA, and I. OGATA, (1975), J. Biol. Chem. 250, 8010 to 8015.

MITANI, F., T. IIZUKA, R. UENO, Y. ISHIMURA, T. KIMURA, S. IZUMI, N. KOMATSU, and K. WATANABE, (1982a), Adv. Enz. Regulation **20**, 213—231.

MITANI, F., T. SHIMIZU, R. UENO, Y. ISHIMURA, S. IZUMI, N. KOMATSU, and K. WATANABE, (1982b), J. Histochem. Cytochem. **30**, 1066—1074.

MITANI, F., T. IIZUKA, H. SHIMADA, R. UENO, and Y. ISHIMURA, (1985), J. Biol. Chem. **260**, 12042—12048.

MOMOI, K., M..OKAMOTO, S. FUJII, C. Y. KIM, Y. MIYAKE, and T. YAMANO, (1983), J. Biol. Chem. **258**, 8855—8860.

MOMOI, K., M. OKAMOTO, and T. YAMANO, (1985), J. Steroid Biochem. **22**, 267—271.

MOROHASHI, K., H. YOSHIOKA, O. GOTOH, Y. OKADA, K. YAMAMOTO, T. MIYATA, K. SOGAWA, Y. FUJII-KURIYAMA, and T. OMURA, (1987), J. Biochem. (Tokyo) **102**, 559 to 568.

MÜLLER, J., (1988), in "Regulation of Aldosterone Biosynthesis". Springer-Verlag, Berlin, p. 1—21.

NABI, N., S. KOMINAMI, S. TAKEMORI, and T. OMURA, (1983), J. Biochem. (Tokyo) **94**, 1517—1527.

NAGAI, K., I. MIYAMORI, M. IKEDA, H. KOSHIDA, R. TAKEDA, K. SUHARA, and M. KATAGIRI, (1986), J. Steroid Biochem. **24**, 321—323.

NEW, M. I. and L. S. LEVINE, (1973), Adv. Human Genet. **4**, 251—326.

OGISHIMA, T., Y. OKADA, S. KOMINAMI, S. TAKEMORI, and T. OMURA, (1983), J. Biochem. (Tokyo) **94**, 1711—1714.

OHNISHI, T., A. WADA, Y. NONAKA, M. OKAMOTO, and T. YAMANO, (1984), Biochem. Int. **9**, 715—723.

OHNISHI, T., A. WADA, Y. NONAKA, T. SUGIYAMA, T. YAMANO, and M. OKAMOTO, (1986), J. Biochem. (Tokyo) **100**, 1065—1076.

OHNISHI, T., A. WADA, M. LAUBER, T. YAMANO, and M. OKAMOTO, (1988), J. Steroid Biochem. **31**, 73—81.

OHTA, M., S. FUJII, A. WADA, T. OHNISHI, T. YAMANO, and M. OKAMOTO, (1987), J. Steroid Biochem. **26**, 73—81.

OHTA, M., S. FUJII, T. OHNISHI, and M. OKAMOTO, (1988), J. Steroid Biochem. **29**, 699 to 707.

OKAMOTO, M., K. MOMOI, S. FUJII, and T. YAMANO, (1982), Biochem. Biophys. Res. Commun. **109**, 236—241.

OKAMOTO, M., M. OHTA, A. WADA, T. OHNISHI, T. SUGIYAMA, and T. YAMANO, (1985), in "Cytochrome P-450, Biochemistry, Biophysics and Induction", (L. VERECZKEY and K. MAGYAR, eds.), Akademiai Kiado, Budapest, p. 497—500.

OMURA, T., R. SATO, D. Y. COOPER, O. ROSENTHAL, and R. W. ESTABROOK, (1965), Fed. Proc. **24**, 1181—1189.

RAMAN, P. B., D. C. SHARMA, and R. I. DORFMAN, (1966), Biochemistry **5**, 1795—1804.

RAMSEYER, J. and B. W. HARDING, (1973), Biochim. Biophys. Acta **315**, 306—316.

RAPP, J. P. and L. K. DAHL, (1972), Endocrinol. **90**, 1435—1446.

RAPP, J. P. and L. K. DAHL, (1976), Biochemistry **15**, 1235—1242.

SATO, H., N. ASHIDA, K. SUHARA, E. ITAGAKI, S. TAKEMORI, and M. KATAGIRI, (1978), Arch. Biochem. Biophys. **190**, 307—314.

SCHLEYER, H., D. Y. COOPER, and O. ROSENTHAL, (1972a), J. Biol. Chem. **247**, 6103 to 6110.

SCHLEYER, H., D. Y. COOPER, S. S. LEVIN, and O. ROSENTHAL, (1972b), in "Biological Hydroxylation Mechanisms". (G. S. BOYD and R. M. S. SMELLIE, eds.), Academic Press, London, p. 187—206.

SHIKITA, M. and P. F. HALL, (1973), J. Biol. Chem. **248**, 5598—5604.

SHIMIZU, T., T. IIZUKA, F. MITANI, Y. ISHIMURA, T. NOZAWA, and M. HATANO, (1981), Biochim. Biophys. Acta **669**, 46—59.

SKELTON, F. R., (1953), Endocrinology **53**, 492−505.

SUGANO, S., T. OHNISHI, N. HATAE, K. ISHIMURA, H. FUJITA, T. YAMANO, and M. OKA-MOTO, (1985), J. Steroid Biochem. **23**, 1013−1021.

SUHARA, K., T. GOMI, H. SATO, E. ITAGAKI, S. TAKEMORI, and M. KATAGIRI, (1978), Arch. Biochem. Biophys. **190**, 290−299.

SUHARA, K., Y. FUJIMURA, M. SHIROO, and M. KATAGIRI, (1984), J. Biol. Chem. **259**, 8729−8736.

SUHARA, K., K. TAKEDA, and M. KATAGIRI, (1986a), Biochem. Biophys. Res. Commun. **136**, 369−375.

SUHARA, K., K. OHASHI, K. TAKEDA, and M. KATAGIRI, (1986b), Biochem. Biophys. Res. Commun. **140**, 530−535.

SUHARA, K., K. FUJII, T. TANI, and M. KATAGIRI, (1987), J. Steroid Biochem. **26**, 113 to 119.

SWEAT, M. L. and M. D. LIPSCOMB, (1955), J. Am. Chem. Soc. **77**, 5185−5187.

SWEAT, M. L., R. A. ALDRICH, C. H. DE BRUIN, W. L. FOWLKS, L. R. HEISELT, and H. S. MASON, (1956), Fed. Proc. **15**, 367.

TAKEMORI, S., K. SUHARA, S. HASHIMOTO, M. HASHIMOTO, H. SATO, T. GOMI, and M. KATAGIRI, (1975a), Biochem. Biophys. Res. Commun. **63**, 588−593.

TAKEMORI, S., H. SATO, T. GOMI, K. SUHARA, and M. KATAGIRI, (1975b), Biochem. Biophys. Res. Commun. **67**, 1151−1157.

WADA, A., M. OKAMOTO, Y. NONAKA, and T. YAMANO, (1984), Biochem. Biophys. Res. Commun. **119**, 365−371.

WADA, A., T. OHNISHI, Y. NONAKA, M. OKAMOTO, and T. YAMANO, (1985), J. Biochem. (Tokyo) **98**, 245−256.

WADA, A., T. OHNISHI, Y. NONAKA, and M. OKAMOTO, (1988), J. Steroid Biochem. **31**, 803−808.

WANG, H.-P. and T. KIMURA, (1976), J. Biol. Chem. **251**, 6068−6074.

WATANUKI, M., B. E. TILLEY, and P. F. HALL, (1978), Biochemistry **17**, 127−130.

WATERMAN, M. R. and E. R. SIMPSON, (1985), Mol. Cell. Endocr. **39**, 81−89.

YAGO, N. and S. ICHII, (1969), J. Biochem. (Tokyo) **65**, 215−224.

YANAGIBASHI, K., M. HANIU, J. E. SHIVELY, W. H. SHEN, and P. HALL, (1986), J. Biol. Chem. **261**, 3556−3562.

Chapter 5
Adrenal Microsomal Cytochrome P-450 Dependent Reactions in Steroidogenesis and Biochemical Properties of the Enzymes Involved therein

S. Takemori and S. Kominami

1. Introduction

Adrenal glands secrete three kinds of steroid hormone: mineralocorticoids, glucocorticoids and androgens. They are responsible for the balance of water and electrolytes in the body, for the overall regulation of the metabolism of protein, carbohydrate and lipid and for the secondary sex characteristics, respectively. Figure 1 shows the biosynthetic pathway of steroid hormones in adrenal cortex cells in which a series of simple and complex cytochrome P-450-dependent monooxygenase reactions are involved. It must be remembered that all the reactions but one are catalyzed by four different species of cytochrome P-450 located both in mitochondria and endoplasmic reticulum in adrenal cortex cells. The exception is 3β-hydroxysteroid dehydrogenase-isomerase which catalyzes the conversion of Δ^5-steroids to Δ^4-steroids. The adrenal mitochondria contain at least two types of cytochrome P-450, that is to say, P-450$_{SCC}$ and P-450$_{11\beta}$. The endoplasmic reticulum also contains two species of cytochrome P-450, P-450$_{C21}$ and P-450$_{17\alpha,lyase}$ (TAKEMORI and KOMINAMI, 1984).

Cholesterol, the source material for steroidogenesis, originates mainly from plasma cholesterol and partly from cholesterol synthesized from acetate in adrenal cells (KRUM et al., 1964; ICHII et al., 1967). Plasma cholesterylesters bound to low density lipoproteins are carried into the adrenal cells by receptor-mediated endocytosis and most of them are stored in lipid droplets in the adrenal gland (BROWN and GOLDSTEIN, 1986). When required, the cholesterol is liberated by the action of cholesterol esterase which is regulated by phosphorylation-dephosphorylation via cAMP-dependent protein kinase (BECKETT and BOYD, 1977) and passes into the mitochondria, where P-450$_{SCC}$ cleaves the C-20—C-22 bond of cholesterol to form pregnenolone. This is the rate-determining step in steroidogenesis and the site of ACTH-dependent acute regulation (see Chapter 3). The pregnenolone passes out of the mitochondria into the endoplasmic reticulum, where various reactions catalyzed by P-450$_{C21}$, P-450$_{17\alpha,lyase}$, and 3β-hydroxysteroid dehydrogenase-isomerase take place. Although there are two competing pathways from pregnenolone to androgen (Δ^4 and Δ^5 pathways), their relative importance is not yet clear. When progesterone is produced from pregnenolone by the action of the dehydrogenase-isomerase, it is either hydroxylated at the 21-position by P-450$_{C21}$ to give 11-deoxycorticosterone or is hydroxylated at the 17α-position by P-450$_{17\alpha,lyase}$. Some of the 17α-hydroxysteroids are successively converted into androgens by C-17,C-20-lyase action (P-450$_{17\alpha,lyase}$) and the rest are hydroxylated at the 21-position to give 11-deoxycortisol. Furthermore, the transfer of 11-deoxycortisol and 11-deoxycorticosterone back into the mitochondria is necessary for 11β-hydroxylation catalyzed by P-450$_{11\beta}$ in the synthesis of cortisol and corticosterone respectively and for further 18-hydroxylation in the synthesis of aldosterone. Therefore, the metabolic intermediates must move back and

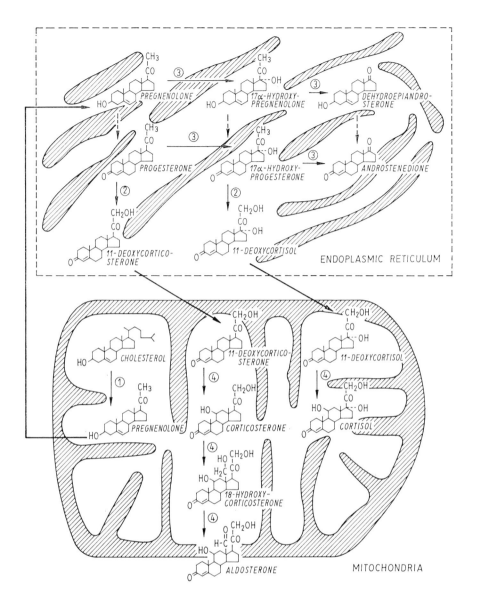

Fig. 1. The biosynthetic route of steroid hormones in adrena lcortex cells. The solid arrows indicate monooxygenase reactions catalyzed by cytochrome P-450: (1), P-450$_{\text{SCC}}$; (2), P-450$_{\text{C21}}$; (3), P-450$_{17\alpha,\text{lyase}}$; (4), P-450$_{11\beta}$. The dotted arrows indicate the reactions catalyzed by 3β-hydroxysteroid dehydrogenase.

forth between subcellular organelles during the series of reactions. Finally, steroid hormones such as corticosterone, cortisol, aldosterone or androgens pass into the blood to be delivered to the target cells.

Two different electron transfer pathways from NADPH to cytochrome P-450 are localized in two organelles of adrenal cortex cells. The electron transfer system of the mitochondria, which is composed of NADPH-adrenodoxin reductase, adrenodoxin and cytochrome P-450 (P-450$_{SCC}$ and P-450$_{11\beta}$), is localized on the matrix side of the mitochondrial inner membrane (MITANI et al., 1982). Both adrenodoxin reductase and adrenodoxin are loosely associated with the membrane, while P-450$_{SCC}$ and P-450$_{11\beta}$ are bound tightly to the membranes. As noted in Chapters 2 and 4, the mitochondrial electron transfer system has been most extensively studied using the purified proteins. With regard to the electron transfer system in the adrenal endoplasmic reticulum, however, until 1980 very little was known about the molecular details of the enzymes participating in the 21- and 17α-hydroxylations as well as the C-17—C-20 bond cleavage of steroid. The 21-hydroxylation of 17α-hydroxyprogesterone in adrenal gland tissues received early the attention of endocrinologists because of its deficiency related closely with congenital adrenal hyperplasia. P-450$_{C21}$ played the role of a pioneer in the history of cytochrome P-450 investigation. It was with this cytochrome in adrenal microsomes that ESTABROOK et al. (1963) first demonstrated the involvement of a cytochrome P-450 in steroid 21-hydroxylation using the photochemical action spectrum technique. Immunochemical studies by MASTERS et al. (1971) showed that the microsomal electron transfer system is composed of NADPH-cytochrome P-450 reductase and cytochrome P-450 but no iron-sulfur protein. However, much less effort was directed towards the isolation and characterization of the microsomal electron transfer system because the low contents of the electron transfer components of adrenal microsomes and their tight association with microsomal membranes hampered their actual isolation. Over the last decade, remarkable progress has been accomplished in the purification and the characterization of the adrenal microsomal electron transfer components. Extensive investigations have been carried out using the reconstituted system composed of the purified enzymes.

In this Chapter, we will describe what is currently known of adrenal microsomal cytochrome P-450-dependent reactions in steroidogenesis and biochemical properties of the enzymes involved therein. We will briefly review aromatase reaction since adrenal cortex is one of estrogen-producing organs in which the amounts secreted are quite small (RAPP, 1986).

S. TAKEMORI; S. KOMINAMI

2. Historical background

Early in the 1950s the conversion of progesterone into the 21- and 17α-hydroxylated derivatives, deoxycorticosterone and 17α-hydroxyprogesterone, was observed after the incubation of progesterone with bovine adrenal tissue homogenates (PLAGER and SAMUELS, 1954). In this study, it was recognized that some cofactors besides the tissue homogenates were required for hydroxylation. RYAN and ENGEL (1957) defined steroid 21-hydroxylation as a reaction in the microsomal fraction in bovine adrenals and showed that the cofactors required for the reaction were molecular oxygen and NADPH. HOFMANN (1960) also concluded that steroid 17α-hydroxylation of the adrenal cortex required NADPH and molecular oxygen. It must be remembered that RYAN and ENGEL recognized the inhibition of 21-hydroxylation by carbon monoxide and the recovery of its inhibition by light illumination, which provided a valuable clue towards the discovery of the physiological function of cytochrome P-450.

In these years, HAYAISHI et al. (1955) and separately MASON et al. (1955) discovered an oxygenase reaction which leads to the direct incorporation of one or two atoms of molecular oxygen per molecule of an organic substrate. The stoichiometry of oxygen and NADPH consumption in steroid 21-hydroxylation was established by COOPER et al. (1963) in which 21-hydroxylation of one mole 17α-hydroxyprogesterone was determined to consume one mole each of molecular oxygen and NADPH in bovine adrenal microsomes. This data confirmed the view that the 21-hydroxylase belongs to the class of enzymes known as monooxygenases (mixed function oxidase). In hepatic microsomes, KLINGENBERG (1958) and GARFINKEL (1958) independently observed unusual absorption of CO-binding pigment. LATER, OMURA and SATO (1964) attributed this pigment to a novel heme protein, cytochrome P-450. Utilizing the photochemical action spectrum method, COOPER et al. (1965) carried out an ingenious experiment to establish the relationship between 21-hydroxylase and cytochrome P-450 in the adrenal cortex microsomes as well as that between the drug metabolism and cytochrome P-450 in hepatic microsomes. The photochemical action spectrum for the light reversal of the carbon monoxide inhibition of 21-hydroxylase activity had a maximum at 450 nm which was just the same as the absorption maximum of the dithionite-reduced carbon-monoxide-difference spectrum of adrenal cortex microsomes. On the basis of these findings, they proposed that cytochrome P-450 in adrenal microsomes is the enzyme catalyzing the 21-hydroxylation.

Adrenal mitochondria had been shown to have 11β-hydroxylase activity for 11-deoxycorticosterone and 11-deoxycortisol (SWEAT and LIPSCOMB, 1955; GRANT and BROWNIE, 1955) which was attributed latterly to cytochrome P-450 in the mitochondria (COOPER et al., 1967). OMURA et al. (1965) found that an iron sulfur protein and some reductase were required for the electron

---

transfer from NADPH to adrenal mitochondrial cytochrome P-450 for 11β-hydroxylation. In that study, their attempt to reduce the mitochondrial cytochrome P-450 by hepatic microsomal NADPH-cytochrome c reductase was not successful. The cholesterol side-chain cleavage system of adrenal mitochondria had been shown by HALKERSTON et al. (1961) to belong to the monooxygenase requiring NADPH and molecular oxygen. Definite proof for the difference of the electron transfer between microsomal and mitochondrial systems was provided by MASTERS et al. (1971) and BARON et al. (1972). They prepared antibodies to NADPH-cytochrome c reductase from pig hepatic microsomes and to adrenodoxin from bovine adrenal mitochondria. The former antibody inhibited ethylmorphine N-demethylase activity mediated by cytochrome P-450 in pig liver microsomes and also the 21-hydroxylation in bovine adrenal microsomes but did not interfere with the reduction of cytochrome P-450 in the bovine adrenal mitochondria. The latter antibody inhibited the mitochondrial cytochrome P-450 reduction but not the microsomal cytochrome P-450 reduction. By reconstitution as well as immunochemical studies, the route of electron transfer for mitochondrial system was established as follows:

NADPH → Flavoprotein (FAD) → Iron-sulfur protein →
Cytochrome P-450

and for the microsomal system:

NADPH → Flavoprotein (FAD, FMN) → Cytochrome P-450.

3.   NADPH-cytochrome P-450 reductase

The hepatic microsomal NADPH-cytochrome P-450 reductase was initially purified as "cytochrome c reductase" from lipase or protease-treated microsomes (WILLIAMS and KAMIN, 1962; PHILLIPS and LANGDON, 1962). However, these preparations were not capable of reconstituting the drug hydroxylation with cytochrome P-450. The reductase has been purified in the native form from the detergent-solubilized microsomes by using an affinity chromatography on 2′,5′-ADP-Sepharose (YASUKOCHI and MASTERS, 1976; VERMILION and COON, 1978) or NADP-Sepharose (DIGNAM and STROBEL, 1977). It is not clear at present whether the species of NADPH-cytochrome P-450 reductase is the same in hepatic and adrenal microsomes. There is evidence that microsomal NADPH-cytochrome P-450 reductase in rat liver and testis can not be distinguished from each other by biochemical and immunological techniques and therefore, are probably the same protein (HALES and BETZ, 1986). Using the antibody against pig liver reductase, DEES et al. (1980) showed that the microsomal reductase in the kidney, liver, and lung of minipigs and rats are all immunochemically similar to pig liver reductase. There is a difference in the content of the reductase between liver and adrenal microsomes. Bovine adrenal microsomes contain a lesser amount (to about one order) of the reductase per mg protein than bovine liver microsomes. The low content of the reductase in bovine adrenal microsomes might be related to the low content of cytochrome P-450 in the microsomes compared with cytochrome P-450 content of the hepatic microsomes.

The purification of NADPH-cytochrome P-450 reductase from bovine adrenal microsomes was carried out by HIWATASHI and ICHIKAWA (1979) and separately by TAKEMORI and KOMINAMI (1980). The reductase was purified from bovine adrenal microsomes by detergent solubilization followed by column chromatographies including 2′,5′-ADP-Sepharose. The reductase has the molecular mass of 78 ± 4 kDa as estimated by SDS polyacrylamide gel electrophoresis and contains about 1 mol each of FMN and FAD as the prosthetic group. The oxidized form of the reductase shows an absorption spectrum of a typical flavoprotein, which has absorption maxima at 380 and 455 nm. An aerobic titration of the reductase with NADPH produced a spectral species with absorption around 600 nm which could be attributed to the air-stable semiquinone radical (KOMINAMI et al., 1982). ESR spectrum was observed at 77 K for this species which was similar to that reported for the hepatic microsomal reductase (IYANAGI and MASON, 1973).

The complete amino acid sequence of the hepatic reductase has been determined by cDNA sequencing (PORTER and KASPER, 1985; KATAGIRI et al., 1986) and also by microsequence analysis (HANIU et al., 1986). It is reasonable to assume that the primary structure of the adrenal reductase is similar to that of the hepatic reductase. When the adrenal reductase incorporated into liposomal membranes was digested by trypsin, 65 kDa water soluble and 13 kDa membrane bound fragments were produced. The 65 kDa fragment contains the flavins and shows NADPH-cytochrome c reductase activity. The reductase at high concentration (e.g. 5 μM) can exist in a stable fashion in an aqueous solution without detergent where the reductase exists as a protein micelle of about 500 kDa as detected by gel filtration (KOMINAMI et al., 1984).

On hepatic reductase, the mechanism of the electron transfer from NADPH to cytochrome c has been studied by KAMIN et al. (1966) who obtained parallel lines in double reciprocal plots of the cytochrome c reduction rate vs. the concentration of oxidized cytochrome c at various concentrations of NADPH. On adrenal microsomal reductase, the double reciprocal plots clearly gave parallel lines, suggesting the electrons from NADPH were transferred to cytochrome c in the mechanism Hexa-Uni-Ping-Pong. The Km values for NADPH and cytochrome c in the steady state electron transfer were 6—8 μM and 9—17 μM, respectively which was not much different from those of hepatic reductase. The cytochrome c reductase activity was quite dependent on the pH and ionic strength of the reaction medium. The maximum molecular activity was estimated to be 6000 mol cytochrome c reduced per min per mol adrenal reductase in 0.3 M potassium phosphate buffer, pH 7.7, at 25 °C (KOMINAMI et al., 1982). The steady state reaction rate of P-450$_{C21}$ reduction by NADPH via the reductase was quite dependent on the concentration of the detergent whereas the reduction of cytochrome c was not (KOMINAMI et al., 1984). The details of the detergent effects on P-450$_{C21}$ reduction and the 21-hydroxylase activity will be discussed in the latter part of this section.

In order to study electron transfer between the reductase and cytochrome P-450 in a reconstituted system resembling the natural membranes, it is necessary for both the proteins to be incorporated into liposomal bilayers. Attempts to incorporate adrenal reductase into liposomal membranes have been made using two different methods (KOMINAMI et al., 1987). One version employed was the usual cholate dialysis method which involved removal of detergents from cholate-solubilized mixtures of adrenal reductase and phospholipids. Eighty percent of the reductase in the membrane incorporated by this method was located at the outer surface of the liposomes, as judged from the susceptibility of the reductase to proteinase treatment. The other method was the spontaneous insertion of the reductase into preformed liposomal vesicles. It was demonstrated from Ficoll density gradient centrifugation and HPLC gel filtration that the former method made the reductase bind tightly to the liposomal membranes, while the latter made the reductase bind loosely to the membranes. The loosely bound reductase was found to be transferable between the vesicles, whereas the tightly bound reductase was not readily transferred. The reductase in both binding mode well supported the steroid 21-hydroxylation. BLACK and COON (1982) proposed two possible binding modes for the hepatic reductase to the membrane of the endoplasmic reticulum as illustrated in Figure 2. The adrenal reductase in the two binding modes might have a similar configuration as proposed by them. It is not yet known which of the above mode structure exists in natural membranes.

4. Steroid 21-hydroxylase reaction

4.1. Purification and molecular properties of P-450$_{C21}$

Earlier investigations on cytochrome P-450 in adrenal cortex microsomes were carried out using microsomal particles or their extract. Attempts at the extraction and purification of cytochrome P-450 were carried out by NARASIM-HULU (1974) and MACKLER et al. (1971), but they were unable to establish the purification procedure. In order to purify cytochrome P-450 for 21-hydroxylation, we must first consider the source of the adrenal microsomes. The adrenals must be easily obtained and the adrenal microsomes must contain

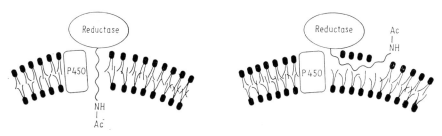

Fig. 2. The two binding modes of NADPH-cytochrome P-450 reductase to the membranes.

Table 1. Cytochrome P-450 content and 21- and 17α-hydroxylase activities of adrenal microsomes in various animals. 21-Hydroxylase activity was measured by the conversion of progesterone to deoxycorticosterone. 17α-Hydroxylase activity was measured by the conversion of progesterone to 17α-hydroxyprogesterone.

Adrenal microsomes	P-450 content	21-Hydroxylase activity	17α-Hydroxylase activity	References
	nmol/mg protein	nmol product/mg protein		
Cow	0.4—0.7	1.4	0.2	KOMINAMI et al. (1983) HIWATASHI and ICHIKAWA (1981) NARASIMHULU and EDDY (1985)
Guinea pig	1.0—2.3	2.4—4.4	5—15	KOMINAMI et al. (1983) MENARD et al. (1976; 1979)
Rabbit	0.46	1.77	0.055	MENARD et al. (1976; 1979)
Dog	1.16—1.14	3.46	7.3	MENARD et al. (1976; 1979)
Rat	0.76—1.12	6.7	N. D.	MENARD et al. (1976; 1979)
Pig	0.86	0.81	0.43	INANO et al. (1969)

considerable amounts of cytochrome P-450 for 21-hydroxylation compared with other cytochrome P-450, where major cytochromes P-450 in adrenal microsomes are those for 17α- and 21-hydroxylations. Cytochrome P-450 contents and 21- and 17α-hydroxylase activities in adrenal microsomes from various sources are summarized in Table 1. The variety in the ratio of 21-hydroxylase activity to 17α-hydroxylase activity is one of the indications that these reactions might be catalyzed by different species of cytochrome P-450. An animal having stronger 17α-hydroxylase activity for progesterone than 21-hydroxylase activity in the adrenals must produce more cortisol than corticosterone since the glucocorticoid and the adrenals having stronger 21-hydroxylase activity for progesterone must secrete predominantly corticosterone. As the source of microsomes for the purification of P-450$_{C21}$ the adrenals of rats and rabbits would seem impractical, since the small size of their adrenal glands makes it very difficult to collect large amounts of microsomes. Bovine adrenals were in fact chosen as the source material of the microsomes for the purification of P-450$_{C21}$.

In general, in the procedure of enzyme purification one can detect the enzyme fraction by the assay of enzyme activities. Since the detergent used for solubilization prevents the interaction of P-450$_{C21}$ with the reductase added externally to the assay mixture (KOMINAMI et al., 1984), it is quite difficult to estimate correctly the activity of each fraction eluted from the chromatography column during purification. P-450$_{C21}$ can be detected by the type I spectral change upon addition of substrate steroid, which was demonstrated by NARASIMHULU (1971) and BRYAN et al. (1974). The microsomes from bovine adrenal cortex can be obtained by the usual differential centrifugations but it is not easy to prepare pure microsomes not contaminated by the mitochondrial particles of less than 10%. The content of cytochrome P-450 in bovine adrenal mitochondria is about 5 times higher than that in the microsomes and the cytochrome P-450 content in the 10% mitochondrial contamination occupies about one third of the total cytochrome P-450 in the microsomal preparation. Using a hydrophobic chromatography on aminooctyl-Sepharose, KOMINAMI et al. (1980) have presented for the first time a procedure for the purification of P-450$_{C21}$ from bovine adrenal microsomes. P-450$_{C21}$ was solubilized with sodium cholate from bovine adrenal microsomes. The chromatogram of the crude extract on an aminooctyl Sepharose column showed two peaks of cytochrome P-450. Upon the addition of 17α-hydroxyprogesterone, the cytochrome P-450 in the first peak did not show any spectral change and that in the second showed the type I spectral change. The latter fraction containing P-450$_{C21}$ was further purified by a second aminooctyl Sepharose chromatography (KOMINAMI et al., 1983). Almost the same purification procedures were applied by HIWATASHI and ICHIKAWA (1981), BUMPS and DUS (1982), YUAN et al. (1983), and NARASIMHULU and EDDY (1985).

The purified preparation was quite stable without substrate in the buffer

---

S. TAKEMORI; S. KOMINAMI

containing 20% glycerol and Emulgen 913 at more than 0.1%. When Emulgen was removed from the sample, one half of $P\text{-}450_{C21}$ precipitated in 30 min at 25 °C. The incubation of the purified preparation in the buffer without glycerol converted 15% of the original $P\text{-}450_{C21}$ into P-420 in 2 hrs at 25 °C. In the reconstituted system consisting of NADPH-cytochrome P-450 reductase and $P\text{-}450_{C21}$ in the presence of 0.005% Emulgen 913, progesterone was converted into deoxycorticosterone and 17α-hydroxyprogesterone into 11-deoxycortisol at the activity of 10 and 19 mol of product per min per mol of $P\text{-}450_{C21}$, respectively. The specific content of the purified $P\text{-}450_{C21}$ was 16—18 nmol per mg of protein. The electrophoretic behaviour of $P\text{-}450_{C21}$ was different from that of $P\text{-}450_{SCC}$ and $P\text{-}450_{11\beta}$ on SDS polyacrylamide in which the band of $P\text{-}450_{C21}$ appeared between those of $P\text{-}450_{SCC}$ and $P\text{-}450_{11\beta}$. The molecular mass was estimated from the amino acid analysis as 54 kDa. The N-terminal amino acid sequence of $P\text{-}450_{C21}$ started from methionine which was definitely different from cytochromes P-450 of mitochondria (OGISHIMA et al., 1983). The difference in the immunochemical property was observed by the use of antibodies with $P\text{-}450_{SCC}$ and $P\text{-}450_{11\beta}$, where the antibodies showed the precipitin lines only with corresponding cytochrome P-450 but did not react with $P\text{-}450_{C21}$. Furthermore, antibody to $P\text{-}450_{17\alpha,lyase}$ purified from guinea pig adrenal microsomes did not react with $P\text{-}450_{C21}$ (SHINZAWA et al., 1985). Recently, the amino acid sequence of bovine adrenal $P\text{-}450_{C21}$ was deduced by YOSHIOKA et al. (1986) from the nucleotide sequence of the cloned cDNA of mRNA which was isolated by using oligonucleotides as synthesized probes based on the partial sequence determined by WHITE et al. (1984). Bovine adrenal $P\text{-}450_{C21}$ is composed of 496 amino acids with a molecular weight of 56,113 Da and has methionine as the N-terminal, which agrees with the earlier observation that $P\text{-}450_{C21}$ is synthesized in its natural form mainly at bound ribosomes (NABI et al., 1983). The amino acid sequence of bovine $P\text{-}450_{C21}$ has high homogeneity with $P\text{-}450_{C21}$ from human and porcine adrenals (HIGASHI et al., 1986; HANIU et al., 1987). A conserved amino acid sequence containing a heme-binding cysteine was found to be near the COOH terminus of $P\text{-}450_{C21}$. Sequence alignment of different cytochromes P-450 indicate that the sequence of adrenal $P\text{-}450_{C21}$ is much closer to those of hepatic microsomal cytochromes P-450 than to those of adrenal mitochondrial $P\text{-}450_{SCC}$ and $P\text{-}450_{11\beta}$ (MOROHASHI et al., 1984; MOROHASHI et al., 1987).

4.2. Interaction of $P\text{-}450_{C21}$ with steroid in detergent solubilized system

As shown in Figure 3, the purified $P\text{-}450_{C21}$ in the presence of Emulgen 913 showed a typical low spin optical absorption spectrum in the oxidized state. The addition of 17α-hydroxyprogesterone to $P\text{-}450_{C21}$ elicited changes in the spectrum, shifting the maximum of the Soret band from 417 nm to 394 nm

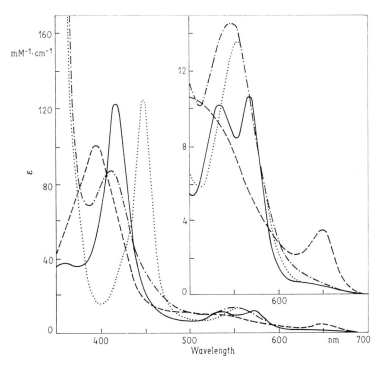

Fig. 3. Absorption spectra of purified P-450$_{C21}$ from bovine adrenal microsomes, taken from KOMINAMI et al. (1980) with permission. (————) oxidized, (—————) 17α-hydroxy progesterone-complex, (·————·) reduced and (·····) reduced CO-complex.

accompanied by a charge transfer band around 650 nm. A typical P-450$_{C21}$-CC spectrum was observed upon the addition of CO to the dithionite-reduced P-450$_{C21}$. The type I spectral changes were also observed upon the addition of steroids such as progesterone, 17α-hydroxyprogesterone, 11β-hydroxyprogesterone, androstenedione and testosterone. The 11β-hydroxysteroids and androgens caused the reverse type I spectral change when they were added to high spin P-450$_{C21}$-substrate complex (KOMINAMI et al., 1980). Steroids such as progesterone, 17α-hydroxyprogesterone, 11β-hydroxyprogesterone and 21-deoxycortisol are hydroxylatable substrates for P-450$_{C21}$, all of which have a double bond at Δ^4 position. MACKLER et al., (1971), however, reported 21-hydroxylation of pregnenolone having a double bond at Δ^5 position in steer adrenal tissues, which might be a special case. The 21-hydroxylation of pregnenolone has not been observed in the adrenals of other animals. HORNSBY (1982) studied the effect of various steroids on the 21-hydroxylation of 21-deoxycortisol using primary cultures of bovine adrenal cortex cells and found that several androgens were quite effective in the inhibition of 21-hydroxylation. This inhibition was attributed to the competition of androgens with the substrate for binding to P-450$_{C21}$. Moreover, using bovine adrenal microsomes, NARASIMHULU (1971) showed that androstenedione caused consumption of NADPH and oxygen without steroid hydroxylation. With the use of the type I spectral change as the criterion for substrate binding, various aspects of binding of steroids to P-450$_{C21}$ have been studied. NARASIMHULU and EDDY (1985) found that the apparent dissociation constant of P-450$_{C21}$-substrate complex was dependent on the cholate concentration and also on the concentration of P-450$_{C21}$ and proposed the view that P-450$_{C21}$ exists in a monomer-dimer equilibrium and that the substrate cannot bind to the dimer. The dissociation constant of 17α-hydroxyprogesterone for P-450$_{C21}$ in the presence of detergent was smaller than that of progesterone, which was also the case in a liposomal P-450$_{C21}$ system (KOMINAMI et al., 1988). The smaller dissociation constant and the higher activity of 21-hydroxylase for 17α-hydroxyprogesterone than progesterone seems to help to simplify the regulation of production either of cortisol or corticosterone for the glucocorticoid in the animal. In the case of higher 21-hydroxylase activity for progesterone, the activity of 17α-hydroxylase might not be the only factor determining which steroid — corticosterone or cortisol — would be produced as the glucocorticoid in the adrenal and the regulation of the hormone biosynthesis would be complicated. As can be seen in Table 1, the cortisol producing animal has a definitely higher 17α-hydroxylase activity. The effects of membrane fluidity of microsomes on the binding of substrate to P-450$_{C21}$ has also been shown (NARASIMHULU, 1977). The fluorescence experiment suggests that typtophans contribute to the substrate-binding site in P-450$_{C21}$ (NARASIMHULU, 1988).

The extinction coefficients (ε max — min) of difference spectra of P-450$_{C21}$ induced upon the addition of a sufficient amount of steroid varied for each steroid, being 113 mM^{-1} cm^{-1} for 17α-hydroxyprogesterone, 86 mM^{-1} cm^{-1} for progesterone and 64 mM^{-1} cm^{-1} for 11β-hydroxyprogesterone in the presence of Emulgen 913 at 25 °C (KOMINAMI et al., 1980). The different values of the extinction coefficients indicate the difference in the content of the high spin form for each steroid P-450$_{C21}$ complex. The spin state conversion with temperature was observed for 17α-hydroxyprogesterone-bound and progesterone-bound P-450$_{C21}$ and also slightly for the substrate-free form (KOMINAMI and TAKEMORI, 1982). These temperature-dependent spin state conversions have also been reported for bacterial P-450$_{cam}$ (SLIGAR, 1976) and for hepatic cytochrome P-450 (RISTAU et al., 1978). These phenomena have been explained as a temperature-dependent equilibrium between high spin and low spin conformations around the heme which is quite sensitive to temperature. On the basis of the temperature dependence, the relative amounts of the high spin form were estimated to be 94% for 17α-hydroxyprogesterone-bound and 78% for progesterone-bound and 25% for the substrate-free P-450$_{C21}$ at 25 °C. The redox potential of cytochrome P-450 was recognized to be dependent on the content of the high spin form by SLIGAR et al. (1979). By following the formation of CO-complex of the reduced P-450$_{C21}$, the effect of spin state on the reduction of P-450$_{C21}$ was examined in the reconstituted system containing P-450$_{C21}$ and NADPH-cytochrome P-450 reductase in the presence of 0.5% Emulgen 913 and 20% glycerol (KOMINAMI and TAKEMORI, 1982). The rate of the reduction was measured in the steady state by varying the concentrations of P-450$_{C21}$ in the presence of a catalytic amount of the reductase with or without steroid substrate. The maximum turnover rate of the reductase for the reduction of P-450$_{C21}$-substrate complex was almost similar to that for substrate-free P-450$_{C21}$, being about 100 mol of P-450$_{C21}$ reduced per min per mol of the reductase at 25 °C. The Km value of the reductase for P-450$_{C21}$ was quite different for the substrate-bound and substrate-free forms being about 50 times larger for the latter, indicating that steroid substrate is required for the interaction of P-450$_{C21}$ with the reductase. The presence of substrate was shown to increase the interaction between hepatic P-450$_{LM2}$ and the reductase (FRENCH et al., 1980). The electron transfer reaction from the reductase to P-450$_{C21}$ was also studied by a stopped flow method (KOMINAMI and TAKEMORI, 1982). Most of P-450$_{C21}$-steroid complexes were reduced within 10 sec but the substrate-free P-450$_{C21}$ was reduced very slowly even though the substrate-free form had 25% high spin content. This experiment indicates that the reduction of P-450$_{C21}$ is not dependent on the content of the high spin form but that the substrate-bound form is essential for the electron transfer from the reductase to P-450$_{C21}$. Adrenal P-450$_{17\alpha,lyase}$ from guinea pigs exists in the low spin form even in the pregnenolone-bound form and is active in pregnenolone hydroxylation (KOMINAMI, 1990), which is one more indication that the high spin form is not essential for the hydroxylation reaction. Furthermore, GUENGERICH (1983) indicated that the spin state did not correlate to the catalytic activity of hepatic cytochromes P-450 in a reconstituted enzyme system.

4.3. Reaction mechanism between P-450$_{C21}$ and NADPH-cytochrome P-450 reductase

The hydroxylase activity of P-450$_{C21}$ in a reconstituted enzyme system is quite dependent on the concentration of Emulgen 913 (KOMINAMI et al., 1984). As illustrated in Figure 4(a), the activity was maximum at an Emulgen concentration of 0.005% and sharply dropped either with the decrease or with the increase of the detergent concentration that figure 0.005%. The decrease of the activity below the optimum could easily be attributed to the rapid precipitation of P-450$_{C21}$ at the lower detergent concentration, which might be due to the exposure of the hydrophobic portion of P-450$_{C21}$ to aqueous solution. It must be remembered here that preincubation of the reconstituted system with detergent is quite important for the measurement of the hydroxylase

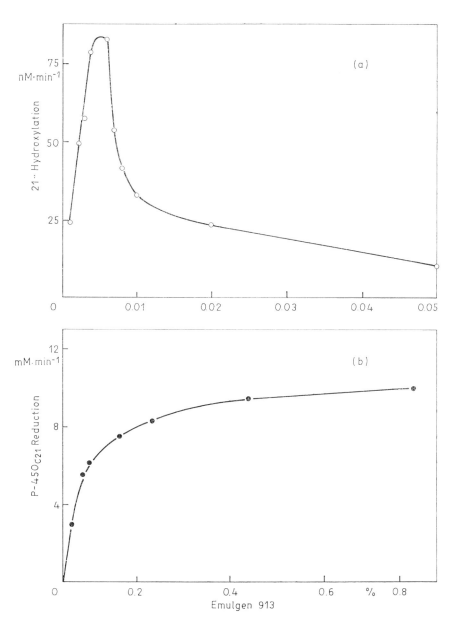

Fig. 4. Effect of Emulgen 913 concentration on progesterone 21-hydroxylase activities (a) and on the initial velocity of $P\text{-}450_{C21}$ reduction (b) in a reconstituted system consisting of $P\text{-}450_{C21}$ and NADPH-cytochrome P-450 reductase, taken from KOMINAMI et al. (1984) with permission. The concentrations of $P\text{-}450_{C21}$ and NADPH-cytochrome P-450 reductase were 8.3 and 16.6 nM (a), and 570 and 0.39 nM (b), respectively.

activity. Usually, the hydroxylase activity was measured after the preincubation of P-450$_{C21}$ and NADPH-cytochrome P-450 reductase with the detergent for more than 10 hr at 0°C. The dependence of the electron transfer reaction from the reductase to P-450$_{C21}$ on Emulgen 913 was investigated in the steady state (KOMINAMI et al., 1984). Surprisingly, the initial velocity of P-450$_{C21}$ reduction in the steady state was enhanced by the increase of the detergent concentration as shown in Fig. 4(b). There is, however, a big difference in the reductase concentration between the measurements for the 21-hydroxylase activity and for the initial reduction rate of P-450$_{C21}$: in the former case, the concentration of the reductase is roughly similar to that of P-450$_{C21}$, while the concentration of the reductase is about 1/1500 of that of P-450$_{C21}$ in the measurement of steady state reduction of P-450$_{C21}$. The hydroxylase activity was found to be dependent on the concentrations of both the enzymes with a fixed concentration of the detergent. The dependence of the hydroxylase activity on the enzyme concentration obeys the mechanism where P-450$_{C21}$ and the reductase form an active complex for the hydroxylation at a molar ratio of 1:1. The equimolar active complex formation between cytochrome P-450 and the reductase in rat liver microsomes and in its reconstituted enzyme system has been reported by MIWA et al. (1978; 1979). The dissociation constant of the active complex between P-450$_{C21}$ and the reductase was calculated to be 1.5 nM at 0.005% Emulgen. The dissociation constant was increased with the increase of the detergent concentration, being 36 nM, 107 nM, 242 nM and 950 nM with the Emulgen concentration of 0.02%, 0.05%, 0.2% and 0.7%, respectively. It is quite apparent that the complex tends to dissociate at high detergent concentration, suggesting the reason why the hydroxylase activity is decreased with the increase of Emulgen 913 concentration.

In order to solve the discrepancy in the detergent dependence between the hydroxylase reaction and the steady state reduction of P-450$_{C21}$, the electron transfer reaction from the reduced reductase to P-450$_{C21}$ was measured at various Emulgen concentrations by the stopped flow method in the presence of CO (KOMINAMI et al., 1984). As shown in Figure 5, the observed reduction process of P-450$_{C21}$ consisted of fast and slow phases and the ratio of the two phases varied with the detergent concentration. The increase of Emulgen concentration decreased the ratio of amount of fast phase to that of slow phase where the rate of the electron transfer in the fast phase was constant but the rate in the slow phase increased with the increase of the Emulgen concentration. The dependence of the hydroxylase activity on the Emulgen concentration is quite similar to the dependence of the amount of the fast phase reduction. The ratio of the amount of P-450$_{C21}$ reduction in the fast phase to that in the slow phase also decreased with the decrease of the concentration of the reductase. The rate of electron transfer in the fast phases did not change with the concentration of the reductase, while the rate of electron transfer in

---

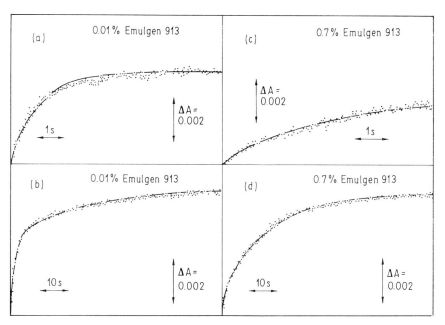

Fig. 5. Typical kinetic traces of P-450$_{C21}$ reduction at 450 nm measured with a stopped flow spectrophotometer, taken from KOMINAMI et al. (1984) with permission. The solution containing 120 nM P-450$_{C21}$, 120 nM NADPH-cytochrome P-450 reductase and 20 μM progesterone was mixed rapidly with an equal volume of 250 μM NADPH solution. Dotted points show the data points used for the curve fittings and solid lines show the simulated curves.

the slow phase increased with the increase of the reductase concentration. The dissociation constant between P-450$_{C21}$ and the reductase could be estimated from the ratio of content of fast phase to that of slow phase when the amount of the fast phase is assumed to be proportional to the amounts of the complex of P-450$_{C21}$ with the reductase, being similar to those obtained from hydroxylase activity. In this sense, the rate of electron transfer in the fast phase is attributed to the electron transfer from the reduced reductase to P-450$_{C21}$ within the stable complex. The rate of reduction in the slow phase must correspond to a random collision between the reduced reductase and P-450$_{C21}$. This dependence upon the Emulgen concentration can be simply explained as follows: at the low Emulgen concentration, the active complex between P-450$_{C21}$ and the reductase at a molar ratio of 1:1 is abundant in the system but the higher detergent concentration interferes with the complex formation. The rate of reduction of P-450$_{C21}$ can be considered to be proportional to the concentration of free reductase which is increased with the increase of the Emulgen concentration. Using gel filtration it was actually detected that the stable complex was composed of several molecules of the two enzyme proteins at an equimolar ratio. Phospholipids were reported to be required for the hydroxylase activity of hepatic microsomal cytochrome P-450, which might implicate the interaction between cytochrome P-450 and the reductase (FRENCH et al., 1980). It must be pointed out that the stable complex formation in the detergent solubilized system does not necessarily reflect the native feature in the microsomal membranes.

4.4. Other properties of adrenal 21-hydroxylase reactions

Cytochrome b_5 is distributed within adrenal cortex microsomes, but its possible role in microsomal electron transfer system is not completely understood. Controversial results have been reported on the effect of cytochrome b_5 on the 21-hydroxylase activity in the reconstituted system involving P-450$_{C21}$. HIWATASHI and ICHIKAWA (1981) have shown no activation upon the addition of cytochrome b_5 to the reconstituted system, but KATAGIRI et al. (1982) have reported the stimulation of the activity by the addition of cytochrome b_5. The role of cytochrome b_5 on the 21-hydroxylation requires more study and must be examined immunochemically using the antibody against cytochrome b_5.

CHASALOW and LIEBERMAN (1979) found that the addition of cytosol from bovine adrenal cortex activated the 21-hydroxylase activity in the microsomes in which the cytosol increased the V_{max} but did not change the Km for 17α-hydroxyprogesterone. The same group has further tried to isolate the activator from the cytosol using acetone precipitation, gel filtration and ion-exchange chromatography (PONTICORVO et al., 1980). The cytosol was separated into eleven protein fractions and all of which had some stimulatory effect on 21-hydroxylase activity in the microsomes. Bovine serum albumin was proved to be active. A low molecular weight compound, subsequently identified as oxidized glutathione, had some stimulatory effect on the activity but ascorbate showed an inhibitory effect (GREENFIELD et al., 1980). Based on these results, they discussed the possible role of glutathione reductase on the regulation of 21-hydroxylase activity in the adrenal cortex.

MENARD et al. (1976) found an impairment of steroid 21-hydroxylase activity in adrenal cortex microsomes by the administration of the mineralocorticoid antagonist, spironolactone, which occurred only in cortisol producing animals such as the guinea pig and dog that had a high activity of 17α-hydroxylase. The decrease of the 21-hydroxylase activity was accompanied by the decrease of cytochrome P-450 content in the adrenal microsomes. In contrast, the administration of spironolactone to corticosterone producing animals such as the rat and rabbit caused an increase in the microsomal 21-hydroxylase activity. In a further study it was found that the sulfur atom located at the carbon 7α-position on the spironolactone molecule was essential for the destruction of the heme and apoprotein of adrenal cytochrome P-450 and suggested that an active metabolite produced from spironolactone, rather than the parent compound itself, was responsible for the decline in microsomal cytochrome P-450 (MENARD et al., 1979). However, GREINER et al., (1978) demonstrated that the actions of spironolactone on adrenal microsomal cytochrome P-450 were not mediated by its major circulating metabolite, canrenone.

Several enzyme activities in hepatic microsomes were decreased by lipid peroxidation induced by Fe^{+2}. In adrenal microsomes, the lipid peroxidation mediated by Fe^{+2} and the cytosol was observed as the enhanced production of malonaldehyde (BROGAN et al., 1983; IMATAKA et al., 1986). In addition, cytochrome P-450 levels, NADPH and NADH cytochrome c reductase activities and the affinity of substrate to cytochrome P-450 decreased as lipid peroxidation progressed. The peroxidation affected not only 21-hydroxylase activity but also the microsomal drug monooxygenase and mitochondrial cytochrome P-450-dependent monooxygenase activities. 17α-Hydroxylase activity was more effectively decreased than that of 21-hydroxylase and the decrease of these enzyme activities was prevented by α-tocopherol (IMATAKA et al., 1986). Administration of $HgCl_2$ to male rats resulted in the decrease of cytochrome P-450 content in adrenal microsomes, which was accompanied by the decrease of the 21-hydroxylase and benzo[a]pyrene hydroxylase activities. In contrast, Hg^{+2} treatment resulted in the increase of cytochrome P-450 content in adrenal mitochondria and elevated 11β-hydroxylase and cholesterol side-chain cleavage activities. The content of adrenal microsomal cytochrome b_5 was increased in Hg^{+2}-treated animals (VELTMAN and MAINES, 1986).

4.5. Molecular biology of 21-hydroxylase deficiency

One of the inborn errors of the metabolism in humans, congenital adrenal hyperplasia, is an autosomal recessive disorder predominantly resulting from a deficiency of 21-hydroxylase in the adrenal cortex. This is to be observed at a high level of frequency, 1 in 5,000—7,000 births. The deficiency of 21-hydroxylase causes a decrease in the production of cortisol and aldosterone and a marked increase in production of adrenal androgen, which results in salt-wasting or virilism (DELAMATER, 1986). DUPONT et al. (1977) found in their genetic trailing of several families that 21-hydroxylase deficiency is closely linked to the histocompatibility complex of human beings, HLA. Since then, research into the molecular genetic basis of 21-hydroxylase deficiency has developed and grown at a rapid rate. WHITE et al. (1984) isolated a cDNA clone for bovine adrenal P-450$_{C21}$ mRNA by using an antiserum against the purified P-450$_{C21}$ and determined the partial sequence of P-450$_{C21}$. The cDNA was used for the isolation of plasmids containing gene fragments of human adrenal genes of P-450$_{C21}$. A series of cosmid clones isolated from a human genomic library by using a probe encoding part of the fourth component of complement, C_4, were found to hybridize with the plasmid containing human P-450$_{C21}$ gene fragments. Restriction mapping and hybridization analysis using the above probes showed that normal individuals have two P-450$_{C21}$ genes, each located near the 3' end of one of the two C_4 genes and the genes were aligned in the sequence of C4A, 21-hydroxylase A, C4B, and 21-hydroxylase B (Fig. 6). It was also found that DNA from a patient with 21-hydroxylase deficiency lacked the 21-hydroxylase B gene but contained the 21-hydroxylase A gene, indicating that in a normal person the B gene is functional but that the A gene is not in normal persons (WHITE et al., 1985).

CARROLL et al. (1985) separately studied the relationship between the genes of HLA complex and 21-hydroxylase. They found that cosmids containing C4A and C4B genes hybridized with human adrenal mRNA and the hybridized sequences were duplicated. Comparison of the duplicated sequences with the cDNA sequence of bovine adrenal P-450$_{C21}$ showed that the genes contained those for human adrenal P-450$_{C21}$. Mapping the genes using a synthetic oligonucleotide based on the bovine cDNA sequence revealed an alignment of the C4A, 21-hydroxylase A, C4B, and 21-hydroxylase B which is the same as that reported by DUPONT et al. (1977). These investigations demonstrated that the linkage between the genes of HLA and 21-hydroxylase results from the structure of the gene alignment in the HLA region of the human chromosome 6 as illustrated in Figure 6 (PHILLIPS and SHEPHARD, 1985). PARKER et al. (1985) found that a similar alignment of the genes of C4A, 21-hydroxylase A, C4B and 21-hydroxylase B lies within the S region of the murine major histocompatibility complex. Using oligonucleotide probes specific to the 21-hydroxylase A and B genes, all mRNA for 21-hydroxylase in murine adrenal glands were shown to be derived from the 21-hydroxylase A gene. Furthermore, they showed that in the mouse the 21-hydroxylase A gene is functional instead of the B gene as found in humans, being based on the results that Y1 adrenal tumor cell transfected with the 21-hydroxylase A gene expressed 21-hydroxylase activity but the cells with 21-hydroxylase B gene did not. The Y1 cells transfected with 21-hydroxylase A gene were used for the studies on the regulation of the expression of 21-hydroxylase (PARKER et al., 1986). The sequence

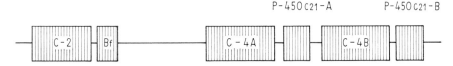

Fig. 6. The HLA-region containing P-450$_{C21}$ genes. C-2, C-4A, C-4B, and factor B (Bf) genes encode complement components.

       S. TAKEMORI; S. KOMINAMI

between 230 and 180 base pair upstream of the transcription initiation was concluded as being required for the expression of murine 21-hydroxylase from the results of several systematic deletion of the sequences. Similar sequences were also present around $200-250$ base pair upstream of bovine and human P-450$_{C21}$, indicating that these are the regulatory regions of the expression of the P-450$_{C21}$.

From these results arose a question about the nonfunctional 21-hydroxylase gene A in humans, whether the 21-hydroxylase A gene itself was a pseudo-faculty gene or if the regulatory region upstream of the gene A is not functional. HIGASHI et al. (1986) isolated the A and B genes from a human genomic library using a P-450$_{C21}$ cDNA and analysed the complete base sequences of both the genes including the introns. Since there are several base replacements in A from B in the regulatory region, it might be possible that the A gene is not expressed effectively. Comparing the two sequences in the coding region, it was found that two genes are highly homologous (98%) but in the A gene there are several alternations from the B gene. There are 8 base pair deletions in the third exon, one base pair insertion in the seventh exon and a transition $(C-T)$ point mutation in the eight exon, which was found to cause premature termination of the translation. This shows clearly the reason why the 21-hydroxylase A gene is nonfunctional. They suggested that the presence of the duplication of highly homologous sets of the C4 and P-450$_{C21}$ genes in a short-length of chromosomal DNA may cause frequent exchange of their DNA sequences by homologous recombination or unequal crossing-over during meiosis. Actually, polymorphisms of 21-hydroxylase genes have been observed in humans (DONOHOUE et al., 1986; JOSPE et al., 1987). In mouse, lethal deletion of the C4 and 21-hydroxylase genes has been reported as being caused by meiotic recombination (SHIROISHI et al., 1987).

The expression of P-450$_{C21}$ genes was observed in bovine adrenal and testis but not in bovine liver, kidney, brain and ovary (CHUNG et al., 1986; JOHN et al., 1986). In mouse liver, mRNA for adrenal P-450$_{C21}$ was detected but the expression was 1/50 in the adrenal gland (AMOR et al., 1985). These expression must be related to the regulatory region of the 21-hydroxylase gene. A cytochrome P-450 having 21-hydroxylase activity for progesterone but not for 17λ-hydroxyprogesterone was purified from rabbit hepatic microsomes and designated P-450I (DIETER et al., 1982). The cDNA sequence of the P-450I exhibits only very limited homology to that of adrenal P-450$_{C21}$ (TUKEY et al., 1985). It is interesting to note that in the congenital adrenal hyperplasia, approximately one-third of the production of deoxycorticosterone can take place in extra-adrenal tissues (ANTONIPILLAI et al., 1980). The 21-hydroxylase in liver is, therefore, likely to be responsible for progesterone metabolism in the case where P-450$_{C21}$ is impaired in the adrenal gland.

5. Steroid 17α-hydroxylase and C-17,20-lyase reactions

5.1. Purification and molecular properties of P-450$_{17\alpha,\text{lyase}}$

In guinea pig and bovine adrenal microsomes, antibody prepared against P-450$_{C21}$ strongly inhibited the 21-hydroxylase activity but had no inhibitory effect on 17α-hydroxylase and lyase activities (KOMINAMI et al., 1983). On the basis of these observations, it was suggested that different P-450 species other than P-450$_{C21}$ should catalyze the steroid 17α-hydroxylation and the cleavage of C17—C20 bond in adrenal microsomes. In order to purify P-450$_{17\alpha,\text{lyase}}$, guinea pig adrenal microsomes were chosen as source materials because they had quite high starting, specific activities for 17α-hydroxylase and lyase as shown in Table 1.

KOMINAMI et al. (1982) have accomplished a purification procedure for P-450$_{17\alpha,\text{lyase}}$ from guinea pig adrenal microsomes. When the microsomes were solubilized with detergents, it was found that P-450$_{17\alpha,\text{lyase}}$ became labile in 20% glycerol alone but the presence of the steroid substrate was required to preserve the P-450 almost completely from the inactivation. This characteristic was very similar to that found with P-450$_{11\beta}$ (TAKEMORI et al., 1975) and P-450$_{\text{cam}}$ (YU et al., 1974) which were markedly stabilized by the substrate. P-450$_{17\alpha,\text{lyase}}$ from guinea pig adrenal microsomes was solubilized with a nonionic detergent, Emulgen 913, in the presence of progesterone as a stabilizing agent and was purified by a series of column chromatography on ω-amino-n-octyl Sepharose, DEAE-Toyo pearl and Matrix gel Red A. At the final stage of the purification, the detergent was removed completely from P-450$_{17\alpha,\text{lyase}}$ using a DEAE-Toyo pearl column. P-450$_{17\alpha,\text{lyase}}$ could be solubilized in a stable form in 20% glycerol without detergent, which was quite different from P-450$_{C21}$. The final preparation recovered 20% of the initial microsomal cytochrome P-450 and had an average specific content of 15 nmol P-450/mg protein. The purified protein exhibited a single polypeptide band with a molecular mass of 52 kDa on SDS polyacrylamide gel electrophoresis. Both progesterone 17α-hydroxylase (8 nmol 17α-hydroxyprogesterone/min/nmol P-450) and 17α-hydroxyprogesterone C-17,20-lyase (12 nmol androstenedione/min/nmol P-450) activities could be reconstituted in the presence of dilauroylphosphatidylcholine upon mixing the purified P-450$_{17\alpha,\text{lyase}}$ with NADPH-cytochrome P-450 reductase. P-450$_{17\alpha,\text{lyase}}$ was also active with pregnenolone as 17α-hydroxylase and with 17α-hydroxypregnenolone as C-17,20-lyase. The absorption spectrum of the oxidized P-450$_{17\alpha,\text{lyase}}$ in the presence of progesterone had maxima at 394 and 650 nm, characteristic of a high spin form. The reduced CO complex of the P-450$_{17\alpha,\text{lyase}}$ showed an absorption maximum at 448 nm.

NAKAJIN et al. (1983) subsequently purified P-450$_{17\alpha,\text{lyase}}$ from porcine adrenal microsomes by sodium cholate solubilization followed by a series of column chromatography on DEAE cellulose, DEAE Sepharose, CM Sepharose and hydroxyapatite. The yield through the procedure was $2-4\%$ of the total microsomal P-450. The P-450$_{17\alpha,lyase}$ was pure on SDS polyacrylamide gel electrophoresis with a molecular mass of 53 kDa and was revealed to be identical with the testicular P-450 species immunochemically. The P-450$_{17\alpha,lyase}$ was active with both Δ^4(progesterone) and Δ^5(pregnenolone) substrates as 17α-hydroxylase and with the corresponding 17α-hydroxysteroids as C-17,20-lyase. The absorption spectrum of P-450$_{17\alpha,lyase}$ in the oxidized form showed the low spin form with an absorption maximum at 417 nm. Upon addition of progesterone, the oxidized P-450$_{17\alpha,lyase}$ showed a high spin type spectrum with absorption peaks at 394, 535, 565 and 650 nm. The reduced CO spectrum showed a peak at 448 nm with a shoulder at 420 nm.

The characteristics of P-450$_{17\alpha,lyase}$ purified from guinea pig and porcine adrenal microsomes are very similar to those of the P-450 species from neonatal pig testicular microsomes (NAKAJIN and HALL, 1981). These three species show both 17α-hydroxylase and C-17,20-lyase activities. The ratio of the activity of 17α-hydroxylase to that of the lyase in the final preparation of adrenal P-450$_{17\alpha,lyase}$ is found to be 2.5 ± 0.3 in the several preparations, which is similar to that of the testicular P-450 species. Their absorption spectra are also similar to each other and the P-450-CO complexes show absorption maxima at 448 nm. There are some differences in their stability. The guinea pig adrenal P-450 is unstable in the absence of progesterone, but the porcine adrenal and testicular P-450's are reported to be stable without steroid substrate.

The homogeneity of guinea pig and porcine P-450$_{17\alpha,lyase}$ was demonstrated by several methods: electrophoresis on SDS polyacrylamide gel, immunoelectrophoresis with anti-P-450$_{17\alpha,lyase}$ IgG and isoelectric focusing (SHINZAWA et al., 1985; NAKAJIN et al., 1984). The guinea pig and porcine P-450's are composed of a single polypeptide having methionine as the NH_2-terminal amino acid (OGISHIMA et al., 1983; NAKAJIN et al., 1984). Linear competitive inhibition of 17α-hydroxyprogesterone C-17,20-lyase reaction by progesterone showed that 17α-hydroxyprogesterone and progesterone are catalyzed at the same active site (SHINZAWA et al., 1983). The Km value for 17α-hydroxyprogesterone was 0.55 μM and the Ki value for progesterone was 0.11 μM. These results for adrenal P-450's agree well with those for testicular P-450 (NAKAJIN et al., 1981).

ZUBER et al. (1986) have cloned cDNA of bovine adrenal P-450$_{17\alpha,lyase}$ and analysed the sequences of two cloned DNA molecules, pB17α-1 and pcD17α-2, which are complementary to mRNA sequences encoding bovine adrenal P-450$_{17\alpha,lyase}$. Clone pcD17α-2 contained an open reading frame coding for the complete amino acid sequence of P-450$_{17\alpha,lyase}$, which corresponds to 509 amino acids and to a molecular weight of 57,251. By Northern blot hybridization analysis, they found the mRNA specific for P-450$_{17\alpha,lyase}$ in bovine adrenal cortex to be a single species, approximately 1850 bases in length. When the amino-terminal sequence up to 16 amino acids of bovine adrenal $P\text{-}450_{17\alpha,\text{lyase}}$ is compared with the corresponding sequence from pig adrenal $P\text{-}450_{17\alpha,\text{lyase}}$, which has been obtained by protein sequence analysis (NAKAJIN et al., 1981; NAKAJIN et al., 1984), both sequences have 75% identity. Likewise, the guinea pig $P\text{-}450_{17\alpha,\text{lyase}}$ sequence is approximately 50% homologous to bovine $P\text{-}450_{17\alpha,\text{lyase}}$. The bovine sequence is more closely related to that of pig than to that of guinea pig. All these amino-terminal sequences have a high content of hydrophobic amino acids, 15 of 16 for the bovine $P\text{-}450_{17\alpha,\text{lyase}}$, 14 of 16 for the pig $P\text{-}450_{17\alpha,\text{lyase}}$ and 12 of 15 for the guinea pig $P\text{-}450_{17\alpha,\text{lyase}}$.

Conclusive evidence for the dual function of $P\text{-}450_{17\alpha,\text{lyase}}$ was provided by expression of bovine $P\text{-}450_{17\alpha,\text{lyase}}$ in COS I (transformed monkey kidney) cells, which normally contain no detectable $P\text{-}450_{17\alpha,\text{lyase}}$ (ZUBER et al., 1986). It was demonstrated that a single polypeptide chain catalyzes both the 17α-hydroxylase and the C-17,20-lyase reaction. This system can catalyze 17α-hydroxylation of pregnenolone and progesterone with equal efficiency but catalyzes about five times as much C-17,20-lyase reaction for 17α-hydroxypregnenolone than that for 17α-hydroxyprogesterone. Furthermore, the identity of adrenal $P\text{-}450_{17\alpha,\text{lyase}}$ with testicular cytochrome P-450 in humans was confirmed by CHUNG et al. (1987), who used porcine $P\text{-}450_{17\alpha,\text{lyase}}$ cDNA to isolate the full-length human adrenal $P\text{-}450_{17\alpha,\text{lyase}}$ cDNA. A human testis P-450 cDNA library contained an identical sequence with adrenal $P\text{-}450_{17\alpha,\text{lyase}}$ cDNA and nuclease SI-protection experiments confirmed that the same $P\text{-}450_{17\alpha,\text{lyase}}$ mRNA exists in human adrenal gland and testis. These results indicate that the testis possess $P\text{-}450_{17\alpha,\text{lyase}}$ identical to that in the adrenal, which might be feasible because testicular Leydig cells and adrenocortical cells share a common embryologic origin (DAHL and BAHN, 1962). This molecular genetic evidence shows clearly that a single P-450 species, termed $P\text{-}450_{17\alpha,\text{lyase}}$, catalyzes both 17α-hydroxylase and C-17,20-lyase reaction in adrenal and testis glands.

5.2. Role of cytochrome b_5 in the function of $P\text{-}450_{17\alpha,\text{lyase}}$

The role of cytochrome b_5 in hepatic microsomal cytochrome P-450-mediated monooxygenase systems has been investigated for many years (HILDEBRANDT and ESTABROOK, 1971; COHEN and ESTABROOK, 1971; CORREIA and MANNERING, 1973). Various forms of cytochrome P-450 showed different requirements for cytochrome b_5 in the reconstituted drug oxidation system (LU et al., 1974; SUGIYAMA et al., 1979; IMAI, 1981). Several studies suggested that cytochrome b_5 may participate in some NADPH-supported cytochrome P-450 dependent reaction as the source of the second electron. The contribution of

S. TAKEMORI; S. KOMINAMI

cytochrome b_5 to cytochrome P-450 dependent reactions related to steroidogenesis will prove very important since cytochrome b_5 is abundantly distributed in adrenal and testicular microsomes (OSHINO, 1978).

COLBY (1980) noted that the addition of pregnenolone to guinea pig adrenal microsomes produces a difference spectrum with peaks at 425 and 557 nm and a trough at 410 nm which is similar to the oxidation-reduction difference spectrum of cytochrome b_5. This phenomenon can be explained as follows: 3β-hydroxysteroid dehydrogenase-isomerase converts pregnenolone into progesterone accompanied with the conversion of NAD^+ into NADH. The NADH reduces cytochrome b_5 by the action of NADH-cytochrome b_5 reductase. If cytochrome b_5 participates in monooxygenase reaction in adrenal microsomes, pregnenolone metabolism and cytochrome b_5 reduction can provide reducing equivalents for cytochrome P-450 reaction. In this case, the linkage between cytochrome b_5 reduction and steroid metabolism in adrenal microsomes may have considerable functional significance.

In testicular steroidogenesis, it has been suggested that cytochrome b_5 may play a physiological role. OHBA et al. (1981) noted that the NADPH-supported C_{21} steroid side-chain cleavage reaction in testicular microsomes is stimulated either by the presence of exogeneous NADH or the oxidation of pregnenolone by 3β-hydroxysteroid dehydrogenase-isomerase and is decreased by addition of the antibody against NADH-cytochrome b_5 reductase. KATAGIRI et al. (1982) noted that the addition of cytochrome b_5 and NADH-cytochrome b_5 reductase to the reconstituted system containing testicular cytochrome P-450 and NADPH-cytochrome P-450 reductase in the presence of NADPH and NADH stimulates the C_{21} steroid side-chain cleavage reaction. Using testicular microsomes and its reconstituted enzyme system, ONODA and HALL (1982) further showed that cytochrome b_5 produces stimulation of lyase activity greater than that on 17α-hydroxylase.

Concerning the role of cytochrome b_5 in the function of P-450$_{17\alpha,\text{lyase}}$ from guinea pig adrenal microsomes, further studies have been performed in the reconstituted enzyme system with dilauroylphosphatidylcholine (SHINZAWA et al., 1985). As illustrated in Figure 7, adrenal P-450$_{17\alpha,\text{lyase}}$ had optimal activities at pH 6.1 both for 17α-hydroxylase and lyase activities and the ratio of these activities in various pH ranges was changed remarkably. The pH dependence of the activities in the presence of cytochrome b_5 was apparently different from those in its absence. The maximum activities for both reactions did not change, but the optimal pH was shifted to 6.6—7.0. Above 6.3—6.5, cytochrome b_5 stimulated both 17α-hydroxylase and lyase activities, but it suppressed these activities below pH 6.3—6.5. The 17α-hydroxylase and lyase activities became 1.7 and 7 times higher than those without cytochrome b_5 at a physiological pH of 7.3, respectively. It is of interest that the dual activities of P-450$_{17\alpha,\text{lyase}}$ are differentialy affected in the presence of cytochrome b_5, suggesting a physiological role for cytochrome b_5 in modulating

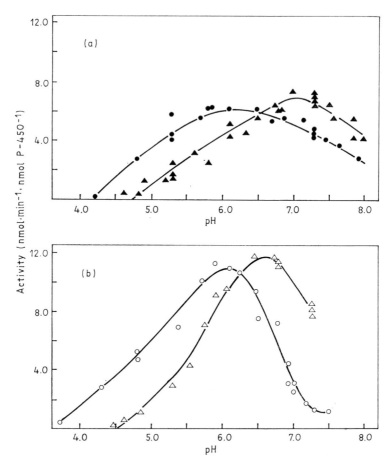

Fig. 7. Effect of pH on 17α-hydroxylase and C-17,20-lyase activities of P-450₁₇α,lyase from guinea pig adrenal microsomes, taken from SHINZAWA et al. (1985) with permission. (a), the 17α-hydroxylase activities without cytochrome b_5 (•——•) and with cytochrome b_5 (▲——▲); (b), C-17,20-lyase activities without cytochrome b_5 (○——○) and with cytochrome b_5 (△——△).

the synthesis of steroid hormone. When NADH-cytochrome b_5 reductase, purified from bovine liver microsomes, was added to the reconstituted system containing cytochrome b_5 in the presence of NADH and NADPH, 17α-hydroxylase and lyase activities were obviously not further stimulated. Furthermore, cytochrome b_5 suppressed the activities at pH 5.3, which is difficult to attribute to the second electron transfer from cytochrome b_5. Taken together, these results suggest that the effect of cytochrome b_5 might be due to the shift of optimal pH rather than the electron transfer from cytochrome b_5 to

S. TAKEMORI; S. KOMINAMI

P-450$_{17\alpha,\text{lyase}}$. However, the possibility that the electron is transferred from NADPH-cytochrome P-450 reductase to cytochrome b$_5$ and further to P-450$_{17\alpha,\text{lyase}}$ cannot be completely excluded.

A variety of responses for the alternative involvement of cytochrome b$_5$ have been reported in liver microsomal cytochromes P-450. P-450$_{\text{LM2}}$ from rabbit liver cannot metabolize prostaglandin unless cytochrome b$_5$ is added, while benzphetamine metabolism by the same cytochrome P-450 is inhibited by addition of cytochrome b$_5$ (OKITA et al., 1981). With regard to the interaction of cytochrome P-450 with cytochrome b$_5$, it has been reported that in the liver microsomal system the complex formation between cytochrome P-450 and cytochrome b$_5$ induced changes in the spin state of cytochrome P-450 and decreased the dissociation constant for substrate (TAMBURINI and GIBSON, 1983; CHIANG, 1981; BÖSTERLING and TRUDELL, 1982). The 17α-hydroxylase and C-17,20-lyase activities of P-450$_{17\alpha,\text{lyase}}$ from pig adrenal microsomes became higher at pH 7.3 with the increase of cytochrome b$_5$ concentration and reached certain levels. Based on kinetic analysis of this data, SHINZAWA et al. (1985) have predicted that P-450$_{17\alpha,\text{lyase}}$ forms a complex with cytochrome b$_5$ at a molar ratio of 1.0. The Kd value of the complex calculated from the activities was estimated to be 0.4 nM at pH 7.3 and 3 nM at pH 5.3. These results were interpreted as meaning that P-450$_{17\alpha,\text{lyase}}$ interacts with cytochrome b$_5$ to form a complex which might induce some conformational change in the P-450 molecule to affect the optimal pH for both activities. The roles of cytochrome b$_5$ as an effector, other than its well known function as an electron carrier, have also been given by the works of LIPSCOMB et al. (1976) and GUENGERICH et al. (1976).

5.3. Regulation of dual activities of P-450$_{17\alpha,\text{lyase}}$

In the synthetic pathway of steroid hormones, P-450$_{17\alpha,\text{lyase}}$ is situated at a branch point between the production of glucocorticoids and adrenal androgens. Both pathways involve 17α-hydroxylation of either pregnenolone or progesterone. In the glucocorticoid (cortisol) production, 17α-hydroxysteroid is further hydroxylated at 21-position by P-450$_{\text{C21}}$. In the androgen synthesis, the product of the 17α-hydroxylation step becomes the substrate for a side-chain cleavage reaction. The 17α-hydroxylation is the predominant reaction in the adrenals which secrete cortisol as a main glucocorticoid and the C17—C20 bond cleavage reaction might be a by-pathway in the adrenal. It is reasonable in testicular steroidogenesis that P-450$_{17\alpha,\text{lyase}}$ catalyzes successively 17α-hydroxylation and the bond cleavage between C-17 and C-20 which is the main metabolic pathway in the production of testosterone. Since P-450$_{17\alpha,\text{lyase}}$ is distributed in the endoplasmic reticulum of endocrine tissues such as the adrenal, testis, ovary and placenta, the regulation of 17α-hydroxylase and lyase activities of P-450$_{17\alpha,\mathrm{lyase}}$ has been of interest to investigators.

SUHARA et al. (1984) attempted to examine the metabolic relation between the production of glucocorticoids and androgens using a reconstituted hybrid enzyme system that contains both testicular P-450 species (P-450$_{17\alpha,\mathrm{lyase}}$) and adrenal P-450$_{\mathrm{C21}}$ as well as NADPH-cytochrome P-450 reductase. When progesterone was reacted in a system with P-450$_{17\alpha,\mathrm{lyase}}$ and P-450$_{\mathrm{C21}}$ at a $1:3$ molar ratio, considerable amounts of deoxycorticosterone and deoxycortisol accumulated with time, whereas formation of androstenedione was limited through the process. It was interpreted that the P-450$_{17\alpha,\mathrm{lyase}}$-mediated formation of androstenedione should be in competition for the common substrate, 17α-hydroxyprogesterone, with P-450$_{\mathrm{C21}}$-mediated formation of 11-deoxycortisol.

SHINZAWA et al. (1985) carried out the following experiments in order to investigate the mechanism of androstenedione production from progesterone. When [^{3}H]-progesterone was incubated with the reconstituted system containing P-450$_{17\alpha,\mathrm{lyase}}$ and NADPH-cytochrome P-450 reductase, radioactivity recovered both 17α-hydroxyprogesterone and androstenedione as products. The addition of 5 times more non-radioactive 17α-hydroxyprogesterone than substrate progesterone decreased the recovered radioactivity in androstenedione to about 50% of that in the absence of non-radioactive 17α-hydroxyprogesterone, although it was expected to be less than 1/5 under such conditions. These results were interpreted as being due to a certain defined portion of 17α-[^{3}H] hydroxyprogesterone, which is produced from [^{3}H] progesterone, being released from P-450$_{17\alpha,\mathrm{lyase}}$ and the other part of the 17α-hydroxyprogesterone not released from the P-450 molecule and being successively subjected to the lyase action in the reaction process from progesterone to androstenedione. It was noted that the presence of cytochrome b$_5$ in the assay system increases the ratio of the recovery of radioactivity in androstenedione to that in the total product. It is most notable that the release of 17α-hydroxyprogesterone from P-450$_{17\alpha,\mathrm{lyase}}$ in a sequential reaction from progesterone to androstenedione is suppressed in the co-presence of cytochrome b$_5$, so that the dual activities of P-450$_{17\alpha,\mathrm{lyase}}$ are differentially affected. These effects of cytochrome b$_5$ might serve as a physiological role in controlling the synthesis of steroid hormones.

Testicular microsomes efficiently convert progesterone to androstenedione. On the other hand, adrenal cortex microsomes catalyze 17α-hydroxylation with a little lyase activity which is also the case even when 21-hydroxylase was completely inhibited by anti-P-450$_{\mathrm{C21}}$-IgG (KOMINAMI et al., 1983). YANAGIHASHI and HALL (1986) have investigated the reason for the low levels of lyase activity in adrenal microsomes as compared with testicular microsomes. The addition of porcine liver NADPH-cytochrome P-450 reductase to adrenal and testicular microsomes increased the activity of lyase relative to 17λ-hydroxylase in both cases. The same effects were observed when the reductase was added to reconstituted enzyme systems using P-450$_{17\lambda,\text{lyase}}$ from adrenal or testis. As the concentration of reductase increased, the lyase activity increased relative to the hydroxylase until the rates of the activities became almost equal. V_{max} was the same for both activities for the two P-450s, but K_m value for the reductase was lower for the hydroxylase activities than that for the lyase activities in the two P-450s. Antibodies to the reductase, when added to testicular microsomes, inhibited both activities but inhibition of the lyase activity was greater than that of the hydroxylase. It was noted that the enzyme activity of the reductase in testicular microsomes is 3—4 times higher than that of adrenal microsomes. These results may account for the greater activity of lyase in testicular compared with adrenal microsomes. DEE et al. (1985) have proposed that the expression of the gene for NADPH-cytochrome P-450 reductase may be coupled to that of cytochrome P-450 in some microsomal systems. Since the expression of the gene for the reductase is different in adrenal and testis, the difference can provide a mechanism for regulating the production of androgen. Finally, it should be remembered that other, unrecognized factors might be responsible for regulating dual activity of P-450$_{17\lambda,\text{lyase}}$.

6. The zonation and function of microsomal cytochrome P-450 in adrenal cortex

The relationship between the anatomical and functional zonation of the adrenal cortex has been investigated for several decades. The adrenal cortex consists of three zones: the zona glomerulosa, the zona fasciculata and the zona reticularis. These zones are differentiated with regard to the steroidogenic activities (LONG, 1975; GOWER, 1984). It is now well established that aldosterone, the main mineralocorticoid, is produced and secreted by cells of the zona glomerulosa, and that these cells are regulated by the renin-angiotensin system (DAVIS, 1967; TAN, 1986). Zonae fasciculata and reticularis produce glucocorticoids and androgens. However, the relative importance between these two zones for the production of glucocorticoids and androgens is not well understood. It is widely accepted that the zona fasciculata is the major site of production of glucocorticoids and that these cells are normally regulated by ACTH (STROTT et al., 1981; HYATT et al., 1983; EACHO and COLBY, 1983). NISHIKAWA and STROTT (1984) evaluated the cortisol production with cells of the zona reticularis of the guinea pig adrenal cortex and suggested that ordinarily little cortisol is produced although the enzyme system is present. Confusion arises concerning the major site of androgen synthesis. DAVISON et al. (1983) reported that the zona reticularis is responsible for androgen secretion from guinea pig adrenal cortex, while HYATT et al. (1983) using cultured adrenal cortical cells demonstrated that the zona fasciculata participates in androgen production more efficiently than the zona reticularis, which is consistent with the results of EACHO and COLBY (1983). In general the adrenal cortex of the guinea pig has been used to examine these problems. The major reason for choosing this animal is that the adrenal cortex is composed of two chromatically distinct zones: a yellow outer zone (zona glomerulosa and zona fasciculata) and an inner brown zone (zona reticularis) (BLACK et al., 1979).

In an attempt to elucidate the functional role of the three zones of adrenal cortex, the localizations of $P\text{-}450_{17\alpha,\text{lyase}}$ and $P\text{-}450_{C21}$, which are necessary for the production of glucocorticoids and androgen, were immunohistochemically studied in the guinea pig adrenal cortex using the direct method of the peroxidase labelled antibody technique (SHINZAWA et al., 1988). The immunocytochemical localizations of these cytochromes were proved to be the smooth surfaced endoplasmic reticulum. This cytochemical evidence is consistent with biochemical observations in which 17α- and 21-hydroxylase activities are located mostly in smooth surfaced microsomal fractions obtained from adrenal gland (INANO et al., 1969). $P\text{-}450_{17\alpha,\text{lyase}}$ was found to be distributed in the zona fasciculata and less in the zona reticularis. The zona glomerulosa was negative to immunohistochemical staining for $P\text{-}450_{17\alpha,\text{lyase}}$. On the other hand, $P\text{-}450_{C21}$ was distributed in all three zones in adrenal cortex, which was consistent with the result by SASANO et al. (1988). HARKINS et al. (1974) reported that microsomes from zona glomerulosa of bovine adrenal cortex had virtually no 17α-hydroxylase activity, while microsomes from zonae fasciculata and reticularis had both 17α- and 21-hydroxylase activities. The above immunohistochemical evidence suggests that the lack of 17α-hydroxylase activity in microsomes from zona glomerulosa is due to the absence of $P\text{-}450_{17\alpha,\text{lyase}}$.

SHINZAWA et al. (1988) further attempted to measure the contents of $P\text{-}450_{17\alpha,\text{lyase}}$ and $P\text{-}450_{C21}$ in guinea pig adrenal microsomes using a highly sensitive ELISA. The ratio of $P\text{-}450_{C21}$ content in the inner zone to that in the outer zone microsomes was estimated to be 1.24. 21-Hydroxylase activity was higher in the inner zone microsomes, which was in good agreement with the relative $P\text{-}450_{C21}$ contents. The ratio of $P\text{-}450_{17\alpha,\text{lyase}}$ content in the inner zone to that in the outer zone microsomes was estimated to be 0.75. 17α-Hydroxylase and C-17,20-lyase activities were 5—6 fold greater in the outer than in the inner zone microsomes, which could not be explained only by the $P\text{-}450_{17\alpha,\text{lyase}}$ content of the two zones. This can be interpreted to mean that some effector might be controlling the activity. As described in the previous section, the addition of cytochrome b_5 to the reconstituted system consisting of $P\text{-}450_{17\alpha,\text{lyase}}$ and NADPH-cytochrome P-450 reductase stimulated the 17α-hydroxylase and C-17,20-lyase activities. However, the ratio of the content of cytochrome b_5 to that of total cytochrome P-450 was about 2.0 in microsomes of both inner and outer zones, showing that the differences in the $P-450_{17\alpha,lyase}$ activity of microsomes in these two zones cannot be attributed to the effect of cytochrome b_5. From the data of enzyme activity and predominant immunohistochemical localization of $P-450_{17\alpha,lyase}$ in zona fasciculata, SHIN-ZAWA et al. (1988) proposed that androgens may be mainly produced in the zona fasciculata in the guinea pig adrenal cortex.

7. Interaction of steroids with adrenal microsomal cytochrome P-450 in membranes

The biosynthesis of steroid hormones takes place in mitochondria and endoplasmic reticulum of adrenal cortex cells. During the process of steroidogenesis, metabolic intermediates must move back and forth between these organellar membranes where the lipid compositions are quite different from each other (NARASIMHULU, 1975; KIMURA, 1986). These metabolic intermediates are all very lipophilic and favor partitioning into the lipid phases of the membranes (FLYNN, 1971; TOMIDA et al., 1978). The steroid metabolizing route seems to be in the direction of increasing the hydrophilicity of the products. PARRY et al. (1976) pointed out the role of the substrate partitioning between the membrane and aqueous phase and also the importance of distinguishing whether the substrate binding site of a membrane protein faces the aqueous or the lipid phase. In order to explore the effect of substrate partitioning and orientation of the substrate binding site of the enzyme in the membranes it will be necessary to incorporate the enzyme into lipid bilayers as a model of natural biomembranes.

$P-450_{17\alpha,lyase}$ and $P-450_{C21}$ have been incorporated into the liposomal membranes by the cholate dialysis method (KOMINAMI et al., 1986; KOMINAMI et al., 1988). The proteoliposomes, with an average diameter of 50 nm, were usually obtained by the dialysis method and more than 80% of the enzymes were found to be located toward the external side of the liposomes, as judged from the reducibility upon the external addition of NADPH-cytochrome P-450 reductase and NADPH. The stabilities of $P-450_{C21}$ and $P-450_{17\alpha,lyase}$ in the liposomes were remarkable: only a small portion of the P-450 was converted into P-420 after the proteoliposomes had been kept at room temperature for a long time without glycerol and substrate, while most of the P-450 in the detergent solubilized state was converted into P-420 in a few hours at $25\,°C$ under the same conditions. The type I spectral change was observed in the $P-450_{17\alpha,lyase}$-proteoliposomes upon the addition of the substrate, where the magnitude of the difference spectra observed under non-saturated condition decreased with the increase of the amount of the $P-450_{17\alpha,lyase}$-free liposomes, as can be seen in Figure 8. The apparent Kd values were increased

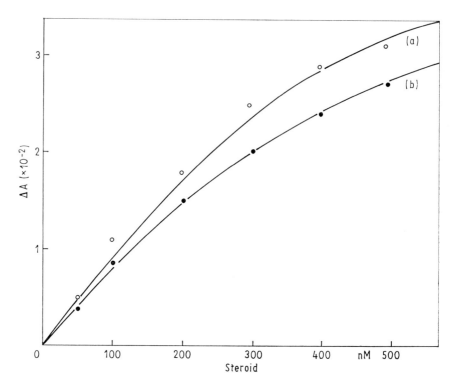

Fig. 8. Effects of phosphatidylcholine (PC) concentration on 17α-hydroxyprogesterone-induced difference spectra of P-450$_{17\alpha,\text{lyase}}$ embedded in PC liposomes, taken from KOMINAMI et al. (1988) with permission. The horizontal axis represents the total concentration of 17α-hydroxyprogesterone in the system. The closed and open circles show the magnitudes of the difference spectra, ΔA (389 nm — 421 nm), observed in the system containing 380 nM P-450$_{17\alpha,\text{lyase}}$ in 1 mg PC/ml and 3 mg PC/ml, respectively. The curves (a) and (b) were generated using Kd values of 40 nM and 95 nM, respectively, which were obtained by the least squares method for the observed points.

with the increase of the phospholipid concentration. The dependence of Kd values of the substrate-P-450 complex on the microsomes concentration has been demonstrated in hepatic microsomes (EBEL et al., 1978).

In general, most steroids can be expected to partition in the membranes. In order to qualify the correct concentration of the steroid in the membrane, the partition coefficient of the steroid, $[SL]_l/[SA]_a$, must be determined, which can be obtained by equilibrium dialysis. The $[SL]_l$ and $[SA]_a$ represent the steroid concentration in the lipid phase and aqueous phase, respectively. These values are defined by the volume of the lipid phase and the aqueous phase. The volume of the aqueous phase is usually almost the same as the total volume of the system since the volume of the membrane lipid is quite small compared with the total volume. The volume of the lipid phase in 1 mg phospholipid in the liposomes can be assumed to occupy 0.001 ml in the solution, which is based on the hydrodynamic experiments that have been performed (HUANG and MASON, 1978). The partition coefficients were determined by employing liposomes composed of different lipid compositions: those composed only of phosphatidylcholine (PC) and those of the phospholipid mixture of phosphatidylcholine, phosphatidylethanolamine and phosphatidylserine at the molar ratio of 5:3:1 (PCES), which is comparable with the composition of bovine adrenal microsomes (NARASIMHULU, 1975; KIMURA, 1986). As summarized in Table 2, the partition coefficients of steroids are apparently larger in PC liposomes than in PCES liposomes.

Table 2. Partition coefficients of steroids between aqueous and lipid phases, taken from KOMINAMI et al. (1988) with permission. The partition coefficients were measured in 50 mM Tris-HCl (pH 7.2) containing 50 mM NaCl and 0.1 mM EDTA at 25 °C by equilibrium dialysis using [^{3}H] steroids.

Steroids	K_p	
	PC liposomes	PCES liposomes
Progesterone	$3\,200 \pm 200$	$2\,300 \pm 200$
17α-Hydroxyprogesterone	$1\,500 \pm 100$	920 ± 50
Androstenedione	440 ± 40	320 ± 20

The rate of the binding and the dissociation of steroids to and from the cytochrome P-450 molecule can be studied by a stopped flow method (KOMINAMI et al., 1986). The binding could be measured by the rapid mixing of the P-450$_{C21}$-proteoliposomes with steroids and the dissociation could be observed by the mixing of proteoliposomes containing P-450$_{C21}$-substrate complex with liposomes not having steroids and P-450$_{C21}$ as in Figure 9(a) and (b). In the latter experiments, a rapid increase in the volume of the lipid phase caused a rapid decrease of steroid concentration around the P-450$_{C21}$ molecule and resulted in the dissociation of a part of the steroid from P-450$_{C21}$. In Figure 9(a) the absorbance at 389 nm and at 421 nm corresponding to the maximum and the minimum of the substrate induced difference spectrum of P-450$_{C21}$ increased and decreased respectively whilst that at 407 nm, the isosbestic point in the difference spectrum did not change. In the dissociation process the reverse optical changes were obtained. The apparent binding rate was

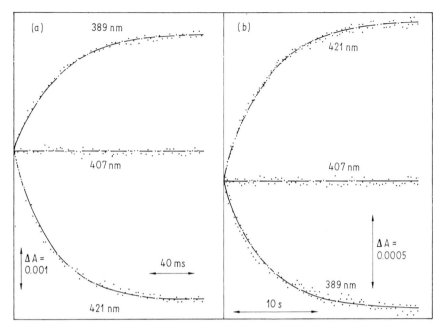

Fig. 9. Stopped flow measurements of the binding (a) and dissociation (b) of substrate to and from P-450$_{C21}$ embedded in liposomes, taken from Kominami et al. (1986) with permission. The dotted points represent the data used for the calculation and the solid lines show the theoretical curves calculated using the best fit parameters. (a) A solution containing 200 nM P-450$_{C21}$ in 0.234 mg/ml PC liposomes was rapidly mixed with one containing 8 µM 17α-hydroxyprogesterone. (b) A solution containing 450 nM 17α-hydroxyprogesterone and 450 nM P-450$_{C21}$ in 1.5 mg/ml PC liposomes was mixed rapidly with one of liposomes of 3.5 mg/ml PC.

increased linearly with the increase of the steroid concentration and was hyperbolically dependent on the concentration of the lipids in the system. The dissociation rates were, however, independent of the lipid concentration. To calculate the kinetic parameters from the apparent parameters, two models, A and L, were considered — as shown in Figure 10. In these models, there are two separate reactions, the partition of the steroid into the membranes and the binding of the steroid with P-450$_{C21}$. For kinetic analysis, it is essential to know which of the reactions is the rate limiting process in the binding process as a whole. The transfer of aromatic molecules between liposomal vesicles has been shown to occur via the aqueous phase and to be controlled by the rate of the release from the liposomal membranes (Doody et al., 1980; Almgren, 1980). Almgren (1980) was able to derive an equation showing the rate of the releasing from the membranes to be proportional to the inverse value of the partition coefficient of the molecule between the liposomal mem-

S. Takemori; S. Kominami

branes and the aqueous solution. Perylene shows the partition coefficient of 10^7 and the rate is in the range of 3 per sec in the PC liposomes. Since the partition coefficients of most steroids except cholesterol are much smaller than that of perylene, the rate of exchange of the steroid between the liposomes is much faster than the binding of the steroid to P-450$_{C21}$. The fast exchange of steroids between liposomal vesicles has been shown in experi-

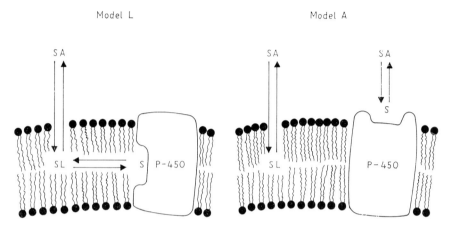

Fig. 10. The interaction between steroid substrates and liposome-bound cytochrome P-450, taken from KOMINAMI et al. (1986) with permission.

ments with the P-450$_{C21}$-liposomal system (KOMINAMI et al., 1986). The detailed analysis of the kinetic parameters suggests that model L is much more reasonable to explain than is model A. Model L seems physiologically reasonable because P-450$_{C21}$ easily reacts with the steroids concentrated into the membranes.

Similar mechanisms for interaction with steroid in liposomal membranes has been proved for P-450$_{17\alpha,lyase}$ (KOMINAMI et al., 1988). The Kd values of P-450$_{17\alpha,lyase}$-steroid complexes were obtained by equilibrium dialysis and optical difference spectrum and were converted into Kd$_L$ and Kd$_A$, where Kd$_L$ and Kd$_A$ represent the binding constants for the case of models L and A (Table 3.). The values of Kd$_L$ in the P-450$_{17\alpha,lyase}$-substrate complex in PC liposomes were equal, within the experimental errors, to those in the PCES liposomes, suggesting that in model L the difference in the lipid composition of the membrane might make little effect on the conformation of P-450$_{17\alpha,lyase}$. This was supported by the results of the extinction coefficients of the substrate-induced difference spectra between those observed in PC and PCES liposomes and which showed little difference. As summarized in Table 3, the values of Kd$_L$ are in

Table 3. Dissociation constants of P-450$_{17\alpha,\text{lyase}}$ — steroid complexes in the liposomes, taken from KOMINAMI et al. (1988) with permission. The dissociation constants were calculated using the experimental results obtained by equilibrium dialysis under the assumption that P-450$_{17\alpha,\text{lyase}}$ interacts only with the steroids in the aqueous phase (Kd$_A$) or with those in the lipid phase of the membrane (Kd$_L$). The equilibrium dialysis was performed in 50 mM Tris-HCl (pH 7.2) containing 50 mM NaCl and 0.1 mM EDTA at 25 °C.

Steroids	Kd$_L$ (μM)		Kd$_A$ (nM)	
	PC liposomes	PCES liposomes	PC liposomes	PCES liposomes
Progesterone	35 $\pm$ 5	40 $\pm$ 8	10 $\pm$ 3	18 $\pm$ 4
17α-Hydroxyprogesterone	26 $\pm$ 6	30 $\pm$ 8	20 $\pm$ 5	32 $\pm$ 8
Androstenedione	600 $\pm$ 80	600 $\pm$ 90	1100 $\pm$ 200	1900 $\pm$ 300

the order of 17α-hydroxyprogesterone, progesterone and androstenedione, which seems physiologically reasonable, because some 17α-hydroxyprogesterone, the metabolic intermediate, must be further metabolized sequentially into androstenedione and the final product, androstenedione, must be readily released from P-450$_{17\alpha,\text{lyase}}$. The agreement of Kd$_L$ values of three different steroids in PC liposomes with those in PCES liposomes within the experimental error cannot be regarded as accidental but must be seen as strong evidence for supporting model L.

Cholesterol-P-450$_{\text{SCC}}$ complex has been reported to release cholesterol when the complex was incorporated into the membrane, indicating that the cholesterol binding site of P-450$_{\text{SCC}}$ faces the lipid phase of the membranes (SEYBERT et al., 1979; DEMEL and DE KRUYFF, 1979). LOMBARDO et al. (1986) provided concrete evidence that for model L the 34 kDa hydrophobic fragment of P-450$_{11\beta}$ embedded in membranes had a partial 11β-hydroxylase activity. For hepatic microsomal cytochrome P-450, several studies suggest that the substrate binding site of cytochrome P-450 might face the lipid phase of the membranes (EBEL et al., 1978; BACKES et al., 1982; TANIGUCHI et al., 1984). The conclusion emerging from these studies is that the four major P-450 species in adrenal cortex are integral membrane proteins embedded in the organelles and that their substrate binding sites face the lipid phase.

8. Xenobiotic metabolism in adrenal microsomes

Extra-hepatic xenobiotic metabolizing activities are distributed in a variety of organs and tissues including the kidneys, the lungs, the skin, the intestines, the adrenal gland, and the spleen, ovary, testis, placenta and brain. Although the levels of these activities are low as compared with those of the liver, their presence may be necessary to metabolize a xenobiotic immediately when it enters the organ. Among these, the xenobiotic metabolisms in endocrine glands have been of particular interest in relation to steroidogenesis. In early studies, it was recognized that a variety of foreign compounds including drugs and carcinogens were rapidly metabolized by adrenocortical tissues. In the human fetus, adrenal aryl hydrocarbon hydroxylase activity was much greater than that of liver aryl hydrocarbon hydroxylase (PELKONEN and KARKI, 1973). Although adrenal xenobiotic metabolism has been found in a number of animal species, activity was particularly high in the human and monkey fetus and in both fetal and adult guinea pigs (JUCHAU et al., 1972; JUCHAU and PEDERSON, 1973; GREINER et al., 1977; COLBY et al., 1982). GUENTHNER et al. (1979) observed that approximately 80% of the total aryl hydrocarbon hydroxylase activity in rat adrenal microsomes exist in the zona fasciculata-reticularis and 20% in the zona glomerulosa.

The relationship between steroidogenic activities and xenobiotic-metaboli-zing activities in adrenal cortex has been analysed from the differences induced with aging, sexual differences, various inhibitors and ACTH. GUENTHNER et al. (1979) observed that the rat adrenal hydroxylase activity was strongly in-hibited by α-naphthoflavone which was quite specific for polycyclic hydro-carbon-induced monooxygenase activities, whereas progesterone 21-hydroxy-lase was not. COLBY et al. (1975) investigated the details of the changes in adrenal microsomal monooxygenase activity which occur with aging in guinea pigs. With aging, the rate of metabolism of xenobiotics by the adrenal micro-somes increased, but steroid 17α- and 21-hydroxylase activities were similar in both the young and the adult animals. In young guinea pigs, the magnitude of the ethylmorphine-induced spectral changes varied dramatically with aging, but steroid-induced spectra by progesterone or 17α-hydroxy-progesterone were not affected by the age of the animal. These results strongly suggest that adrenal xenobiotic and steroidogenic activities are independently controlled by aging.

Using cultured bovine adrenal cortical cells, DIBARTOLOMEIS and JEFCOATE (1984) examined the interrelationship between adrenal steroidogenesis and polycyclic aromatic hydrocarbon metabolism. ACTH selectively induced steroidogenic cytochrome P-450-dependent enzyme activities in cell cultures. Benzoanthracene suppressed steroidogenesis as an inhibitor of 17α-hydroxylase but induced benzo[a]pyrene metabolism several times above control levels. Their results indicated that cytochrome P-450 isozymes involved in steroidogenesis and polycyclic aromatic hydrocarbon metabolism are separately controlled. GREINER et al. (1977) performed comparisons of the effects of several physiological variables on the adrenal microsomal drug and the steroid metabolism in guinea pig. The rate of adrenal ethylmorphine metabolism increased with maturation in males but not in females, resulting in a sex difference in adrenal enzyme activ ityin adult guinea pigs. 21-Hydroxylase activity, in contrast, was noted to be similar in adrenals from males and females. ACTH administration decreased ethylmorphine demethylase activity but did not affect 21-hydroxylation. Testosterone, when given to female guinea pigs increased the rate of ethylmorphine metabolism and decreased 21-hydroxylase activity. Simple aromatic hydrocarbons, in contrast, inhibited 21-hydroxylation but did not affect ethylmorphine metabolism. These studies support the contention that adrenal drug and steroid metabolisms are independently regulated. These investigations strongly suggest that there may be some drug monooxygenase in the microsomes other than the steroidogenic cytochrome P-450, but the existence of such monooxygenase has not yet been experimentally confirmed.

MOCHIZUKI et al. (1988) measured the steroid and benzo[a]pyrene hydroxylase activities of $P-450_{17\alpha,lyase}$ and $P-450_{C21}$ using a reconstituted system containing NADPH-cytochrome P-450 reductase in the presence of dilauroylphosphatidylcholine. In the reconstituted system, $P-450_{C21}$ catalyzed only progesterone 21-hydroxylation, whereas $P-450_{17\alpha,lyase}$ catalyzed benzo[a]pyrene hydroxylation as well as progesterone 17α-hydroxylation. This may be interpreted as an indication that all the benzo[a]pyrene hydroxylase activity in the microsomes can be attributed to the action of $P-450_{17\alpha,lyase}$. This possibility was examined from enzyme inhibition in guinea pig adrenal microsomes by anti-$P-450_{17\alpha,lyase}$ IgG. As illustrated in Figure 11(a), when the benzo[a]pyrene hydroxylase activity of the microsomes without the IgG was defined as 100%, only about 30% inhibition of the benzo[a]pyrene hydroxylase activities was observed in the microsomes at the maximum addition of the IgG. On the other hand, 21-hydroxylase activity increased rather with addition of IgG. The addition of anti-$P-450_{17\alpha,lyase}$ IgG to the reconstituted system containing $P-450_{17\alpha,lyase}$ could completely inhibit the both activities of progesterone 17α-hydroxylation and benzo[a]pyrene hydroxylation (Fig. 11(b)), suggesting that partial inhibition of benzo[a]pyrene hydroxylase in the microsomes by anti-$P-450_{17\alpha,lyase}$ IgG is not due to a partial inhibitory effect of IgG on $P-450_{17\alpha,lyase}$. This data indicates that $P-450_{17\alpha,lyase}$ (but not $P-450_{C21}$) participates to a certain degree in the benzo[a]pyrene hydroxylase activity of adrenal microsomes. HORNSBY et al. (1985) reported an unusual example in that the benzo[a]pyrene metabolism in cultured fetal human adrenal cells was stimulated by ACTH, which suggests strongly that $P-450_{17\alpha,lyase}$ might participate in the metabolism of a carcinogen because the synthesis of $P-450_{17\alpha,lyase}$ in adrenal cells is regulated by ACTH. $P-450_{17\alpha,lyase}$ exhibited fairly high xenobiotic-

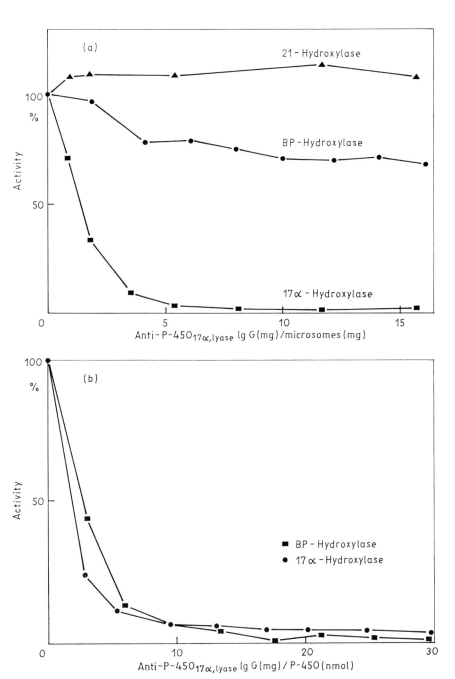

Fig. 11. Effect of anti-P-450$_{17\alpha,\text{lyase}}$-IgG on steroid and benzo[a]pyrene hydroxylase activities of guinea pig adrenal microsomes (a) and the reconstituted system consisting of P-450$_{17\alpha,\text{lyase}}$ and NADPH-cytochrome P-450 reductase (b), taken from MOCHIZUKI et al. (1988) with permission.

metabolizing activity in the reconstituted system. SUHARA et al. (1984) reported that the P-450 species from pig testis microsomes, which can be considered as an identical species to P-450$_{17\alpha,\text{lyase}}$ in the guinea pig adrenal microsomes, shows various xenobiotic-metabolizing activities in addition to steroid 17α-hydroxylase and C-17,20-lyase activities. It is of interest that P-450$_{17\alpha,\text{lyase}}$ has an apparently broader substrate specificity than P-450$_{C21}$.

A form of cytochrome P-450 metabolizing polycyclic aromatic hydrocarbons, distinct from steroidogenic P-450, has been partially purified from rat adrenal microsomes and has been suggested as being a protein with a molecular mass of 57 kDa, which is similar to the MC-inducible hepatic enzyme (GUENTHNER and NEBERT, 1979; MONTELIUS and RYDSTROM, 1982). In an attempt to clarify the existence of a novel enzyme having benzo[a]pyrene hydroxylase activity other than P-450$_{17\alpha,\text{lyase}}$ in guinea pig adrenal microsomes, MOCHIZUKI et al. (1988) removed P-450$_{17\alpha,\text{lyase}}$ from the detergent-solubilized microsomes using an immobilized anti-P-450$_{17\alpha,\text{lyase}}$ IgG column. The isolated benzo[a]pyrene hydroxylase fraction contained 0.12 nmol cytochrome P-450 per mg protein and did not show any steroid hydroxylase activity even in the presence of excess amounts of additional NADPH-cytochrome P-450 reductase. It was demonstrated from enzyme-linked immunosorbent assay using anti-P-450$_{17\alpha,\text{lyase}}$ IgG that more than 98% of P-450$_{17\alpha,\text{lyase}}$ in the solubilized microsomal preparation was removed by treatment of the immobilized IgG column. The benzo[a]pyrene hydroxylase activity of the benzo[a]-pyrene hydroxylase fraction was noted to be 1.1 nmol product per nmol of cytochrome P-450 per min and was demonstrated to have an obligatory requirement for molecular oxygen, NADPH and NADPH-cytochrome P-450 reductase. Moreover, benzo[a]pyrene hydroxylase activity in the presence of 20% O_2 and 80% CO decreased to about 20% of that measured aerobically. These results unambiguously indicate that certain cytochrome P-450 in benzo[a]-pyrene hydroxylase fraction is capable of catalyzing benzo[a]pyrene hydroxylation. The amount of the cytochrome P-450 specific for xenobiotic metabolism as a percentage of total cytochrome P-450 in the microsomes is likely to be 10—20%. The xenobiotic metabolism in the adrenal microsomes seems to be carried out mostly by the cytochrome P-450 in the benzo[a]pyrene hydroxylase fraction, partially by P-450$_{17\alpha,\text{lyase}}$ and much less by P-450$_{C21}$.

Differential regulation between xenobiotic metabolism and steroidogenesis activities by various treatments including ACTH and inhibitors may be explained by the characteristics of the cytochromes P-450 in benzo[a]pyrene hydroxylase fraction which differ from those of the steroidogenic cytochrome P-450. The change in the xenobiotic metabolizing activity caused by physiological variables including age or sex may indicate that the genetic expression of the cytochromes P-450 in the benzo[a]pyrene hydroxylase fraction might be controlled for some physiological purpose.

S. TAKEMORI; S. KOMINAMI

9.   3β-Hydroxy-Δ⁵-steroid dehydrogenase-Δ⁵-isomerase

The conversion of 5-ene-3β-hydroxysteroids to the 4-ene-3-oxo configuration represents an essential step in the biosynthetic route of steroid hormones. The enzyme responsible for this step is the NAD^+-dependent 3β-hydroxy-$Δ^5$-steroid dehydrogenase which oxidizes the 3β-hydroxy group to a 3 keto group and 3-ketosteroid $Δ^4,Δ^5$-isomerase which is a membrane-bound protein found in endocrine glands such as the adrenal, ovary, testis and placenta. Although BEYER and SAMUELS (1956) concluded that in rat adrenal cortex 3β-hydroxysteroid dehydrogenase-isomerase was a constituent of the endoplasmic reticulum only, other investigators have reported results which suggest that mitochondria do indeed contain this enzyme activity (McCUNE et al., 1970; BASCH and FINEGOLD, 1971; HOCHBERG et al., 1974; CHAPMAN and SAUER, 1979). Additionally, when rat adrenal cortex tissue slices were cytochemically stained for 3β-hydroxysteroid dehydrogenase activity, the product of the reaction was found in the mitochondria as well as in the smooth endoplasmic reticulum (BERCHTOLD, 1977). These observations suggest that 3β-hydroxysteroid dehydrogenase-isomerase has a dual intracellular location in the adrenal cortex. The role of the mitochondrial enzyme in vivo and its relationship to the microsomal enzyme are unknown. Mitochondrial and micro-somal enzymes may be considered as isozymes and it will be of great interest to determine whether the two enzymes are identical or different.

The question as to whether the dehydrogenase and the isomerase activities are present in the same protein has often been discussed. There has been considerable debate about the multiplicity of 3β-hydroxysteroid dehydro-genase-isomerase. Concerning the substrate specificity of each enzymatic reac-tion, a number of hypothesis have been proposed for kinetics with inhibitors, partial solubilization or studies of pathological cases. The separation of the dehydrogenase and the isomerase from bovine adrenal microsomes has been carried out by GALLAY et al. (1978). Using a mixture of a substituted betaine and sodium cholate, separation of the two enzyme activities was successfully accomplished by chromatography on a DEAE-Biogel A column. Dehydro-genase activity was eluted in the initial wash, while elution of the isomerase required the increase of the ionic strength. This separation of the activities strongly suggested that the dehydrogenase and the isomerase can be con-sidered as distinct protein entities. Moreover, two dehydrogenase peaks were observed on the column chromatography: in the first, dehydrogenase activity toward dehydroepiandrosterone, pregnenolone and 17Λ-hydroxypregnenolone was present and in the second the oxidation of dehydroepiandrosterone and pregnenolone only was detectable. The data supported the hypothesis that several substrate specific dehydrogenase might exist. HIWATASHI et al. (1985) reported the purification of 3β-hydroxysteroid dehydrogenase from bovine adrenal microsomes. The enzyme was solubilized with cholate and purified using various chromatographies in the presence of Emulgen 913. The purified enzyme gave a single homogeneous protein band with a molecular mass of 41 kDa on SDS polyacrylamide gel electrophoresis and showed no steroid isomerase activity.

Unlike the works of GALLARY et al. (1978) and HIWATASHI et al. (1985), ISHII-OHBA et al. (1986) have co-purified 3β-hydroxysteroid dehydrogenase and steroid isomerase after solubilization of rat adrenal microsomes with sodium cholate. Its molecular mass was noted to be 46 kDa by SDS polyacrylamide gel electrophoresis and 91 kDa by gel filtration. During the purification, 3β-hydroxysteroid dehydrogenase activity could not be separated from the steroid isomerase activity as suggested by the earlier work of FORD and ENGEL (1974). Finally, they obtained a pure enzyme protein which exhibited both activities. Their data strongly supports the contention that dehydrogenation of 3β-hydroxy-5-ene-steroid and isomerization of 3-oxo-5-ene-steroid are catalyzed by a single enzyme. Concerning the substrate specificity of the purified enzyme it was found that 3β-hydroxy-5-ane-steroids are better substrates than pregnenolone or dehydroepiandrosterone (3β-hydroxy-5-ene-steroid) although the physiological significance of the 3β-hydroxy-5-ane-steroid dehydrogenase activity of this enzyme is unclear. A partially purified 3β-hydroxysteroid dehydrogenase from microsomes of rat testicular Leydig cells catalyzed oxidation of 3β-hydroxy group of both 5-ane- and 5-ene-steroids whose molecular mass was estimated as 35 kDa by SDS polyacrylamide gel electrophoresis and by a gel filtration (LARNER and WIEBE, 1983). The adrenal enzyme seems to be distinct from that in Leydig cells with regard to molecular mass. The reaction mechanism of steroid isomerase purified from *Pseudomonas testosteroni* which had no 3β-hydroxysteroid dehydrogenase activity had already been studied in detail (BATZOLD et al., 1976). In the case of 3β-hydroxysteroid dehydrogenase-isomerase from mammalian sources, there are many ambiguous points concerning the enzyme characterizations and further investigations are required.

10. Biosynthesis of estrogens from androgens (aromatase reaction)

An enzyme known as aromatase is responsible for converting the androgens to the corresponding estrogens. The aromatase catalyzes a complex reaction involving removal of the angular C_{19} methyl group and aromatization of the A ring as indicated in Figure 12. The principal sites of estrogen formation are the granulosa cells of the ovary, the placenta, adipose tissue, Sertoli and Leydig cells of the testis and multiple sites in the brain. Adrenal cortex is one of estrogen-producing glands in which the amounts secreted are quite

S. TAKEMORI; S. KOMINAMI

small and do not contribute significantly to the estrogen pool in the body (LAWRENCE and GRIFFITHS, 1966; BAIRD et al., 1969). It is generally believed that the aromatase activity is located in the endoplasmic reticulum. In addition, aromatase activity is also found in mitochondrial preparations of placenta and of other tissues (ALLEN et al., 1970; CANICK and RYAN, 1978). However, when the total mitochondrial fraction was separated into sub-fractions, most of the aromatase activity was detected in light mitochondria and little activity

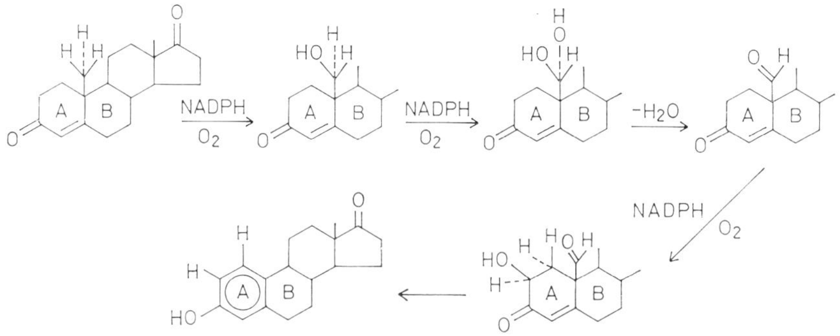

Fig. 12. Enzymatic conversion of androgen to estrogen.

was found in heavy mitochondria. These findings suggest that aromatase activity in the mitochondrial fraction may be due to contamination by microsomes (MOORTHY and MEIGS, 1978). SUHARA et al. (1986) presented unusual data showing that adrenal mitochondrial P-450$_{11\beta}$ can mediate the formation of estrogen, suggesting a possibility that such a mechanism may play an important role in generating adrenal estrogens.

The placental aromatase was first identified as a microsomal enzyme by RYAN (1959), who demonstrated that NADPH and O_2 are required for the conversion of androstenedione to estrone. THOMPSON and SIITERI (1974) showed that cytochrome P-450 was involved in this reaction and determined that 3 mol each of NADPH and O_2 are consumed for each mole of estrogen formed. This data suggests that three hydroxylation stages are involved in this reaction sequence. The first two stages of the hydroxylation take place on the C_{19} methyl group, giving rise successively to the 19-hydroxyandrostenedione (MEYER, 1955; WILCO and ENGEL, 1965) and 19-dihydroxyandrostenedione which dehydrates to the 19-aldehyde structure (AKHTAR and SKINNER, 1968; SKINNER and AKHTAR, 1969). The site of the third hydroxylation at the final stage of the transformation from androgen to estrogen remains unknown. Different mechanisms have been proposed that either 1β-hydroxylation

(Townsley and Brodie, 1968), 2β-hydroxylation (Hahn and Fishman, 1984) or 19-peroxidation (Akhtar et al., 1982) would form an intermediate which would rapidly and nonenzymatically collapse into an estrogen. The most consistent evidence would imply the 2β-position as the final hydroxylation step in estrogen biosynthesis.

Cytochrome P-450 catalyzing the aromatization, referred to as P-450$_{arom}$, has received much attention because of this interesting reaction mechanism, the role of estrogens in endocrine physiology and estrogen-dependent diseases. Several research groups have attempted the isolation of P-450$_{arom}$ from the human placental microsomes (Osawa and Higashiyama, 1980; Mendelson et al., 1985; Nakajin et al., 1986; Tan and Muto, 1986; Hagerman, 1987). Despite their efforts, however, P-450$_{arom}$ has so far resisted all attempts at purification to homogeneity. These failures are due not only to its membrane-bound nature, but also because the solubilized enzyme is very unstable. Therefore, until recently, the characteristics of the P-450$_{arom}$ have not been so well understood as those of other steroidogenic P-450 species. Kellis and Vickery (1987) have succeeded in purifying P-450$_{arom}$ from human placental microsomes to a specific content of 11.5 nmol of P-450 per mg of protein using chromatographies on phenyl Sepharose, DE 52 and hydroxyapatite in the presence of androstenedione as a stabilizer. The purified enzyme, with a molecular mass of 55 kDa, displayed spectroscopic properties typical of cytochrome P-450. The androstenedione complex of P-450$_{arom}$ exhibited a Soret maximum at 393 nm and the charge transfer band at 642 nm, was strongly suggestive of high spin form. The low spin type spectrum, which had a Soret peak at 417 nm, was generated from the high spin form by adding a hydroxy-α-naphthoflavone, a potent inhibitor of P-450$_{arom}$. Aromatase activity could be reconstituted with rabbit liver microsomal NADPH-cytochrome P-450 reductase. P-450$_{arom}$ catalyzed the aromatization of androstenedione (maximal turnover number: 6), 19-hydroxyandrostenedione (12) and 19-oxoandrostene-dione (21). Testosterone and 16α-hydroxytestosterone were also aromatized at maximal rates similar to androstenedione. Recently, the purification of human placental P-450$_{arom}$ has been reported by Harada (1988). The reported preparation had a molecular mass of 51 kDa and showed the high aromatase activity.

Since estrogens function both in normal endocrine process and in certain hormone-dependent cancers, attention has been directed towards developing inhibitors to control estrogen biosynthesis. Several different classes of P-450$_{arom}$ inhibitors have been developed, namely, competitive substrate analogs, substrate derivatives possessing functional groups capable of direct covalent reactions, mechanism-based suicide substrates and non-steroid compounds which coordinate with the P-450$_{arom}$ heme group (Ortiz de Montellano and Reich, 1986). Novel antitumour agents that act by selective inhibition of P-450$_{arom}$ may be developed in the future.

Acknowledgements

The authors deeply appreciate the kindness of Dr. YAMAZAKI for discussing the contexts of this review and of Mr. IKUSHIRO for help in the drawing of the figures.

11. References

AKHTAR, M. and S. J. M. SKINNER, (1968), Biochem. J. **109**, 318—321.
AKHTAR, M., M. R. CALDER, D. L. CORINA, and J. N. WRIGHT, (1982), Biochem. J. **201**, 569—580.
ALLEN, F. A., E. VALDIVIA, and A. E. COLAS, (1970), Gynecol. Invest. **1**, 277—287.
ALMGREN, M., (1980), J. Am. Chem. Soc. **109**, 7882—7887.
AMOR, M., M. TOSI, C. DUPONCHEL, M. STEINMETZ, and T. MEO, (1985), Proc. Natl. Acad. Sci. U.S.A. **87**, 4453—4457.
ANTONIPILLAI, I., E. MOGHISST, S. D. FRASIER, and R. HORTON, (1980), J. Clin. Endocrinol. Meta. **57**, 580—584.
BACKES, W. L., M. HOGABOOM, and W. J. CANADY, (1982), J. Biol. Chem. **257**, 4063 to 4070.
BAIRD, D. T., A. UNO, and J. C. MELBY, (1969), J. Endocr. **45**, 135—136.
BARON, J., W. E. TAYLOR, and B. S. S. MASTERS, (1972), Arch. Biochem. Biophys. **150**, 105—115.
BASCH, R. S. and M. J. FINEGOLD, (1971), Biochem. J. **125**, 983—989.
BATZOLD, F. H., A. M. BENSON, D. F. COVEY, C. H. ROBINSON, and P. TALALAY, (1976), Adv. Enz. Regul. **14**, 243—267.
BECKETT, G. J. and G. S. BOYD, (1977), Eur. J. Biochem. **72**, 223—233.
BERCHTOLD, J. P., (1977), Histochemistry **50**, 175—190.
BEYER, K. P. and L. T. SAMUELS, (1956), J. Biol. Chem. **219**, 69—76.
BLACK, V. H., E. ROBBINS, N. MCNAMARA, and T. HUIMA, (1979), Amer. J. Anat. **156**, 453—504.
BLACK, S. D. and M. J. COON, (1982), J. Biol. Chem. **257**, 5929—5938.
BOSTERLING, B. and J. R. TRUDELL, (1982), J. Biol. Chem. **257**, 4783—4787.
BROGAN, W. C., P. R. MILES, and H. D. COLBY, (1983), Biochim. Biophys. Acta **758**, 114—120.
BROWN, M. S. and J. L. GOLDSTEIN, (1986), Science **232**, 34—47.
BRYAN, G. T., A. M. LEWIS, J. B. HARKINS, S. F. MICHELETTI, and G. S. BOYD, (1974), Steroid **23**, 185—201.
BUMPUS, J. A. and K. M. DUS, (1982), J. Biol. Chem. **257**, 12696—12704.
CANICK, J. A., and K. J. RYAN, (1978), Steroid **32**, 499—509.
CARROLL, M. C., R. D. CAMBELL, and R. R. PORTER, (1985), Proc. Natl. Acad. Sci. U.S.A. **82**, 521—525.
CHAPMAN, J. C. and L. A. SAUER, (1979), J. Biol. Chem. **254**, 6624—6630.
CHASALOW, F. I. and S. LIEBERMAN, (1979), J. Biol. Chem. **254**, 3777—3782.
CHIANG, J. Y. L., (1981), Arch. Biochem. Biophys. **211**, 662—673.
CHUNG, B.-C., J. PICADO-LEONARD, M. HANIU, M. BIENKOWSKI, P. F. HALL, J. E. SHIVELY, and W. L. MILLER, (1978), Proc. Natl. Acad. Sci. U.S.A. **84**, 407—411.
CHUNG, B.-H., K. J. MATTESON, and W. L. MILLER, (1986), Proc. Natl. Acad. Sci. U.S.A. **83**, 4243—4247.
COHEN, B. S. and R. W. ESTABROOK, (1971), Arch. Biochem. Biophys. **143**, 37—45.
COLBY, H. D., R. C. RUMBAUGH, and R. E. STITZEL, (1975), J. Biol. Chem. **107**, 1359 to 1363.

COLBY, H. D., (1980), J. Steroid Biochem. **13**, 861—867.
COLBY, H. D., P. B. JOHNSON, M. R. POPE, and J. S. ZULKOSKI, (1982), Biochem. Pharmacol. **31**, 639—646.
COOPER, D. Y., R. W. ESTABROOK, and O. ROSENTHAL, (1963), J. Biol. Chem. **238**, 1320—1323.
COOPER, D. Y., S. LEVIN, S. NARASIMHULU, O. ROSENTHAL, and R. W. ESTABROOK, (1965), Science **147**, 400—402.
COOPER, D. Y., B. NOVACK, O. FOROFF, A. SLADE, E. SAUDERS, S. NARASIMHULU, and O. ROSENTHAL, (1967), Fed. Proc. **26**, 341.
CORREIA, M. A. and G. J. MANNERING, (1973), Mol. Pharmacol. **9**, 455—469.
DAHL, E. V. and R. C. BAHN, (1962), Am. J. Pathol. **40**, 587—598.
DAVIS, J. O., (1967), in: The Adrenal Cortex, (A. B. EISENSTEIN, ed.), Little, Brown and Co., Boston, pp. 203—247.
DAVISON, B., D. M. LARGE, D. C. ANDERSON, and W. R. ROBERTSON, (1983), J. Steroid Biochem. **18**, 285—290.
DEE, A., G. CARLSON, C. SMITH, B. S. S. MASTERS, and M. R. WATERMAN, (1985), Biochem. Biophys. Res. Commun. **128**, 650—656.
DEES, J. H., L. D. COE, Y. YASUKOCHI, and B. S. S. MASTERS, (1980), Science **208**, 1473—1475.
DELAMATER, P. V., (1986), in: The Adrenal Gland, (P. J. MULROW, ed.)., Elsevier Science Publishing Co., New York, pp. 363—382.
DEMEL, R. A. and B. DE KRUYFF, (1976), Biochim. Biophys. Acta **457**, 109—132.
DIBARTOLOMEIS, M. J. and C. R. JEFCOATE, (1984), Mol. Pharmacol. **25**, 476—486.
DIETER, H. H., U. MULLER-EBERHARD, and E. F. JOHNSON, (1982), Biochem. Biophys. Res. Commun. **105**, 515—520.
DIGNAM, J. D. and H. W. STROBEL, (1977), Biochemistry **16**, 1116—1123.
DONOHOUE, P. A., N. JOSPE, C. J. MIGEON, R. H. MCLEAN, W. B. BIAS, P. C. WHITE, and C. VAN DOP, (1986), Biochem. Biophys. Res. Commun. **136**, 722—729.
DOODY, M. C., H. J. POWNALL, Y. J. KAO, and L. C. SMITH, (1980), Biochemistry **19**, 108—116.
DUPONT, B., S. E. OBERFIELD, E. M. SMITHWICK, T. D. LEE, and L. S. LEVINE, (1977), Lancent **2**, 1309—1311.
EACHO, P. I. and H. D. COLBY, (1983), Life Sci. **32**, 1119—1127.
EACHO, P. I. and H. D. COLBY, (1985), Endocrinology **116**, 536—541.
EBEL, R. E., D. H. O'KEEFFE, and J. A. PETERSON, (1978), J. Biol. Chem. **253**, 3888 to 3897.
ESTABROOK, R. W., D. Y. COOPER, and O. ROSENTHAL, (1963), Biochem. Z. **338**, 741 to 755.
FLYNN, G. L., (1971), J. Pharm. Sci. **60**, 345—353.
FORD, H. C. and L. L. ENGEL, (1974), J. Biol. Chem. **249**, 1363—1368.
FRENCH, J. S., F. P. GUENGERICH, and M. J. COON, (1980), J. Biol. Chem. **255**, 4112 to 4119.
GALLAY, J., M. VINCENT, C. DE PAILLERETS, and A. ALFSEN, (1978), Biochim. Biophys. Acta **529**, 79—87.
GARFINKEL, D., (1958), Arch. Biochem. Biophys. **77**, 493—509.
GOWER, D. B., (1984), in: Biochemistry of Steroid Hormones, (H. L. J. MARKIN, ed.), Blackwell Scientific Publication, Oxford, pp. 117—169.
GRANT, J. K. and A. C. BROWNIE, (1955), Biochim. Biophys. Acta **18**, 433.
GREENFIELD, N., L. PONTICORVO, F. CHASALOW, and S. LIEBERMAN, (1980), Arch. Biochem. Biophys. **200**, 232—244.
GREINER, J. W., R. E. KRAMER, R. C. RUMBAUGH, and H. D. COLBY, (1977), Life Sci. **20**, 1017—1026.

         S. TAKEMORI; S. KOMINAMI

GREINER, J. W., R. C. RUMBAUGH, R. E. KRAMER, and H. D. COLBY, (1978), Endocrinology **103**, 1313—1320.
GUENGRICH, F. P., D. P. BALLOU, and M. J. COON, (1976), Biochem. Biophys. Res. Commun. **70**, 951—956.
GUENGERICH, F. P., (1983), Biochemistry **22**, 2811—2820.
GUENTHER, T. M., D. W. NEBERT, and R. H. MENARD, (1979), Mol. Pharmacol. **15**, 719—728.
HAGERMAN, D. D., (1987), J. Biol. Chem. **262**, 2398—2400.
HAHN, E. F. and J. FISHMAN, (1984), J. Biol. Chem. **259**, 1689—1694.
HALES, D. B. and G. BETZ, (1986), Endocrinology **119**, 811—818.
HALKERSTON, I. D., K., J. EICHHORN, and O. HECHTER, (1961), J. Biol. Chem. **236**, 374—380.
HANIU, M., T. IYANAGI, P. MILLER, T. D. LEE, and J. E. SHIVELY, (1986), Biochemistry **25**, 7906—7911.
HANIU, M., K. YANAGIHASHI, P. F. HALL, and J. E. SHIVELY, (1987), Arch. Biochem. Biophys. **254**, 380—384.
HARADA, N., (1988), J. Biochem. **103**, 106—113.
HARKINS, J. B., E. B. NELSON, B. S. S. MASTERS, and G. T. BRYAN, (1974), Endocrinology **94**, 897—902.
HAYAISHI, O., M. KATAGIRI, and S. ROTHBERG, (1955), J. Am. Chem. Soc. **77**, 5450 to 5451.
HIGASHI, Y., H. YOSHIOKA, M. YAMANE, O. GOTOH, and Y. FUJII-KURIYAMA, (1986), Proc. Natl. Acad. Sci. U.S.A. **83**, 2841—2845.
HILDEBRANDT, A. and R. W. ESTABROOK, (1971), Arch. Biochem. Biophys. **143**, 66—79.
HIWATASHI, A. and Y. ICHIKAWA, (1979), Biochim. Biophys. Acta **580**, 44—63.
HIWATASHI, A. and Y. ICHIKAWA, (1981), Biochim. Biophys. Acta **664**, 33—48.
HIWATASHI, A., I. HAMAMOTO, and Y. ICHIKAWA, (1985), J. Biochem. **98**, 1519—1526.
HOCHBERG, R. S., S. LADANY, M. WELCH, and S. LIEBERMAN, (1974), Biochemistry **13**, 1938—1945.
HOFMANN, F. G., (1960), Biochim. Biophys. Acta **37**, 566—567.
HORNSBY, P. J., (1982), Endocrinology **111**, 1092—1101.
HORNSBY, P. J., S. E. HARRIS, and K. A. ALDERN, (1985), Biochem. Biophys. Res. Commun. **131**, 167—173.
HUANG, C. and J. T. MASON, (1978), Proc. Natl. Acad. Sci. U.S.A. **75**, 308—310.
HYATT, P. J., J. B. G. BELL, K. BHATT, and J. F. TAIT, (1983), J. Endocr. **96**, 1—14.
ICHII, S., S. KOBAYASHI, N. YAGO, and S. OMATA, (1967), Endocrinologica Japonica **14**, 138—142.
IMAI, Y., (1981), J. Biochem. **89**, 351—362.
IMATAKA, H., K. SUZUKI, and B. TAMAOKI, (1985), Biochem. Biophys. Res. Commun. **128**, 657—663.
INANO, H., A. INANO, and B. TAMAOKI, (1969), Biochim. Biophys. Acta **191**, 251—271.
ISHII-OHBA, H., N. SAIKI, H. INANO, and B. TAMAOKI, (1986), J. Steroid Biochem. **24**, 753—760.
IYANAGI, T. and H. S. MASON, (1973), Biochemistry **12**, 2297—2308.
JOHN, M. E., T. OKAMURA, A. DEE, B. ADLER, M. C. JOHN, P. C. WHITE, E. R. SIMPSON, and M. R. WATERMAN, (1986), Biochemistry **25**, 2846—2853.
JOSPE, N., P. A. DONOHOUE, C. VAN DOP, R. H. MCLEAN, W. B. BIAS, and C. J. MIGEON, (1987), Biochem. Biophys. Res. Commun. **142**, 798—804.
JUCHAU, M. R., M. G. PEDERSEN, and K. G. SYMMS, (1972), Biochem. Pharmacol. **21**, 2269—2272.
JUCHAU, M. R. and M. G. PEDERSEN, (1973), Life Sci. **12**, 193—204.

KAMIN, H., B. S. S. MASTERS, and Q. H. GIBSON, (1966), in: Flavins and Flavoproteins, (E. C. SLATER, ed.), Elsevier Publishing Co., Amsterdam, pp. 306—324.

KATAGIRI, M., K. SUHARA, M. SHIROO, and Y. FUJIMURA, (1982), Biochem. Biophys. Res. Commun. **108**, 379—384.

KATAGIRI, M., H. MURAKAMI, Y. YABUSAKI, T. SUGIYAMA, M. OKAMOTO, T. YAMANO, and H. OHKAWA, (1986), J. Biochem. **100**, 945—954.

KELLIS, J. T. Jr. and L. E. VICKERY, (1987), J. Biol. Chem. **262**, 4413—4420.

KIMURA, T., (1986), J. Steroid Biochem. **25**, 711—716.

KLINGENBERG, M., (1958), Arch. Biochem. Biophys. **75**, 376—386.

KOMINAMI, S., H. OCHI, Y. KOBAYASHI, and S. TAKEMORI, (1980), J. Biol. Chem. **255**, 3386—3394.

KOMINAMI, S., K. SHINZAWA, and S. TAKEMORI, (1982), Biochem. Biophys. Res. Commun. **109**, 916—921.

KOMINAMI, S., T. OGISHIMA, and S. TAKEMORI, (1982), in: Flavins and Flavoproteins, (V. MASSEY, and C. H. WILLIAMS, Jr., eds.), Elsevier North Holland Inc., Amsterdam, pp. 715—718.

KOMINAMI, S. and S. TAKEMORI, (1982), Biochim. Biophys. Acta **709**, 147—153.

KOMINAMI, S., K. SHINZAWA, and S. TAKEMORI, (1982), Biochem. Biophys. Res. Commun. **109**, 916—921.

KOMINAMI, S., K. SHINZAWA, and S. TAKEMORI, (1983), Biochim. Biophys. Acta **755**, 163—169.

KOMINAMI, S., H. HARA, T. OGISHIMA, and S. TAKEMORI, (1984), J. Biol. Chem. **259**, 2991—2999.

KOMINAMI, S., Y. ITO, and S. TAKEMORI, (1986), J. Biol. Chem. **261**, 2077—2083.

KOMINAMI, S., S. IKUSHIRRO, and S. TAKEMORI, (1987), Biochim. Biophys. Acta **905**, 143—150.

KOMINAMI, S., A. HIGUCHI, and S. TAKEMORE, (1988), Biochim. Biophys. Acta **937**, 177—183.

KOMINAMI, S., (1990), personal communication.

KRUM, A. A., M. D. MORRIS, and L. L. BENNETT, (1964), Endocrinology **74**, 543—547.

LARNER, J. M. and J. P. WIEBE, (1983), J. Steroid Biochem. **18**, 541—550.

LAWRENCE, J. R. and GRIFFITHS, (1966), Biochem. J. **99**, 27C—28C.

LIPSCOMB, J. D., S. C. SLIGAR, M. J. NAMTVEDT, and I. C. GUNSALUS, (1976), J. Biol. Chem. **251**, 1116—1124.

LOMBARDO, A., M. LAINE, G. DEFAYE, N. MONNIER, C. GUIDICELLI, and E. M. CHAMBAZ, (1986), Biochim. Biophys. Acta **863**, 71—81.

LONG, J. A., (1975), in: Handbook of Physiology, (R. O. GREEP and R. B. ASTWOOD, eds.), American Physiological Society, Washington DC, vol. 6, pp. 13—24.

LU, A. Y. H., S. B. WEST, M. VORE, D. RYAN, and W. LEVIN, (1974), J. Biol. Chem. **249**, 6701—6709.

MACKLER, B., B. HAYNES, D. S. TATTONI, D. F. TIPPIT, and V. C. KELLEY, (1971), Arch. Biochem. Biophys. **145**, 194—198.

MASON, H. S., W. L. FOWLKS, and E. PETERSON, (1955), J. Am. Chem. Soc. **77**, 2914 to 2915.

MASTERS, B. S. S., J. BARON, W. E. TAYLOR, E. L. ISAACSON, and J. LoSPALLUTO, (1971), J. Biol. Chem. **246**, 4143—4150.

McCUNE, R. W., S. ROBERTS, and P. L. YOUNG, (1970), J. Biol. Chem. **245**, 3859—3867.

MENARD, R. H., F. C. BARTTER, and J. R. GILLETE, (1976), Arch. Biochem. Biophys. **173**, 395—402.

MENARD, R. H., T. H. GUENTHNER, H. KON, and J. R. GILLETTE, (1979), J. Biol. Chem. **254**, 1726—1733.

MENDELSON, C. R., E. E. WRIGHT, C. T. EVANS, J. C. PORTER, and E. R. SIMPSON, (1985), Arch. Biochem. Biophys. **243**, 480—491.

MEYER, A. S., (1955), Biochim. Biophys. Acta **17**, 441—442.

MITANI, F., T. IIZUKA, R. UENO, Y. ISHIMURA, T. KIMURA, S. IZUMI, N. KOMATSU, and K. WATANABE, (1982), in: Advances in Enzyme Regulation, (G. WEBER, ed.), Pergamon Press, Oxford and New York, pp. 213—231.

MIWA, G. T., S. B. WEST, and A. Y. H. LU, (1978), J. Biol. Chem. **253**, 1921—1929.

MIWA, G. T., S. B. WEST, M.-T. HUANG, and A. Y. H. LU, (1979), J. Biol. Chem. **254**, 5695—5700.

MOCHIZUKI, H., S. KOMINAMI, and S. TAKEMORI, (1988), Biochim. Biophys. Acta **964**, 83—89.

MONTELIUS, J. and J. RYDSTROM, (1982), in: Cytochrome P-450 Biochemistry, Biophysics and Environmental Implications, (M. HIETANEN, M. LAITINEN, and O. HANNINEN, eds.), Elsevier Biomedical Press, Amsterdam, pp. 349—352.

MOORTHY, K. B. and R. A. MEIGS, (1978), Biochim. Biophys. Acta **528**, 222—229.

MOROHASHI, K., Y. FUJII-KURIYAMA, Y. OKADA, K. SOGAWA, T. HIROSE, S. INAYAMA, and T. OMURA, (1984), Proc. Natl. Acad. Sci. U.S.A. **81**, 4647—4651.

MOROHASHI, K., H. YOSHIOKA, O. GOTOH, Y. OKADA, K. YAMAMOTO, T. MIYATA, K. SOGAWA, Y. FUJII-KURIYAMA, and T. OMURA, (1987), J. Biochem. **102**, 559—568.

NABI, N., S. KOMINAMI, S. TAKEMORI, and T. OMURA, (1983), J. Biochem. **94**, 1517 to 1527.

NAKAJIN, S., and P. F. HALL, (1981), J. Biol. Chem. **256**, 3871—3876.

NAKAJIN, S., P. F. HALL, and M. ONODA, (1981), J. Biol. Chem. **256**, 6134—6139.

NAKAJIN, S., J. E. SHIVELY, P.-M. YUAN, and P. F. HALL, (1981), Biochemistry **20**, 4037 to 4042.

NAKAJIN, S., M. SHINODA, and P. F. HALL, (1983), Biochem. Biophys. Res. Commun. **111**, 512—517.

NAKAJIN, S., M. SHINODA, M. HANIU, J. E. SHIVELY, and P. F. HALL, (1984), J. Biol. Chem. **259**, 3971—3976.

NAKAJIN, S., M. SHINODA, and P. F. HALL, (1986), Biochem. Biophys. Res. Commun. **134**, 704—710.

NARASIMHULU, S., (1971), Arch. Biochem. Biophys. **147**, 391—404.

NARASIMHULU, S., (1974), in: Cytochromes P-450 and b$_5$, (D. Y. COOPER, O. ROSENTHAL, R. SNYDER, and C. WITMER, eds.), Plenum Press, New York, pp. 271—286.

NARASIMHULU, S., (1975), Adv. Exp. Med. Biol. **58**, 371—386.

NARASIMHULU, S., (1977), Biochim. Biophys. Acta **487**, 378—387.

NARASIMHULU, S. and C. R. EDDY, (1985), Biochemistry **24**, 4287—4294.

NARASIMHULU, S., (1988), Biochemistry **27**, 1147—1153.

NISHIKAWA, T. and C. A. STROTT, (1984), Endocrinology **114**, 486—491.

OGISHIMA, T., Y. OKADA, S. KOMINAMI, S. TAKEMORI, and T. OMURA, (1983), J. Biochem. **94**, 1711—1714.

OHBA, H., H. INANO, and B. TAMAOKI, (1981), Biochem. Biophys. Res. Commun. **103**, 1273—1280.

OKITA, B. T., L. K. PARKHILL, V. YASUKOCHI, B. S. S. MASTERS, A. D. THEOHARIDES, and D. KUPFER, (1981), J. Biol. Chem. **256**, 5961—5964.

OMURA, T. and R. SATO, (1964), J. Biol. Chem. **239**, 2370—2378.

OMURA, T., R. SATO, D. Y. COOPER, O. ROSENTHAL, and R. W. ESTABROOK, (1965), Fed. Proc. **24**, 1181—1189.

ONODA, M. and P. F. HALL, (1982), Biochem. Biophys. Res. Commun. **108**, 454—460.

ORTIZ DE MONTELLANO, P. R. and N. O. REICH, (1986), in: Cytochrome P-450, (P. R. ORITIZ DE MONTELLANO, ed.), Plenum Press, New York, pp. 273—314.

Osawa, Y. and T. Higashiyama, (1980), in: Microsomes, Drug Oxidation and Chemical Carcinogenesis, (M. J. Coon, A. H. Conney and R. W. Estabrook, eds.), Academic Press, Orlando, vol. 1, pp. 225—228.

Oshino, N., (1978), Pharmac Ther. A2, 477—515.

Parker, K. L., D. D. Chaplin, M. Wong, J. G. Seidman, J. A. Smith, and B. P. Schimmer, (1985), Proc. Natl. Acad. Sci., U.S.A. 82, 7860—7864.

Parker, K. L., B. P. Schimmer, D. D. Chaplin, and J. G. Seidman, (1986), J. Biol. Chem. 261, 15353—15355.

Parry, G., D. N. Palmer, and D. J. Williams, (1976), FEBS Lett. 67, 123—129.

Pelkonen, O. and N. T. Karki, (1973), Life Sci. 13, 1163—1180.

Phillips, A. H. and R. G. Langdon, (1962), J. Biol. Chem. 237, 2652—2660.

Phillips, I. R. and E. A. Shephard, (1985), Nature 314, 130—131.

Plager, J. E. and L. T. Samuels, (1954), J. Biol. Chem. 211, 21—29.

Ponticorvo, L., N. Greenfield, A. Wolfson, F. Charrsalow, and S. Lieberman, (1980), Arch. Biochem. Biophys. 200, 223—231.

Porter, T. D. and C. B. Kasper, (1985), Proc. Natl. Acad. Sci. U.S.A. 82, 973—977.

Rapp, J. P., (1986), in: The Adrenal Gland, (P. J. Mulrow, ed.), Elsevier Science Publishers B.V., New York, pp. 65—78.

Ristau, O., H. Rein, G. R. Janig, and K. Ruckpaul, (1978), Biochim. Biophys. Acta 536, 226—234.

Ryan, K. J. and L. L. Engel, (1957), J. Biol. Chem. 225, 103—114.

Ryan, K. J., (1959), J. Biol. Chem. 234, 268—272.

Sasano, H., P. C. White, M. I. New, and N. Sasano, (1988), Endocrinology 122, 291—295.

Seybert, D. W., J. R. Lancaster, Jr., J. D. Lambeth, and H. Kamin, (1979), J. Biol. Chem. 254, 12088—12098.

Shinzawa, K., S. Kominami, and S. Takemori, (1985), Biochim. Biophys. Acta 833, 151—160.

Shinzawa, K., S. Ishibashi, M. Murakoshi, K. Watanabe, S. Kominami, A. Kawahara, and S. Takemori, (1988), J. Endocrinol., 119, 191—200.

Shiroishi, T., T. Sagai, S. Natsume-Sakai, and K. Moriwaki, (1987), Proc. Natl. Acad. Sci. U.S.A. 84, 2819—2823.

Skinner, S. J. M. and M. Akhtar, (1969), Biochem. J. 114, 75—81.

Sligar, S. G., (1976), Biochemistry 15, 5399—5406.

Sligar, S. G., D. L. Cinti, G. G. Gibson, and J. B. Schenkman, (1979), Biochem. Biophys. Res. Commun. 90, 925—932.

Strott, C. A., A. K. Goff, and C. D. Lyons, (1981), Endocrinology 109, 2249—2251.

Sugiyama, T., N. Miki, and T. Yamano, (1980), J. Biochem. 87, 1457—1467.

Suhara, K., Y. Fujimura, M. Shiroo, and M. Katagiri, (1984), J. Biol. Chem. 259, 8729—8736.

Suhara, K., K. Ohashi, K. Takeda, and M. Katagiri, (1986), Biochem. Biophys. Res. Commun. 140, 530—535.

Sweat, M. L. and M. D. Lipscomb, (1955), J. Am. Chem. Soc. 77, 5185—5187.

Takemori, S., H. Sato, T. Gomi, K. Suhara, and M. Katagiri, (1975), Biochem. Biophys. Res. Commun. 67, 1151—1157.

Takemori, S. and S. Kominami, (1982), in: Oxygenases and Oxygen Metabolism, (M. Nozaki, S. Yamamoto, Y. Ishimura, M. J. Coon, L. Ernster, and R. W. Estabrook, eds.), Academic Press, New York, pp. 403—408.

Takemori, S. and S. Kominami, (1984), Trends in Biochem. Sci. 9, 393—396.

Tamburini, P. P. and G. G. Gibson, (1983), J. Biol. Chem. 258, 13444—13452.

Tan, S. Y., (1986), in: The Adrenal Gland (P. J. Mulrow, ed.), Elsevier Science Co., New York, pp. 153—167.

TAN, L. and N. MUTO, (1986), Eur. J. Biochem. **156**, 243—250.
TANIGUCHI, H., Y. IMAI, and R. SATO, (1984), Biochem. Biophys. Res. Commun. **118**, 916—922.
THOMPSON, E. A., Jr. and P. K. SIITERI, (1974), J. Biol. Chem. **249**, 5364—5378.
TOMIDA, H., T. YOTSUYANAGI, and K. IKEDA, (1978), Chem. Pharm. Bull. (Tokyo) **26**, 2832—2837.
TOWNSLEY, J. D. and H. J. BRODIE, (1968), Biochemistry **7**, 33—40.
TUKEY, R. H., S. OKINO, H. BARNES, K. F. GRIFFIN, and E. F. JOHNSON. (1985), J. Biol. Chem. **260**, 13347—13354.
VELTMAN, J. D. and M. D. MAINES, (1980), Arch. Biochem. Biophys. **248**, 467—478.
VERMILION, J. L. and M. J. COON, (1978), J. Biol. Chem. **253**, 2694—2704.
WHITE, P. C., M. I. NEW, and B. DUPONT, (1984), Proc. Natl. Acad. Sci. U.S.A. **81**, 1986—1990.
WHITE, P. C., D. GROSSBERGER, B. J. ONUFER, D. D. CHAPLIN, M. I. NEW, B. DUPONT. and J. L. STROMINGER, (1985), Proc. Natl. Acad. Sci. U.S.A. **82**, 1089—1093.
WILCOX, R. S. and L. L. ENGEL, (1965), Steroid **6**, (suppl. 2) 249.
WILLIAMS, C. H., Jr. and H. KAMIN, (1962), J. Biol. Chem. **237**, 587—595.
YANAGIHASHI, K. and P. F. HALL, (1986), J. Biol. Chem. **261**, 8429—8433.
YASUKOCHI, Y. and B. S. S. MASTERS, (1976), J. Biol. Chem. **251**, 5337—5344.
YOSHIOKA, H., K. MOROHASHI, K. SOGAWA, M. YAMANE, S. KOMINAMI, S. TAKEMORI. Y. OKADA, T. OMURA, and Y. FUJII-KURIYAMA, (1986), J. Biol. Chem. **261**, 4106 to 4109.
YU, C.-A., I. C. GUNSALUS, M. KATAGIRI, K. SUHARA, and S. TAKEMORI, (1974), J. Biol. Chem. **249**, 94—101.
YUAN, P. M., S. NAKAJIN, M. HANIU, M. SHINODA, P. F. HALL, and T. E. SHIVELY, (1983), Biochemistry **22**, 143—149.
ZUBER, M. X., E. E. JOHN, T. OKAMURA, E. R. SIMPSON, and M. R. WATERMAN, (1986), J. Biol. Chem. **261**, 2475—2482.
ZUBER, M. X., E. R. SIMPSON, and M. R. WATERMAN, (1986), Science **234**, 1258—1261.

Chapter 6

Microbial Steroid Hydroxylating Enzymes in Glucocorticoid Production

R. MEGGES, M. MÜLLER-FROHNE, D. PFEIL, and K. RUCKPAUL

1. Introduction

Biotechnological aspects of steroid biotransformation have to be understood as part of the rapidly developing biotechnology in general. This remarkable process is based on at least two facts. (i) Basic research in biological sciences such as biochemistry, biophysics, molecular biology, microbiology and toxicology has accumulated an enormous amount of data .(ii) Molecular biology, and in particular genetic engineering, has provided the necessary techniques for application. Both these facts make it possible to develop biotechnological processes applicable to industrial utilization.

Any attempt to describe the development of biotechnological application chronologically is hampered by parallel and overlapping developments, the use and application of different methods and forms of the biocatalysts and, finally, the production of a great variety of compounds. Using a simplifying approach, four general lines of application may be differentiated.

(1) Biocatalytically active cells after phenotypical or genotypical optimization without engineered genetic changes are used as biocatalysts producing or converting either low molecular substances (e.g. organic acids, amino acids, steroids, sugars) or providing the source of high molecular substances such as enzymes or simply by use of their capability to produce proteins.

(2) Microbial, plant or mammalian cells are used as natural producers of practically useful compounds after genetic manipulation. Thus, after cloning of genetic material from different sources the genes are expressed in appropriate hosts to produce the compounds wanted. This category comprises the production of insulin, human growth hormone, interferon, interleukin, chymosin, tissue plasminogen activator and others but is also aimed at amplifying the production of natural compounds by genetic engineering.

(3) Biocatalysts can also be used for biotechnological purposes in a cellfree form as isolated soluble or matrix bound enzymes.

(4) Finally, it is also possible to make the biosynthesizing machinery synthesizing in-vitro constructed proteins which are neither constituents of the host nor naturally existing molecules using such cells either as the source of artificial protein constructs or as biocatalysts with improved properties.

All these possibilities have been practiced already, are being extended now and will be in the future. In this article some aspects of steroid biotransformation will be discussed.

Steroid hormones regulate essential biological functions in human and animal organisms, acting as male or female sex hormones (androgens, estrogens and gestagens), as glucocorticoids and mineralocorticoids. The natural compounds and/or their derivatives show anti-inflammatory, antirheumatic, anabolic, sedative, cytostatic, contraceptive or antihormonal activities (APPLEZWEIG, 1978). Their use as drugs has therefore increased remarkably within the last four decades.

This challenge could be tackled only by partial synthesis of steroid hormones in which chemical synthesis is combined with enzymatic transformation starting from naturally occurring steroid raw materials. In 1978 about 190 t of steroid drugs were produced from about 980 t raw materials (WIECHERT, 1979). Microbial transformations have been used commercially in the large scale production of steroids to economize complicated synthetic steps due to the capability of microbial enzyme systems to catalyze reactions with high regio- and stereospecificity. Microorganisms are superior to cell cultures from endocrine glands because of their rapid cell division, growth metabolic intensity which offers favourable conditions for genetic selection. Therefore, as yet cell cultures have not gained any practical importance for steroid production, although numerous transformation reactions, e.g. from cholesterol into pregnane derivatives or into androgens, have become known. Consequently, this article deals mainly with the application of microorganisms in biotechnological steps of the production of glucocorticoids.

The earliest interest in microbial transformations of steroids developed at the end of the 19th century, starting with the detection of the hydrogenation of the double bond of cholesterol in the gut (BONDZYNSKI and HUMNICKI, 1896). Further studies on the metabolism of bile acids (BONDI, 1908) and on the catabolic oxidation of sterols by various mycobacteria (SÖHNGEN, 1913) expanded the first observations. These studies, however, did not originate extended systematic investigations although since 1913 it has been known that numerous microbial species such as *Nocardia, Pseudomonas, Mycobacterium, Corynebacterium* and *Arthrobacter* are able to grow on sterols such as cholesterol and β-sitosterol as unique carbon source.

The first systematic research into a possible application of microorganisms for selected transformations of steroids was initiated by MAMOLI and VERCELLONE in 1937 and 1938. These authors described different stereospecific redox conversions of C3- and/or C17-ketones in alcohols of C18 and C19 steroids by yeasts. Similar transformations of C21 steroids and cholic acids (C24) were performed by KIM (1937, 1939). MAMOLI and VERCELLONE (1938) further observed specific steroid transformations by bacteria which inadvertently had contaminated yeast cultures.

These early investigations established microbial interconversions of ketones and alcohols, ester hydrolysis, double bond isomerizations and reductions and formed the basis for further studies in the 40s, where for the first time two microbial transformations of continuing biotechnological importance were observed. With cholesterol as substrate both a 7-dehydrogenation by *Azotobacter* sp. (HORVATH and KRAMLI, 1947) and a 7-hydroxylation by *Proactino-*

myces roseus (KRAMLI and HORVATH, 1948, 1949) were described. As a result of these observations the path was cleared for the exploitation of microorganisms for the hydroxylation of steroids. This progress was of special significance, because due to the discovery of the anti-inflammatory activity of cortisone by HENCH et al. (1949) and the identification of cortisone and cortisol* as steroid hormones of the adrenal gland the industrial production of these compounds has become of high economic importance.

The isolation of naturally occuring corticosteroids practiced in the very beginning, particularly that of cortisone and cortisol, by extraction of animal adrenals is nowadays of no practical importance and has been replaced by partial synthetic procedures. The first biotechnological synthesis of a corticosteroid hormone was the application of the finding of HECHTER et al. (1949) that perfused adrenal glands are capable of converting cortexon into corticosterone by enzymatic 11β-hydroxylation. This procedure was used by industry (G. D. Searle & Co., cf. Sebek and Perlman, 1979).

But within a short time these developments were out of date due to the pioneering results obtained by PETERSON and MURRAY (1952) and HANSON et al. (1953) who found that by microbial 11α- and 11β-hydroxylation of C21 steroids, natural corticosteroid hormones can be produced. Together with the findings of TURFITT (1943, 1944) on the microbial decay of cholesterol and β-sitosterol into C19 steroid intermediates, the biotechnological basis for the commercial production of glucocorticosteroids was established.

The remarkable breakthrough in the production of glucocorticosteroids by combining chemical reactions with essential reaction steps catalyzed by microorganisms was initiated by the Upjohn Company. It was immediately followed by similar procedures developed by other pharmaceutical companies such as Syntex, Searle, Squibb, Merck, Pfizer, Schering and others. The various procedures (cf. FIESER and FIESER, 1959; ONKEN, 1971; ERHART and RUSCHING, 1972; LANGECKER et al., 1977; SMITH, 1984) will be discussed in section 5.

The consequence of the development was an enormous increase in the application of glucocorticoids as anti-inflammatory and antirheumatic drugs and at the same time a dramatic decrease in prices. In 1950 (1949 Merck first introduced cortisone) cortical hormone sales were $ 13 million; by 1954 they were $ 50 million and they reached $ 120 million in 1959. In 1965 domestic sales in the United States reached about $ 160 million. The sales for sex steroids and progestational hormones increased from $ 12 million in 1950 up to about $ 150 million in 1965 after detection of their properties as antifertility agents (MURRAY, 1976). The value of steroid drugs produced worldwide reached 2 billion dollars in 1980 and can be assumed to have significantly increased since that time. This sum is distributed between different steroid

---

* cortisol corresponds to hydrocortisone

classes: 76% glucocorticoids and contraceptives, 13% estrogens, spirono-lactone and cardiotonic steroid lactones, and 20% others (NOMINE, 1980). The increased sales originate from an almost unlimited availability of these new drugs based on the use of microorganisms in the partial synthesis of these hormones. The introduction of these new techniques was accompanied by significantly diminished prices. Cortisone was priced at about $ 30 per gram in 1949 and bulk prices for cortisone or cortisol were listed as being less than $ 1.00 per gram in 1967. The price of progesterone as a basis for commercial production of steroid hormones of about $ 15 per gram in 1949 decreased to under $ 0.15 per gram in 1967 (MURRAY, 1976).

Despite these remarkable developments, which have been of importance for a long time, the expensive synthesis of cortisone from cholic acid using only chemical transformations has gained in interest again recently after improving the synthesis by diminishing the number of reaction steps from 36 to 11. The reason for this is due to limited production and a rise of prices for diosgenine, the most used starting material from vegetable sources.

2.   Structure of steroids

Steroids are derivatives of cyclopentanoperhydrophenanthrene or gonane (1*, Fig. 1; cf. FIESER and FIESER, 1959; nomenclature: IUPAC, 1972). The gonane molecule is not plain. The six C-atoms connecting the four rings ABCD are chiral. Therefore, $2^6 = 64$ stereoisomers of gonane are possible. Only three of them (2—4) are realized in naturally occurring steroids. They differ markedly in shape. The arrangement of the rings ABCD in 2 is nearly "plain" (AB trans, BC trans, CD trans), in 4 it is strongly bent (AB cis, BC trans, CD cis), and 3 shows an intermediate form (AB cis, BC trans, CD trans). Compound 2 is the structural basis of steroid hormones, from compound 3 the bile acids and from compound 4 the cardioactive steroid lactones are derived. Substitution at C-atoms 10 and/or 13 and 17 of 2 or 3 leads to the structural basis of different classes of steroids: 5—9 (Fig. 1).

5—7 are the basic structures of hormonal steroids and most of their derivatives. 8 represents the structural basis of bile acids and 9 that of sterols and sapogenins, used as raw materials in the partial synthesis of steroid drugs. Introduction of different substituents (hydroxy- and/or keto-, carboxygroups, double bonds etc.) at different positions of 5—9 leads to the individual steroids (Fig. 2).

The glucocorticoids are derivatives of pregnane (7) and especially of 3,20-diketo-21-hydroxy-4-pregnene (= cortexone) with additional substituents at C11 and/or C17 as shown in Table 1.

* The structural formulas in this chapter are denoted by underlined numbers.

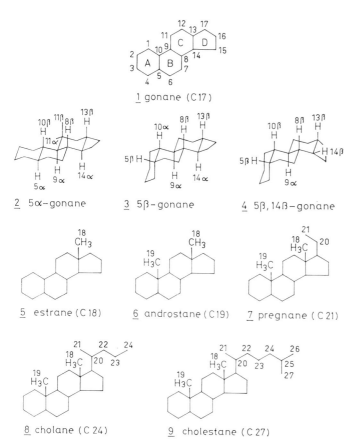

1 gonane (C17)

2 5α-gonane   **3** 5β-gonane   **4** 5β,14β-gonane

5 estrane (C18)   **6** androstane (C19)   **7** pregnane (C21)

8 cholane (C24)   **9** cholestane (C27)

Fig. 1. Basic structures of steroids.

From these structures numerous synthetic analogues are derived with a more or less improved biological activity profile by introduction of a double bond and/or additional substituents at C6, C9, C16 (Table 2).

3. Biochemical principles of microbial steroid transformations

Scientific progress since the first discovery of a microbial steroid transformation (BONDZYNSKI and HUMNICKI, 1896) is characterized mainly by practical considerations and commercial aspects (MAHATO and MUKHERJEE, 1984; MAHATO and BANERJEE, 1985). The basic principles of reaction mechanisms, reaction pathways, their thermodynamics and kinetics on the other hand originate predominantly from isolated and purified mammalian steroid trans-

Table 1. Structures of naturally occuring glucocorticoids as derivatives of 3,20-diketo-21-hydroxy-4-pregnene (= cortexone)

CH$_2$OH
|
CO

(11) (17)

O

	Substituents other than H at	
Glucocorticoid	C11	C17
Cortexone	—	—
Corticosterone	OH (ß)	—
Cortexolone (A)	—	OH (α)
Cortisol	OH (ß)	OH (α)
Cortisone	O	OH (α)

(A) = Reichstein's Substance S

forming enzyme systems. Only a little data exists about corresponding enzymes of microbial origin.

Sterols and steroids are usually not physiological substrates in the metabolic pathway of microorganisms, except ergosterol in the case of some yeasts or cholesterol as a constituent of some bacterial membranes. Therefore, steroids like other biological substances are catabolized ultimately to CO$_2$ and H$_2$O on attack by microorganisms. This process consists of different individual steps. Ring structure and side chains are independently attacked and the reactions do not follow any predetermined sequence as known from mammalian systems where the side chain degradation of a sterol proceeds by cleavage at the C20—C22-bond with intermediate formation of pregnane derivatives and subsequent splitting of the C17—C20-bond leading to 17-ketosteroids. The microbial side chain cleavage of sterols, however, proceeds by a mechanism similar to the β-oxidation of fatty acids. After C26 hydroxylation and oxidation of the alcohol group into a carboxy group, propionic acid, acetic

R. Megges et al.

Table 2. Synthetic glucocorticoids: derivatives of prednisolone (31)

Steroid	Substituents at the steroid skeleton		
	C6α	**C9α**	**C16α**
Prednisolone	H	H	H
Medrol	CH_3	H	H
Prednyliden	H	H	CH_2
Triamcinolone	H	F	OH
Dexamethasone	H	F	CH_3
Paramethasone	F	H	CH_3
Fluocinolone	F	F	OH
Fluocortolone	F	H	CH_3 (17-deoxy)
Deoxymethasone	H	F	CH_3 (17-deoxy)

acid and again propionic acid are split off. Thereby C24- and C22-carboxylic acids, without the formation of pregnane derivatives, are generated as intermediates. Usually, the microbial sterol side chain cleavage reaction is accompanied by the oxidation of the C3-hydroxyl group, leading finally to the formation of 3,17-diketo compounds.

The steps following in the microbial catabolic pathway are reactions which modify the reactivity of the steroid skeleton by introducing substituents, thereby changing the redox potential or the polarity of the molecule. Besides 9α-hydroxylation, the introduction of the double bond in ring A by many microorganisms is the first degradative step of the steroid ring system being used as carbon source. Hydroxylation of steroid molecules by microorganisms may be understood as a detoxifying reaction (MARSHEK, 1971). Oxidations and reductions of steroids are probably catalyzed by enzymes which are likewise involved in reactions of cellular metabolism and energy production and therefore represent relatively unspecific reactions (MARSHECK, 1971). Some of these enzymes involved therein are inducible, either by the sterols or by their metabolites (SZENTIRMAI, 1988).

Several microbial catabolic reactions are applicable to the biotechnological synthesis of steroid compounds. For the production of glucocorticoid hormones the following reactions are of importance (Table 3): oxidative side chain cleavage, formation of the 3-keto group, double bond shifting (delta 5 to delta 4), dehydrogenation for site specific introduction of a double bond in ring A, stereo- and regiospecific hydroxylations.

Hydroxylations are of special importance for glucocorticoid production and therefore they are dealt with in the following sections in more detail.

3.1. Mechanisms and specificities of microbial hydroxylations

The most important biotransformations leading to glucocorticoids are hydroxylations. Steroid hydroxylating enzymes of mammalian cells catalyze the substitution of the steroid molecule only at few distinct positions, whereas microbial hydroxylations are known to occur in almost all possible positions in the steroid nucleus at both angular methyl groups and e.g. at C21 and C26 in the side chains (MAHATO and MUKHERJEE, 1984; MAHATO and BANERJEE, 1985). Therefore, microbial hydroxylations give access to parts of the steroid molecule which can be attacked chemically only with great difficulty (HOLLAND, 1982). They proceed via stereospecific abstraction of the hydrogen atom at the site to be substituted. In addition to monohydroxylations di- or polyhydroxylations of substrates by microorganisms are also possible. Several characteristic patterns such as 6β, 11α- and 12β, 15α-dihydroxylations, are frequently observed. Dihydroxylations may occur as the result of two distinct individual hydroxylases acting independently or in a fixed sequence, but may result also in characteristic patterns catalyzed by one enzyme with diminished regiospecificity.

We still lack a satisfactory understanding of the microbial steroid hydroxylation including its reaction mechanisms. Our knowledge about enzymatically catalyzed hydroxylations of steroids originates mainly from investigations of steroid hydroxylases of mammalian origin (see chapters 1—4 of this volume) and the 15α-hydroxylases of *Bacillus megaterium* (BERG et al., 1979). The available data suggests, irrespective of the source, a cytochrome P-450 dependence of the steroid hydroxylating enzymes (HOLLAND, 1982). Cytochrome P-450 as terminal oxidase, utilizes molecular oxygen and requires an electron transferring system linked to NADPH-dependent dehydrogenases (WOLTERS, 1985). The general characteristics of the catalytic cycle of cytochrome

Table 3. Examples of steroid transforming microorganisms of biotechnical relevance

Organisms	Reaction	Substrates	Products
Eubacteria:			
Arthrobacter simplex	Side chain degradation*)	Cholesterol	Androst-1,4-diene-3,17-dione
Mycobacteria:			
Mycobacterium fortuitum	1,2-Dehydrogenation	Cortisol	Prednisolone
		Cortisone	Prednisone
		6α-Fluoro-16α-hydroxycortisole	Triamcinolone
		6α-Fluoro-16α-methylcortisol	Dexamethasone
	Side chain degradation*)	β-Sitosterol	Androst-1,4-diene-3,17-dione
	9α-Hydroxylation		9α-Hydroxy-androst-4-ene-3,17-dione
Actinomycetes:			
Steptomyces roseochromogenes	16a-Hydroxylation	9a-Fluoro-prednisolone	Triamcinolone
Steptomyces lavandulae	Side chain degradation	Progesterone	Androst-1,4-diene-3,17-dione
Phycomycetes:			
Cunninghamella blakesleeana	11β-Hydroxylation	11-Deoxycortisole	Cortisol
elegans	11α-Hydroxylation	11-Deoxycortisole	11α, 17α, 21-Tri-hydroxypregn-4-ene-3,20-dione
Rhizopus nigricans	11α-Hydroxylation	Progesterone	11α-Hydroxy-progesterone
Ascomycetes:			
Aspergillus ochraceus	11α-Hydroxylation	Progesterone	11α-Hydroxy-progesterone
Fungi imperfecti:			
Curvularia lunata	11β-Hydroxylation	11-Deoxy-cortisol	Cortisol
Fusarium solani	Side chain degradation	Progesterone	Androst-1,4-diene-3,17-dione
	Oxidation at C17		
	Dehydrogenation		

*) also transformation of the 3β-OH-delta-5 system into the 3-keto-delta-4 system

P-450 dependent monooxygenases appear to be independent of the source of the enzyme. The necessary reducing equivalents are transferred to the heme either via a NADPH or NADH dependent flavoprotein alone or together with an iron sulfur protein (ferredoxin) acting as electron shuttle between the reductase and the terminal oxygenase depending on the source of the enzyme (BERG, et al., 1975; BRESKVAR and HUDNIK-PLEVNIK, 1977).

The microbial side chain cleavage reaction of sterols differs from the corresponding mammalian cytochrome P-450 catalyzed reaction from cholesterol to pregnenolone via the (20 S) or (22R) 20- or 22-monohydroxy- and the (20S, 22R) 20, 22-dihydroxy-derivatives. The microbial side chain degradation by oxidative processes leads from sterols to C22- or C19-steroids or from progesterone to testololactone derivatives. This carbon-carbon splitting reaction in microorganisms appears to run via oxygen insertion of a BAYER-VILLIGER type (CARLSTRÖM, 1973). So far, cytochrome P-450 dependent hydroxylation reactions have not been detected in sterol side chain cleavage of microorganisms (KIESLICH, 1980a).

Corresponding to known mammalian cytochrome P-450 dependent hydroxylation reactions the nature of the microbial oxidizing species seems to be clear, but factors which control the binding and possible activation of the substrate have not been identified and the detailed mechanism of the hydroxylating step is unclear as yet (cf. HOLLAND, 1982).

It has been proposed that steroidal substrates are bound, via a condensation reaction between the keto group and a primary amino group, to the protein as imines or via an addition of a thiol group to the 3-keto-delta-4-ketosteroid as thioethers (HOLLAND, 1982). A specific relationship may exist between the position of the substituents of the substrates and the site of hydroxylation. Investigating the hydroxylation of saturated cyclic and polycyclic compounds as model substrates by *Sporotrichum sulfurescens*, MURRAY et al. (1976) postulated an enzyme-substrate complex in which hydroxylation occurs at a methylene group about 5.5 Å away from an electron enriched substituent of the substrate. More recent work with the same microorganism using bridged bicyclic and polycyclic aralkylamide substrates as model compounds, has shown that hydroxylations do not occur systematically 5.5 Å away from the carbonyl oxygens. These findings suggest that either the amide nitrogen or the lipophilic aromatic ring system rather than the carbonyl oxygen may determine the regio-specific hydroxylation of these substrates. A relationship between the position of the substituent of the steroid substrates and the site of hydroxylation has been established for several microorganisms by employing steroid substrates with systematically varied oxygen substituents (hydroxy- and/or carbonyl-groups) in defined positions (JONES et al., 1973).

The presence of a carbonyl or hydroxyl group in the A or D ring exerts a dominant directing influence which results in the introduction of two equatorial hydroxy groups at a distance of about 4 Å from each other. The site of

R. MEGGES et al.

hydroxylation has a distance of 6.5 Å and 7.5 Å, respectively, from the position of the directing substituent.

The presence of an oxygen substituent at a site close (about 4 Å) to the predicted hydroxylation position appears to inhibit hydroxylation at this position. Oxygen carrying substituents such as enol, ethers and acetals may also exert a directing influence on the hydroxylation, but in general with lowered yield and specificity.

The directing influence of halogen substituents on the specificity of hydroxylations has been studied. Presence and position of a halogen substituent have little directing influence on hydroxylations catalyzed by *Calonectria decora* and *Rhizopus nigricans*. 11α-hydroxylation by *Aspergillus ochraceus* is unaffected by the presence of halogen only if the latter is not located close to C11 (HOLLAND and THOMAS, 1982).

Several other fungi show a pattern of hydroxylation products similar to that deduced from *Calonectria decora*. In fungi a definite geometrical relationship between substrate substituents and the site of hydroxylation has not been clearly shown as yet. *Aspergillus ochraceus* shows a predilection for 11α-hydroxylation of a wide range of substrates in both pregnane and androstane series, irrespective of the location of substituents in the molecules. *Rhizopus nigricans* catalyzes predominantly 11α-hydroxylation of pregnanes, but exhibits a substituent directed pattern of hydroxylation with 5α-androstane substrates. Using a series of monoketo, diketo, and ketoalcohol substrates three sites on the enzyme were deduced which are capable either of binding or hydroxylation. Monoketo substrates are dihydroxylated after binding of the substrate carbonyl to one of these sites, dioxygenated substrates on the other hand are monohydroxylated at the third site after binding of the substrate to the other two sites. The substrate may be bound either in "normal" and/or "reverse" fashion (BROWNE et al., 1973).

The fungus *Rhizopus arrhizus* shows a hydroxylation pattern of 5α-androstanes similar to that of *Rhizopus nigricans*, but no clear relationship was observed for hydroxylations performed by *Rhizopus circinnas*.

In hydroxylation reactions of steroids by fungi no specific activation of the substrate's C—H bond by the enzyme has been found, but rather the position of hydroxylation is controlled by the geometry of the active site of the enzyme. Because not all steroid C—H bonds are equally reactive this also may have influence on determining the site of hydroxylation. For example CRABB et al. (1982) have demonstrated by dry ozonation of steroids that selectivity in oxidations at C14 and hydroxylation at allylic positions of the steroids is common.

Investigations on the 6β-hydroxylation of 3-keto-delta-4-ketosteroids by *Rhizopus arrhizus* have shown that the reaction proceeds via binding of the substrate to the enzyme as delta-3,5-dienol, which stereoelectronically controls the enzymic oxidation by axial addition to the dienol (HOLLAND, 1975). For the conformationally constrained steroid only the β-oriented product is obtained.

The structure of steroids which are used as substrates or inducers in microbial transformations strongly influences the kinetics and specificity of the hydroxylation reaction which may lead to unwanted by-products especially in large scale production in industry. However, side reactions can be diminished or completely suppressed by modification or derivatization of the steroidal substrate.

The fungus *Curvularia lunata*, introduced by SHULL and KITA (1955) as 11β-hydroxylating microorganism for the production of cortisol from Reichstein's compound S, forms up to 35% by-products, mainly 14α-hydroxy and $7\alpha,14\alpha$-dihydroxy derivatives. Substitution of the substrate at position 17α- or 16α-increases the yield of the 11β-hydroxy product up to 90%. Moreover, 14a-hydroxylation can be suppressed by a constant pH-value < 6.5 during fermentation. SIH (1962) found that the 1-dehydrogenation and the 9α-hydroxylation of *Pseudomonas* sp. and *Nocardia* sp. could be inhibited specifically to yield the product wanted. When the 9α-position of sterols and C21 or C19 steroids was blocked by a fluorine atom the 1-dehydrogenated derivative was obtained. When KCN was added, the 9α-hydroxy derivative accumulated. By introducing an oxygen atom at C19, SIH et al. (1968) succeeded in preventing bacterial degradation of the steroid ring system to CO_2 and H_2O during fermentation. NAGASAWA et al. (1970) studied the transformation of 167 steroids into androsta-1,4-diene-3,17-dione by *Arthrobacter simplex* and could prevent steroid nucleus degradation by addition of chelating agents, redox dyes or metal ions. Chelating of Fe^{2+} or Cu^{2+} prevents breakdown of the steroid skeleton by microorganisms degrading side chains of sterols (DE FLINES, 1969).

Oxygen-containing substituents, such as carbonyl-, hydroxy-, enolether- or acetal-groups, also have a direct influence on the hydroxylase activity. They reduce the yield and specificity of the reaction. The 11β-hydroxylation by certain fungi can be blocked by replacing the 18-methyl group by an ethyl group (MARSHEK, 1971).

The influence of changes in the A- or B-ring system of pregnane-20-ketones on the 11α-hydroxylation by *Aspergillus ochraceus* (NRRL 405) was studied by TAN and SMITH (1968). Subtle changes in the A ring, like the removal of an oxygen function at C3, did not prevent the steroid from being attacked at its "normal" site. Movement of the ketone in the A ring prevents the hydroxylation at 6-position in the B ring by *Calonectria decora*. Both microorganisms display the ability to hydroxylate 3-deoxy as well as 3-keto steroids.

Hydroxylation of progesterone at C11 was the subject of a study of 53 strains of streptomyces, yeasts and fungi (MONEM et al., 1969). The results show that all strains of *Rhizopus nigricans* and of *Aspergillus* tested were 11α-hydroxylators. No 11α-hydroxylation was observed among *Fusaria, Alteria, Streptomyces* and Yeast. All strains of *Aspergillus* produced 11α-hydroxy- and

R. MEGGES et al.

6β,11α-dihydroxy-progesterone. *Cladosporium cladosporides, Cunninghamella blakesleeana, Aspergillus niger* and *Rhizopus nigricans* show 11α-, 17α-, 21-hydroxylating activity toward progesterone. *Cladosporium cladosporides* produced 11β,17a,21-trihydroxyprogesterone (cortisol), too (EL-REFAI et al., 1970).

Hydroxylation at the C16 position is not confined to certain species of *Streptomyces*, but has been reported also for *Rhizopus nigricans* when the fungi were grown in the presence of 5a-androstan-3-one and its derivatives. Unsubstituted 11α- or 11β-hydroxycompounds of 5α-androstan-3-one are hydroxylated in 16β-position by *Rhizopus nigricans* with yields of 18—56%, while the 16β-hydroxy compound is hydroxylated to the 11α,16β-dihydroxy (60% yield) and 3β,11α,16β-trihydroxy derivatives (23% yield) (ELIN and KOGAN, 1966).

4. Microbial steroid transforming enzyme systems

Both biosynthesis and the biotransformation of steroids in the last three decades have been characterized rather by pragmatic use of microorganisms than by the application of isolated enzymes. The reason for this may be due to well known procedures in classical genetics and phenotypical optimization as compared with risky application of isolated multienzyme systems, including the unresolved problem of cofactor regeneration. Consequently, only few papers dealing with isolation, characterization and purification of steroid transforming enzyme systems of microbial origin have appeared as yet (Table 4).

Among microbial enzymes active in steroid biotransformation which have so far been investigated in cellfree systems are mainly those which catalyze hydroxylation (cf. SMITH, 1984).

4.1. Cellfree systems from eukaryotic microorganisms

In contrast to the rapid expansion of biotechnology the enzymological exploration of microorganisms in general and that of microbial steroid converting enzyme systems in particular has developed relatively slowly. The resulting discrepancy may be explained at least in part by biochemical reasons: difficult accessibility of the respective enzymes due to their low concentration, lability of the isolated proteins, necessity of cofactor regeneration. But by far the most important reason for this lack of progress was that there was neither a real scientific interest nor a commercial need in the period before genetic engineering techniques became available. The promising new techniques which develop along with molecular biology has shown the necessity of knowing more about microbial enzymes and their mechanism of action and thus initi-

Table 4. Microbial cellfree steroid hydroxylases

Microorganisms	Substrate	Positions	References
Eucaryotes: *Aspergillus niger*	Progesterone	11α —	NGUYEN-DANG et al., 1971
Aspergillus niger 12 Y	Progesterone	11α — 11β — 21	ABDEL-FATTAH and BADAWI, 1975a, 1975b
Aspergillus ochraceus NRRL 405	Progesterone	11α —	SHIBAHARA et al., 1970 TAN and FALARDEAU, 1970[a] JAYANTHI et al., 1982
	19-Nortestosterone	11α —	SHIBAHARA et al., 1970
Aspergillus ochraceus TS	Progesterone	11α —	GHOSH and SAMANTA, 1981
Curvularia lunata NRRL 2380	Reichstein's compound S	11β —	ZUIDWEG et al., 1962[b] ZUIDWEG, 1968
	19-Nortestosterone	10β — 11β — 14α —	ZUIDWEG, 1968[c] LIN and SMITH, 1970b
Cochliobolus lunatus m 118	Reichstein's compound S	11β — 12β —	HÖRHOLD et al., 1986
Rhizopus nigricans REF 129	Progesterone	11α — 6β — 17α — 21	SALLAM et al., 1971
Rhizopus nigricans ATCC 6227b	Progesterone	11α — 6β —	BRESKVAR and HUDNIK-PLEVNIK 1977a, b
Procaryotes: *Streptomyces roseochromogenes* ATCC 3347	Progesterone	11α —	ELIN and KOGAN, 1966[b] ELIN et al., 1970
Bacillus megaterium KM ATCC 13632	Deoxycorticosterone	15β — 15α — 6β — 11α —	WILSON and VESTLING, 1965

Table 4. (continued)

Mikroorganisms	Substrate	Positions	References
Bacillus megaterium ATCC 13 368	Progesterone	15β —	BERG et al., 1975, 1976, 1979 BERG and RAFTER, 1981
Nocardia restrictus	Progesterone	9α —	CHANG and SIH, 1964[d])

[a]) 17/20-Lyase observed; [b]) 20-Ketoreductase observed, too; [c]) 17β-Alcoholdehydrogenase observed; [d]) 9α-11α-Epoxidase present, too

ated, aside from cellular studies, those on cellfree systems and on isolated proteins.

As for eukaryotic microorganisms the characterization of cellfree hydroxylases is rendered difficult due to their membrane bound character. At cell disruption the instability in general increases with the degree of disintegration. The utilization of isolated specific steroid hydroxylases for biotechnological purposes has not proved advantageous because the isolated enzyme system is also not free of side reactions. In all known cases, the cellfree system transformed steroid substrates to the same products as the intact cell. New or unexpected reactions were not observed. Substrate optimization (cf. section 3.1) has been found favourable using the advantage of stereo- and regioselective enzymatic catalysis. Studies with cellfree hydroxylases and other steroid-converting enzymes have revealed the complexity of the reaction mechanisms. Its elucidation is of interest not only for basic research but may also provide necessary prerequisities for pheno- and genotypical optimization of cell cultures and for genetic engineering as well.

Steroid transforming monooxygenatic systems are assumed to contain cytochrome P-450 as an essential component. Inhibition by metyrapone, SKF525A and cytochrome c to almost 60%, inhibition by CO and p-chloro-mercuribenzoate to more than 80% as well as the consumption of oxygen and the dependence on pyridine nucleotides have been shown to exist in cytochrome P-450 dependent steroid transforming hydroxylases (BRESKVAR and HUDNIK-PLEVNIK, 1977a; MADYASTHA et al., 1984; MÜLLER-FROHNE et al., 1990).

4.1.1. 11α-Hydroxylase

The *Rhizopus* strains *arrhizus* and *nigricans* are the sources from where most of the biochemical data on microbial 11α-hydroxylases are derived. Already in 1977, BRESKVAR and HUDNIK-PLEVNIK found that the 11α-hydroxylase in *Rhizopus nigricans* is cytochrome P-450 dependent by showing the characteristic CO-difference spectrum with a maximum at 450 nm and the inhibition of enzymatic activity in the presence of CO (BRESKVAR and HUDNIK-PLEVNIK 1977a, b). Its functional linkage with 11α-hydroxylation was further shown by the difference spectrum of the enzyme and its substrate progesterone, which means that the cytochrome P-450 containing fraction from *Rhizopus nigricans* exhibits the modified type II difference spectrum, as defined by SCHENKMAN et al. (1967) with a maximum at 410 nm and a minimum at 395 nm. Although this difference spectrum is not typical for an enzyme substrate interaction which is characterized by a type I difference spectrum (ORRENIUS et al., 1972), it nevertheless indicates an interaction between cytochrome P-450 and progesterone (BRESKVAR and HUDNIK-PLEVNIK, 1977a).

The 11α-hydroxylase, together with the corresponding flavoprotein have been shown by differential centrifugation to be localized in the microsomal fraction (BRESKVAR and HUDNIK-PLEVNIK, 1977, 1981). The flavoprotein was found to be only loosely bound to the membrane. The specific activity of the NADPH dependent reductase in the microsomal fraction of *Rhizopus nigricans* is inducible by progesterone, starting from as low as 0.1 up to 10—80 nmol reduced cytochrome c/min/mg protein, that is about ten times lower than e.g. in rat liver microsomes with 400 nmol reduced cytochrome c/min/mg protein after phenobarbital induction (YASUKOCHI and MASTERS, 1976).

The content of cytochrome P-450 in the noninduced microsomal fraction of *Rhizopus nigricans* is lower than 0.01 nmol/mg protein. It increases during induction by progesterone up to 0.2—1.0 nmol/mg protein. This inducing factor is remarkably high in comparison with a 1.5—2-fold increase of P-450 after phenobarbital induction in rat liver microsomes (RUTTEN et al., 1987) but is in the same range as compared to other microorganisms. In microsomes of *Candida tropicalis* a specific P-450 content of 0.04 up to 0.7 nmol/mg protein has been obtained, dependent on the growth conditions of the yeast, either on glucose or alkane with carbon or oxygen limitation (KÄPPELI, 1986). Obviously microorganisms such as yeast and fungi react more intensively on inducers than mammalian P-450 systems.

The solubilization of the microsomal fraction proved to be extremely difficult (BRESKVAR, 1983). However in 1985, ČREŠNAR et al. found the electron transmitting system to consist of a FAD containing flavoprotein (rhizoporedoxin reductase) and a nonheme iron protein (rhizoporedoxin) (ČREŠNAR et al., 1985). Recently after stabilising the components, an active electron transport chain in the completely reconstituted 11α-hydroxylase system was

R. MEGGES et al.

found (BRESKVAR et al., 1987). Fungi of the strains *Aspergillus, Curvularia* and *Cunninghamella* are further effective in 11α-hydroxylation. 11α-hydroxylation of progesterone in a cellfree system of *Aspergillus ochraceus* has been described by several authors (SHIBAHARA et al., 1970; GOSH and SAMANTA, 1981; JAYANTHI et al., 1882). All authors agree that the 11α-hydroxylase is an inducible enzyme. The substrate progesterone proved a most effective inducer, whereas 11α-hydroxyprogesterone as reaction product induces a 6β-hydroxylase in *Aspergillus ochraceus* proceeding as a side reaction (SHIBAHARA et al., 1970).

Microsomal fractions from progesterone treated *Aspergillus ochraceus* cultures failed to show the characteristic CO-difference spectrum (DUTTA et al., 1983). Rather, an intensive negative peak appeared at 442—444 nm and a maximum at 427—430 nm. According to JAYANTHI et al. (1982), the NADPH-dependent cytochrome c-reductase as protein component of the 11α-hydroxylase is also induced in fungal cultures by steroids. Despite some information about the 11α-hydroxylase activity of the fungi *Rhizopus nigricans* and *Aspergillus ochraceus* there exist far less data on the NADPH-dependent reductase activity, the amount of cytochrome P-450 and the resulting 11α-hydroxylase activity in a reconstituted system. Therefore, reliable conclusions on the specific 11α-hydroxylase activity of these fungi cannot be drawn as yet. Moreover, knowledge is lacking about the composition of the essential components of the 11α-hydroxylase and about their localization in the cell as well.

Isolation of the 11α-hydroxylase in *Cunninghamella elegans* has not been described so far. The microorganism was used to study the role of the cell wall in metabolic processes (SEDLACZEK et al., 1984). For this purpose fungal protoplasts were prepared, and the transport and conversion of steroidal substances investigated. As compared to complete cells, the steroid transformation has been found to increase fourfold in protoplasts with an unchanged pattern of metabolites, indicating the cell wall passage as a rate limiting step. The permeation of steroids can be increased only if distinct alterations of the cell wall structure are possible or if the steroid structure can be modified. So far the use of protoplasts has presented a useful approach for isolating procedures but it is of no importance in large scale processes because of economical reasons caused by laborious preparation.

4.1.2. 11β-Hydroxylase

Glucocorticoids as important antiinflammatory drugs require 11β-hydroxylation as a functionally essential group. Therefore, interest has been focused on 11β-hydroxylating microorganisms. Studies by ZUIDWEG (1962, 1968) on the isolation and characterization of 11β-hydroxylase of *Curvularia lunata* were based on the early finding of 11β-hydroxylation activity in that fungus by SHULL and KITA (1955). Zuidweg found that the 11β-hydroxylation requires NADPH in addition to oxygen. He was unable to find a heme or flavine carrying enzyme. The enzyme is extremely labile and has to be stabilized at disintegration by large amounts of EDTA and SH-group protecting agents. The high proteolytic activity of cellfree extracts certainly contributes to the lability of the enzyme. The cellfree fractions possess still measurable hydroxylase activity, which can be inhibited by metyrapon and p-chloromercuribenzoate.

Topological studies on *Curvularia lunata* by LIN and SMITH (1970) revealed that the 11β-hydroxylase activity remained in the cytosolic supernatant, probably due to the detergent Tween-80 used for extraction. The authors used the supernatant of 25.000 and 60.000 x g fractions free of mitochondria. Higher g-values did not lead to separation of 11β-hydroxylase activity. In addition to 11β-hydroxylation, 10β- and 14α-hydroxylated products of 19-nortestosterone were found in high yield in a cellfree extract from *Curvularia lunata*. Identical K_m values for all three reactions suggested one enzyme rather than different ones with three individual reactions.

Fungi of the *Cochliobolus* genus have so far rarely been used for steroid transformation. According to HÖRHOLD et al., (1981, 1986), *Cochliobolus lunatus* hydroxylates Reichstein's substance S (17α,21-dihydroxy-4-pregnen-3,20-dion) exclusively in 11β-position. However (20S)-20-carboxy-1,4-pregnadien-3-on is hydroxylated in 12β-position too, which in the authors opinion has to be attributed to the additional double bond. This fungal 11β-hydroxylase could be characterized as a membrane bound cytochrome P-450-dependent enzyme system, which as second essential component contains a NADPH-reductase (MÜLLER-FROHNE et al., 1990).

4.1.3. Other Steroid monooxygenases

Studies on progesterone metabolism have shown that a variety of fungi is able to attack the steroid skeleton, such as *Fusarium*, *Penicillium*, *Aspergillus*, *Mucor* and *Gliocladium* sp. (RAHIM and SIH, 1966). The steroid monooxygenase from *Cylindrocarpon radiculata* has been purified by use of affinity chromatography on pregnenolone-Sepharose and characterized in more detail (ITAGAKI, 1986). It turned out to be a flavoprotein consisting of two subunits (1 FAD per subunit) with a molecular weight of 115.000 for the dimer. The enzyme is able to degrade steroids; it forms dehydrotestololactone from progesterone via testosterone acetate. The side-chain degrading reaction needs NADPH and oxygen, and thus represents a typical monooxygenase reaction.

4.2. Cellfree systems from prokaryotic microorganisms

Steroid-converting enzymes have been isolated and examined both from prokaryotic and eukaryotic microorganisms. As compared with the membrane bound hydroxylases from eukaryotes those from procaryotes have proved superior for studies on cellfree systems due to its soluble state. Some examples of prokaryotic cellfree systems are listed in Table 4.

4.2.1. Monooxygenases

Two different cytochrome P-450 dependent monooxygenases have been isolated from *Bacillus megaterium*. BERG et al. have described a soluble cytochrome P-450 from *Bacillus megaterium* (ATCC 13368), which catalyzes 15β-hydroxylation of 3-keto-delta-4-steroids (BERG et al., 1975, 1976, 1979, 1981, 1982). To a lesser extent, a 6β-hydroxylation proceeds as side reaction. The reaction requires NADPH and O_2. Responsible for the electron transport are a FMN-containing reductase and an iron-sulphur-redoxin.

After reconstitution cytochrome P-450$_{meg}$ is able to 15β-hydroxylate progesterone (0.8 nMol progesterone/min/nMol P-450). Heterologous systems with electron transport components from other sources (rabbit liver NADPH-cytochrome c-reductase, bovine adrenal adrenodoxin and its reductase) also show 15β-hydroxylase activity.

In some parameters cytochrome P-450$_{meg}$ resembles P-450$_{CAM}$ from *Pseudomonas putida* and that from *Rhizobium japonicum* (see Table 5) (YU and GUNSALUS, 1974; DUS et al., 1976; BERG et al., 1979).

FULCO and coworkers (STEVENSON et al., 1983; SCHWALB et al., 1985; NARHI and FULCO, 1986; FULCO and RUETTINGER, 1987) investigating *Bacillus mega-*

Table 5. Molecular weight and isoelectric point of different bacterial cytochromes P-450

Form of P-450	Molecular weight	Isoelectric point
P-450 CAM	46,000	4.5
P-450$_{meg}$	52,000	4.9
P-450$_{Rh.jap.}$	50,000	4.9

terium (ATCC 14581) found a quite extraordinary phenobarbital-inducible form of cytochrome P-450, with a molecular weight of roughly 119 kd. In the presence of NADPH and O_2 this cytochrome P-450$_{BM3}$ in native state is able to hydroxylate long-chain fatty acids.

Limited trypsin proteolysis in the presence of a substrate cleaves P-450$_{BM3}$ into two polypeptides of about 66.000 and 55.000 dalton. The 66.000 domain contains both 1 FAD and 1 FMN per polypeptide chain but no heme, it reduces cytochrome c in the presence of NADPH and is derived from the C-terminal of the polypeptide chain portion. The smaller domain of 55.000 dalton is actually a mixture of three discrete peptides. They contain heme and show the characteristic P-450 absorption peak in their reduced CO-bound form. Some of these domains bind other substrates independent of the presence of a few N-terminal amino acids which are essential for substrate binding. Reconstitution of the peptides after proteolysis did not result in an enzymatically active system. Therefore this P-450$_{BM3}$ is self-consistent and represents a fused protein composed of P-450 and reductase.

Two smaller forms of cytochrome P-450, the BM1 and BM2 forms (mol. wt. 46.000 and 38.000) are less active but still able to hydroxylate fatty acids.

Steroid hydroxylase activities could not be demonstrated in cellfree preparations of *Bacillus megaterium* (ATCC 14581) whereas, conversely, P-450$_{meg}$ isolated from *Bacillus megaterium* (ATCC 13368) was unable to hydroxylate fatty acids. Obviously, no relationship exists between the two cytochromes.

4.2.2. Hydroxylases and dehydrogenases

Cellfree 1-hydrogenases and hydroxylases from *Pseudomonas testosteroni* and *Nocardia restrictus* have been investigated (SIH and LAVAL, 1962; CHANG and SIH, 1964; DAVIDSON and TALALAY, 1966; STRIJEVSKY, 1982; ZEILLINGER and SPONA, 1986). CHANG and SIH (1964) described a 9α-hydroxylase from *Nocardia restrictus*. Among the substrates used, progesterone proved the best. Depending on which steroid *Nocardia restrictus* was cultured, the organism synthesized either an epoxidase in the presence of androsta-4,9(11)-diene-3,17-dione or the 9α-hydroxylase in the presence of progesterone. The hydroxylase activity exhibits an unusual optimum at pH 9.5 in the cellfree system. At pH 7.4, similar to other hydroxylases, only 40% of the activity was found. The enzymatic activities are stimulated by cofactors such as NADH or NADPH but they are not essential.

The authors were not able to decide without ambiguity, whether the activity originates from a single enzyme with two activities (9α-hydroxylation and epoxidation) or from two enzymes with distinct activities. In 1982, STRIJEWSKY provided evidence that the 9α-hydroxylase of the *Nocardia* species M117 did not represent a cytochrome P-450-dependent enzyme system. The isolated and

R. MEGGES et al.

purified proteins are a NADH-dependent flavoprotein reductase and two iron-sulfur proteins. All spectral characteristics suggest the absence of a CO-binding hemoprotein. The successful reconstitution of the hydroxylase activity of the cytochrome-free enzyme preparation excludes any participation of cytochrome P-450 in this hydroxylase system. An optimum in activity at pH 9.5 for 9x-hydroxylase from *Nocardia* sp. M117 corresponds with the data derived by CHANG and SIH (1964) from *Nocardia restrictus*.

DAVIDSON and TALALAY (1966) described a soluble inducible delta-4-5β-dehydrogenase from *Pseudomonas testosteroni* (ACTC 11196). This enzyme has been purified approximately 100-fold by ammoniumsulfate precipitation and ion exchange chromatography. Similar cellfree enzymes have been obtained from the commercially useful *Bacillus sphaericus*, *Bacterium cyclo-oxydans*, *Mycobacterium globiforme*, and *Arthrobacter simplex* (SMITH, 1984). The delta-1-dehydrogenases are inducible and competent in the 1-dehydrogenation of substrates, precursors and intermediates in the synthesis of glucocorticoids. The application of isolated enzymes for producing glucocorticoids on a large scale has not been described as yet.

5. Commercially important steroid transformations in glucocorticoid production

The development of technically useful microbial processes in steroid production started at Schering in 1937 with the synthesis of testosterone. Since 1952 this area has developed rapidly, especially with the introduction of microbially catalyzed reaction steps into the synthesis of glucocorticoids (Table 6).

A variety of hydroxylations, hydrogenations, dehydrogenations and the splitting of carbon-carbon-bonds have been carried out using different microorganisms for producing certain derivatives which otherwise are difficult to synthesize. However, only a few reactions have been applied on an industrial scale. 11x-, 11β- and 16x-hydroxylations, 1-dehydrogenation, oxidation of 3-hydroxy-delta-5-steroids to 3-oxo-delta-4-steroids and side chain degradation of sterols have been found to be the most important ones. Moreover, there are coupled reactions in which several reaction steps are combined depending on the enzyme machinery of the microorganisms used. The necessary transformation of the 3-OH-delta-5-system of sterols into the 3-keto-delta-4-system is often linked with the side chain degradation, 1-dehydrogenation and/or 9x-hydroxylation in microorganisms.

Despite the many advantages of using biocatalysts microbial reactions are also not free of problems which in large scale production are of special importance. During transformations, side products may be formed — for instance by hydroxylation at other positions than C11, or additional hydroxylation or

Table 6. Technical useful microbial reactions in corticosteroid production

Reaction	Transformation	Position at steroid	Year of publication
Carbon-carbon bond scission	side chain splitting	17	1965
Hydroxylation	C—H → C—OH	9α	1965
		11α	1952
		11β	1953
		16α	1957
		21	1977
Oxidation/ isomerization	C—OH → C=O delta-5 → delta-4	3/4,5	1965
Dehydrogenation	saturated system → delta-1 double bond	1,2	1955
Hydrolysis	—OAcetyl → —OH	3,21	1966

reduction of the 20-keto group due to the presence of different enzymes or due to reduced specificity of one operating enzyme. These side reactions decrease the yield and render the purification more expensive by necessary subsequent separation procedures. They may be overcome more or less by the inhibition of unwanted or the induction of wanted activities, by species and strain selection or mutation or by structurally changing the starting material. Yield variations may further originate from biological variability due to different batches of living cells.

Contaminating microorganisms have to be avoided by maintaining sterile conditions to carefully exclude strains with human pathogenicity. A high volume-time efficiency may be impaired by the low solubility of steroids in aqueous solutions.

5.1. Raw materials

As necessary prerequisites for an appropriate raw material the following properties are required.

(a) A stereochemical suitable steroid skeleton (cf. Fig. 1).

(b) Availability of the raw material in sufficient amounts with an acceptable price. The requirements for raw material in 1978 amounted up to 1000 tons and is rising strongly (WIECHERT, 1979; KIESLICH, 1980b; SMITH, 1984).

(c) Transformation to necessary intermediates and to the final product must be as simple as possible.

These prerequisites are best met, but not always to the same degree, by the compounds listed in Figure 2, which represent the most important raw materials for glucocorticoid synthesis.

Some other compounds are less important raw materials including ergosterol, eburoic acid, smilagenin, sarsapogenin, solasodine, tomatidine due to either limited availability or more complicated transformations as compared to the raw material in use now.

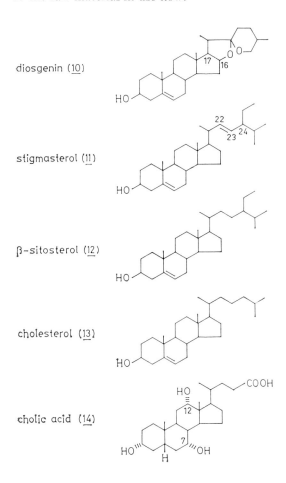

diosgenin (<u>10</u>)

stigmasterol (<u>11</u>)

β-sitosterol (<u>12</u>)

cholesterol (<u>13</u>)

cholic acid (<u>14</u>)

Fig. 2. Main raw materials from plants (<u>10</u>−<u>12</u>) and animal sources (<u>13</u>, <u>14</u>) for partial synthesis of corticosteroids.

The main glucocorticoides are produced by partial synthesis according to four procedures which are listed corresponding to the order of development:

1. From cholic acid (14, Fig. 2)
2. From stigmasterol (11) via progesterone (17, Fig. 4)
3. From diosgenine (10) via 16-dehydropregnenolone-3-acetate (16, Fig. 3)
4. From sterols (11—13, Fig. 2) via 17-keto androstenones or 20-carboxy-pregnenes (18, 19, 21 or 20, 23; Fig. 5)

Procedure 1 is the oldest and is based only on chemical transformations. Remarkable progress was achieved by the combination of chemical and microbial transformation steps. Microbial side chain cleavage and hydroxylation of the nonactivated 11α- or 11β-position (section 5.3 or 5.4), which is difficult to achieve by chemical means, enabled the utilization of new raw materials for glucocorticoid synthesis: stigmasterol (11), diosgenin (10) β-sitosterol (12) and other sterols such as cholesterol (13) or even naturally occurring mixtures of various sterols which can be used without any separation. Besides the advantageous microbial introduction of the hydroxy group in 11-position, the microbial 1-dehydrogenation — necessary for the production of prednisolone — also proved superior when compared to chemical methods (cf. section 5.6.).

In the Peoples Republic of China, 30—50 t hecogenin are isolated per year from the pulp of the sisal fibre production from the leaves of *Agave sisalana*. Hecogenin is the 5,6-dihydro-12-keto derivative of diosgenin (10). Introduction of the delta-1 or delta-1,4-system in the saturated ring A is achieved by microbial transformation with *Arthrobacter* or *Nocardia strains* (FA, 1988).

5.1.1. Diosgenin

Diosgenin (10) belongs to the group of steroid sapogenins. It is mainly isolated from the Mexican Barbasco plant (*Dioscorea composita*). Its side chain degradation was elaborated in the 40s by MARKER et al. (cf. FIESER and FIESER, 1959, p. 548 ff.) and leads in three steps to the important intermediate 3β-acetoxy-16-dehydro-pregnenolone (16, Fig. 3) with a yield of about 45%.

The chemical side chain degradation of 10 (sapogenin) is more effective than that of sterol side chains caused by the lacking functionalization of the side chain in the latter. Due to this fact diosgenin (10) was from the beginning of partial synthesis of glucocorticoids the raw material used most on a large scale. But since 1976, production of (10) in Mexico has not any longer covered the requirements for partial synthesis, this being due to limited resources and export restrictions which led to more than 20-fold higher prices (cf. NOMINE, 1980). The shortage of diosgenin favoured the use of domestic

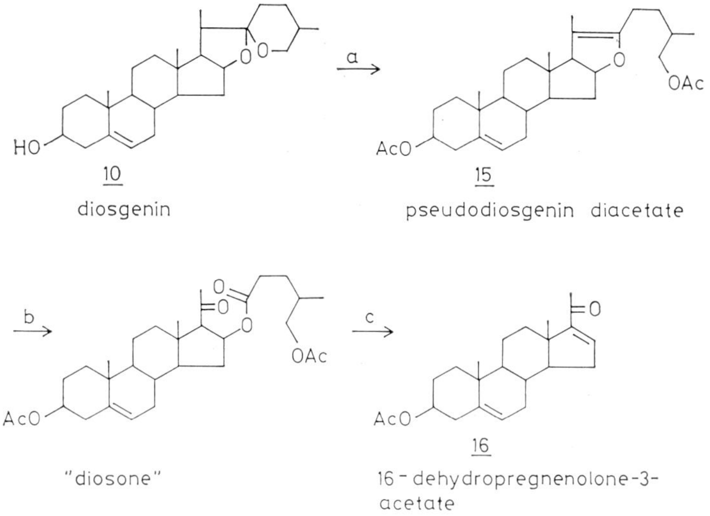

Fig. 3. MARKER degradation of diosgenin (10)-
a acetanhydride, 200°C
b CrO₃, acetic acid
c acetic acid, 140°C

raw materials in steroid producing countries outside of Mexico as the main producer of diosgenin (10). The Peoples Republic of China produces about 200 t diosgenin per year from wild growing *Dioscorea*, thus covering the domestic need. Cultivation of *Dioscorea* decreases the content of diosgenin (FA, 1988).

5.1.2. Stigmasterol

Stigmasterol (11) (12—25%) together with β-sitosterol (12) are components of the unsaponifiable part of soya oil. The isolation and purification of 11 proved difficult. However, the easy degradation of the side chain, which contains a double bond at C22 proved its suitability for the production of progesterone (17) as another important intermediate in glucocorticoid synthesis, via two pathways with 40 or 60% yield (MEYSTRE and MIESCHER, 1949; FIESER and FIESER, 1959, p. 554; see Fig. 4A, B).

The introduction of microbial side chain degradation has economized the degradation process by using a mixture of 11 and 12 or other sterol mixtures (see section 5.2.).

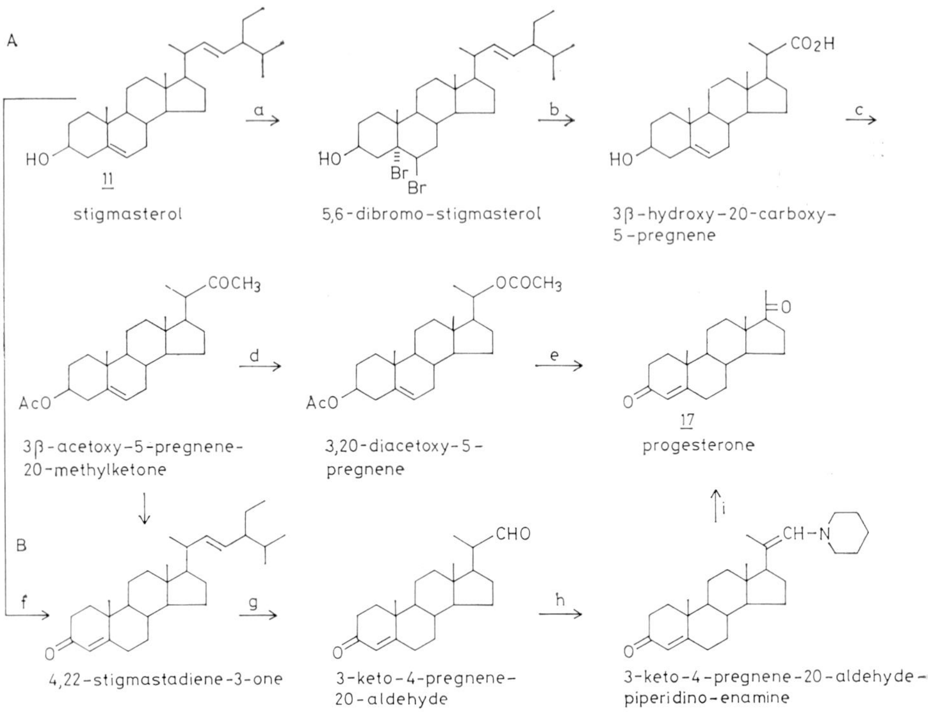

Fig. 4. Two pathways (A, B) from stigmasterol (**11**) to progesterone (**17**).

A a Br_2
   b CrO_3; Zn
   c Ac_2O; $SOCl_2$; $Cd(Me)_2$
   d Br_2; peracid; Zn
   e hydrolysis; OPPENAUER
      oxidation; CrO_3

B f OPPENAUER oxidation
   g O_3
   h piperidine
   i $Na_2Cr_2O_7$, Ac_2O, C_6H_6

5.1.3. β-Sitosterol

β-Sitosterol (**12**) as a component of soya oil, wheat germ oil, tall oil and sugar cane wax is available in the largest amounts as compared to the other compounds of Figure 2. For instance, in 1974 tall oil as a side product of cellulose production from wood was produced in the USA, which contained 20000 tons of **12** (CONNER et al., 1976). Its final purification may be achieved by recrystallization (cf. OTTO et al., 1987). Despite the fact, that **12** is available in sufficient amounts, it became of interest as a starting material only when its microbial side chain degradation was developed in the 60s (see section 5.2.). By this process β-sitosterol (**12**) recently developed into the most important raw material, followed by diosgenin (**10**).

R. MEGGES et al.

5.1.4. Cholesterol

Cholesterol (13) is widely distributed in animal tissues. It is produced from brain, spinal cord and animal fats (the content of 13 in wool grease amounts up to 15%). The purification of 13 is complicated and expensive. One of the most recent methods is the extraction from saponified oils of fats with super-critical CO_2 (KUBOTA, 1987). Cholesterol like β-sitosterol is transformed by microbial side chain cleavage to a C19 or C22 intermediate which are used for the production of glucocorticoids. As compared to diosgenin (10) and β-sitosterol (12) cholesterol (13) is of minor biotechnological importance.

5.1.5. Cholic acid

Cholic acid (14) belongs to the group of bile acids and is mainly isolated from ox bile which contains 25 g/litre (cf. FIESER and FIESER, 1959, p. 53). Cortisone (Table 1) was synthesized from 14 in 36 steps (SARETT, 1946; McKENZIE et al., 1948). This synthesis was improved by some groups, resulting in the reduction to 11 steps (cf. ERHART and RUSCHING, 1972). Due to the shortage of diosgenin (10) this chemical synthesis of glucocorticoids has become of technical interest again.

5.2. Microbial side chain degradation of sterols and coupled reactions

Glucocorticoid production requires side chain degradation of the starting sterols as the initiating step. Before the discovery of microbial side chain cleavage only cholic acid, diosgenin and stigmasterol were used as starting compounds for glucocorticoid production because its side chain could chemical-ly be split off in relatively good yields. The development of microbial side chain cleaving procedures opened the way for using sterols with saturated side chains for partial synthesis of glucocorticoids, such as β-sitosterol which is available in the largest amount compared to all other raw materials (cf. MARTIN, 1984). Technically useful processes were patented in three countries at nearly the same time (ARIMA et al., 1965; Kon. Nederland, Gist-en Spiritus-fab., 1966, 1968; WIX et al., 1966) and improved after that (cf. for instance GROSSE et al., 1987; BÖHME et al., 1987). This development opened a new area in raw material supply substituting the shortage of diosgenin. Contrary to side chain cleavage in mammals which leads to C21 steroids microbial side chain cleavage leads to C19 (18, 19, 21, Fig. 5) or C22 (20, 23) steroids. C19 and C22-steroids likewise are useful intermediates for glucocorticoid pro-duction, but the synthesis through C19 intermediates is the more developed one.

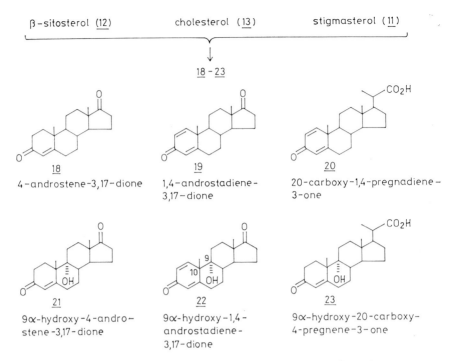

β-sitosterol (12)   cholesterol (13)   stigmasterol (11)

18 - 23

18
4-androstene-3,17-dione

19
1,4-androstadiene-3,17-dione

20
20-carboxy-1,4-pregnadiene-3-one

21
9α-hydroxy-4-andro-stene-3,17-dione

22
9α-hydroxy-1,4-androstadiene-3,17-dione

23
9α-hydroxy-20-carboxy-4-pregnene-3-one

Fig. 5. Important products of microbial side chain degradation of sterols.

C19 steroids obtained in this way required chemical introduction of C20 and C21, for which numerous procedures are available. The classical methods are the Wittig reaction, the pathway via 17-cyanohydrins, the addition of acetylenes, and the Reformatzky reaction (cf. FRIED and EDWARDS, 1972). Three other, more recent, methods are the addition (i) of α-methoxyvinyl-lithium (BALDWIN et al., 1974), (ii) of acetylene and phenylsulfenyl-chloride (VAN RHEENEN and SHEPHARD, 1979), and (iii) of methyl lithiomethoxy-acetate (NEEF et al., 1980). Procedures for splitting one carbon from C22 steroids are also available (NEULAND et al., 1979a, b).

Microbial side chain splitting is advantageously coupled with the transformation of the sterols from 3β-OH-delta-5 compounds to 3-keto-delta-4 steroids required for the synthesis of glucocorticoids. Further reactions such as (i) 1-dehydrogenation to delta-1 compounds (19, 20, Fig. 5) and/or (ii) 9a-hydroxylation leading to 21 or 23 (Fig. 5) have been observed concomitant with side chain cleavage.

Both these reactions are advantageous. The delta-1 double bond is essential for highly active glucocorticoids. 9α-OH is useful for the introduction of the delta-9α (11) double bond, leading to 11β-OH or 11β-OH-9α-halogenosteroids.

R. MEGGES et al.

Simultaneous 1-dehydrogenation and 9α-hydroxylation has to be prevented, because such steroids (as 22, Fig. 5) are unstable. They represent the starting compounds of steroid skeleton degradation by splitting the C9—C10 bond. Type and strain of microorganism, reaction conditions and inhibitors determine the kind of resulting reaction product(s).

With strains of *Mycobacterium* up to 80% of starting sterols (11—13, Fig. 2) are converted to 4-androstene-3,17-dione (18) and 1,4-androsta-diene-3,17-dione (19) (cf. MARTIN, 1984) with substrate concentrations of several grams per litre and incubation times of 3—4 days. The proportion of 18 and 19 is regulated by strain selection or by reaction conditions: by anaerobic post-fermentation, the delta-1 double bond may be hydrogenated again, for instance with formation of 18 from 19 (HÖRHOLD et al., 1979).

The unwanted 9α-hydroxylation is prevented by use of inhibitors (cf. 3.1) or strain selection (cf. MARTIN, 1984). 9α-OH derivatives without a 1,2-double bond are stable and accessible by use of *Mycobacterium* mutants devoid of 1,2-dehydrogenase (cf. ATRAT et al., 1987, SEIDEL et al., 1987). The C22 intermediate 20 is available in good yields with *Nocardia erythropolis* from cholesterol (82%) and β-sitosterol (73%) (SCHUBERT et al., 1976). Also useful is a *Corynebacterium* mutant, leading to different yields depending on the starting sterol: cholesterol (80%), stigmasterol (35%), β-sitosterol (27%) (HILL et al., 1982, IIDA, 1986) or lithocholic acid* (IIDA, 1987). 23 may be produced from β-sitosterol (11) and other sterols beside 21 with *Mycobacterium fortuitum* strains in "good yield" (ANTOSZ et al., 1977).

5.3. 11α-Hydroxylation

PETERSON and MURRAY (1952) found that progesterone (17) is hydroxylated in 11α-position by *Rhizopus arrhizus* or *Rhizopus elegans* in good yield (Fig. 6).

As in all commercially useful microbial transformations, many screening experiments had to be performed before suitable species were obtained. 475

* = 7,12-dideoxy cholic acid, cf. 14, Fig. 2

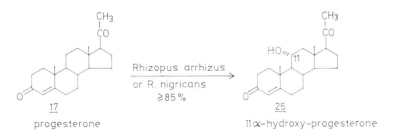

Fig. 6. 11α-Hydroxylation.

strains from 38 species of *Aspergillus* and 475 strains from *Penicillium* had to be screened for their 11α-hydroxylating capacity towards progesterone (17) as usual substrate by DULANY (1955). Numerous other reports describe screening operations with hundreds of species and strains, and even more strains have been selected by individual steroid manufactures. An example is *Aspergillus ochraceus*, which allows the application of substrate concentration up to 50 g/l. Besides 11α-hydroxylation, small amounts of 6β, 11α-dihydroxylated products are formed (WEAVER et al., 1960). Not only progesterone (17) but many other intermediates may be 11α-hydroxylated with high regio- and stereospecificity (LANGECKER et al., 1977; KIESLICH, 1980a).

This microbial reaction was an important breakthrough in partial synthesis of glucocorticoids because it was now possible for the first time to functionalize the nonactivated C11. This allows the use of starting materials without a substituent at C12, such as diosgenine (10) or stigmasterol (11), leading to progesterone (17, cf. Fig. 4). Prior to this discovery functionalization at C12 — as in cholic acid (14) or hecogenine (= 12-keto-4,5-dihydro-diosgenine 10) — was a necessary prerequisite for the introduction of 11β-OH by chemical means.

The chemical transformation of the microbially introduced 11α-OH into the 11β-configuration — required for glucocorticoid activity — is performed by chromic acid oxidation to the 11-keto group after protection of the keto groups in 3- and 20-position and its reduction with sodium borohydride.

Furthermore, 11α-OH derivatives are of interest as intermediates for the synthesis of 9α-substituted glucocorticoids (cf. Tab. 2) via delta-9(11) intermediates, accessible by dehydration of 11α-OH, or 11β-OH, 9α-OH.

5.4. 11β-Hydroxylation

Microbial 11β-hydroxylation is a further milestone in the partial synthesis of glucocorticoids, because it allows immediate hydroxylation at C11 in that configuration which is required for glucocorticoid activity. Reaction of Reichstein's substance S (26, Fig. 7), which is accessible from diosgenin (10) or

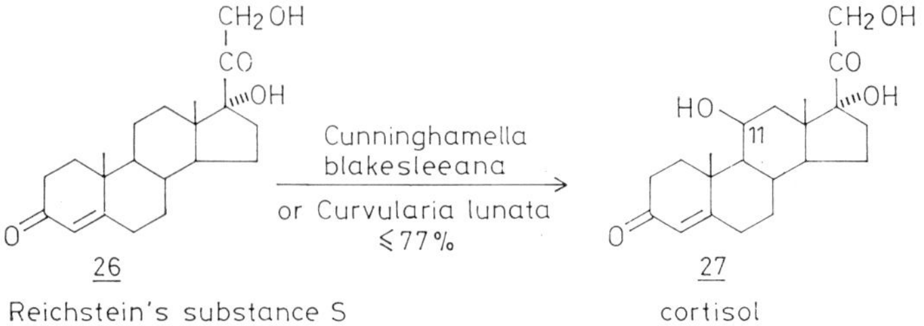

Fig. 7. 11β-Hydroxylation.

sterols (11–13), with the fungi *Cunninghamella blakesleeana* or *Curvularia lunata* leads to the introduction of an 11β-OH group, thus forming cortisol (27) (Hanson et al., 1953; Fried et al., 1955).

The 11β-position (cf. 2, Fig. 1) is axial and therefore more strongly hindered by 1,3-diaxial interactions with 18- and 19-CH₃ and 8β-H than the equatorial 11α-position. This is one possible reason why 11β-hydroxylation proceeds with lower yields and more side reactions as compared with the 11α-hydroxylation (see section 5.3.). With *Cunninghamella blakesleeana* 6β-hydroxylation and 11-ketone formation and with *Curvularia lunata* 7-, 9α-, and 14α-hydroxylations were also observed. However, neither strain selection nor changes of

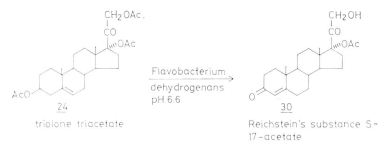

Fig. 8. Ester hydrolysis.

the incubation medium nor variation of fermentation conditions proved appropriate to gain high yields of the product. Instead, variations of the substrate improved formation of 11β-hydroxylated products. Side reactions at the α-side have been evidenced to be suppressed by voluminous substituents at the α-side of the steroid nucleus. Acetylation of 17α-OH group (cf. 24, Fig. 8) has been found quite effective in this respect and increases the yield in 11β-hydroxylation up to 90% (de Flines, 1966; Kieslich, 1969; Smith, 1984).

Comparative studies on 11β-hydroxylase activity proved *Streptomyces fradia* and *Cunninghamella blackesleeana* less effective than *Curvularia lunata* (Smith, 1984).

5.5. 16α-Hydroxylation

The microbial 16α-hydroxylation by *Streptomyces roseochromogenes* has proved appropriate in the production of the steroid hormone analogue Triamcinolone (29, Fig. 9) and its 6α-fluoro derivative fluocinolone (Thoma et al., 1957; Smith, 1984).

Both compounds are highly active glucocorticoids. 16α-hydroxylation of

9α-fluoro-hydrocortisol by *Streptomyces roseochromogenes* with subsequent 1-dehydrogenation constitutes an alternative manufacturing process of Triamcinolone. This two-stage process replaces a chemical and microbiological process with 14 steps (1% yield) starting from Reichstein's substance S (BERNSTEIN et al., 1959). Several *Streptomyces roseochromogenes* strains, as well as other species, are effective in 16α-hydroxylation of 9α-fluorohydrocortisone but further strain selections seem to be necessary because the 2β-hydroxylating capacity of the known strains leads to unwanted by-products (GOODMAN and SMITH, 1961).

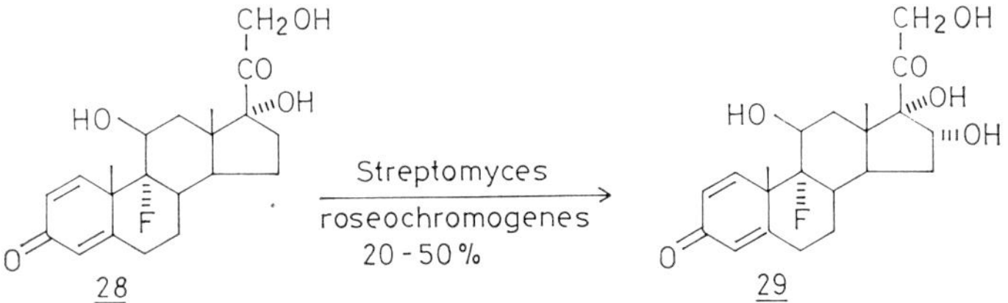

Fig. 9. 16α-Hydroxylation.

The 16α-hydroxylation of the retrosteroid 9β,10α-pregna-4,6-diene-3,20-dione byre sting cells of *Sepedonium ampullosporum* has been demonstrated. The 16α-hydroxylase does not appear to require induction and thus differs from the 11β-hydroxylase of *Curvularia lunata* (McGREGOR et al., 1972).

Resting cells of all important 11- and 16α-hydroxylating microorganisms (*Curvularia lunata*, *Aspergillus ochraceus*, *Rhizopus nigricans*, and *Streptomyces roseochromogenes*) can be used for steroid hydroxylation without nutrient media. This alternative process has not become of technical importance as yet, but commercial interest in this direction is continuing due to the following reasons. Of considerable advantage is the time shortening for the hydroxylating process. Moreover, the resting cells can be charged with substrate several times for product formation, whereas endogenous metabolites and cofactors are being consumed (SMITH, 1984).

The 16α-hydroxylation of *Streptomyces roseochromogenes* (NRRL b-1233), which was immobilized in photopolymerized polyethyleneglycol diacrylate gel has also been described. Incubation of the immobilized mycelium up to 60 h resulted in enhanced yields of 16α-hydroxylated substrate in comparison with vegetative cells. Immobilized cells could be used for three batch transformations before yields began to decrease (CHUN et al., 1981).

R. MEGGES et al.

5.6. 1-Dehydrogenation

Introduction of the 1,2-double bond into steroids is of special importance because it leads to a great number of corticoid derivatives with increased anti-inflammatory and antirheumatic activity (see Table 2).

With regard to its effectiveness, chemical dehydrogenation cannot compete with corresponding microbial transformations. Dehydrogenations are quite common among bacteria with delta-4-3-ketones, saturated ketones and delta-5-3β-alcohols as substrates. The occurrence of 1-dehydrogenase is unique in microorganisms and is not found in mammals. Most ring A dehydrogenating microbes contain both delta-1- and delta-4-dehydrogenating activity (MAR-SHEK, 1971).

For *Pseudomonas testosteroni* and *Bacterium sphaericus* HAYANO et al. (1961) have shown that 1-dehydrogenation proceeds via hydroxylated intermediates and enolization of the 3-ketone toward the 2-position which is followed by

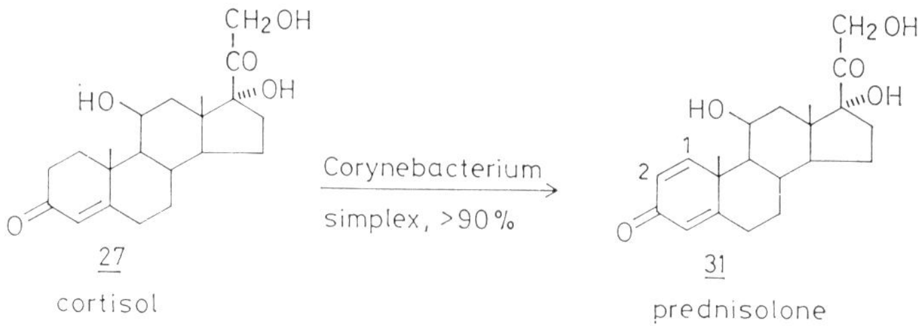

Fig. 10. 1,2-Dehydrogenation.

loss of the activated hydrogen at C1. Several 1-dehydrogenating micro-organisms exhibit a strong tendency to degrade steroid substrates. By addition of quinoid compounds to the medium such as quinone, phenazines, thiazines and oxaquines, the so-called 'steroid destructase system' is inhibited without decrease of the activity of 1-dehydrogenase (WIX et al., 1968).

Since the finding of NOBILE et al., (1955) that *Corynebacterium simplex* contains 1-dehydrogenase activity this microorganism has been mostly used for the 1-dehydrogenation of cortisol (27, Fig. 10).

In a special process termed 'pseudo crystallofermentation' up to 500 g cortisol/l are incubated with *Corynebacterium simplex* for 5 days resulting in more than 90% conversion to prednisolon (31) (KONDO, 1961). The 1-dehydrogenation of cortisol 17-acetate by *Bacterium lentus* leads to prednisone 17-acetate (CEJKA, 1976). Contrary to hydroxylating microbial strains improvement of dehydrogenating microorganisms proved less effective. High yields and minimal by-products are obtained by careful control of fermentation conditions.

---

6. Concluding remarks

Since the discovery of the antiinflammatory and antirheumatic actions of glucocorticoids by HENCH et al. (1949), this group of steroid hormones underwent an enormous development during the following decades. The detection and utilization of microorganisms for partial synthesis of glucocorticoids and its consequences from a dramatic decline in prices to a general availability for medical purposes is one of the early examples of the effectiveness of biotechnology. Molecular biology, and in particular genetic engineering, have opened promising new pathways for the application and further development of biotransforming processes.

It is the aim of this article to summarize our knowledge in this field. The biochemical data available so far and the knowledge about the physiological importance of microbial enzyme systems which are involved in the bioconversion of steroids are as yet very limited. A much deeper understanding about the chemical nature, its composition and its topology in the cell, requires major efforts in basic research to provide a level of accumulated data which would be sufficient for applied use.

But despite these challenging gaps, remarkable new results with potential value for biotechnology have been elaborated in the last 5 or 6 years and may stimulate further studies: the biosynthetic machinery for proteins has been shown to be species independent and pro- and eucaryotic microorganisms are for instance capable of expression of eucaryotic genetic information of mammals. We have learned that rate limited biocatalysis by induction dependent biosynthesis of the donor can be overcome by selection of appropriate promotor systems in the host. We are on the threshold of understanding how insertion into distinct membranes is regulated. We have begun to construct new molecules with desired specified properties. Taking into account these possibilities, biotechnology in general — and biotransformation of steroids in particular — have entered a new area of investigation the results of which may contribute to the well-being of mankind.

7. References

ABDEL-FATTAH, A. F. and M. A. BADAWI, (1975a), J. Gen. Appl. Microbiol. **21**, 217 to 223.
ABDEL-FATTAH, A. F. and M. A. BADAWI, (1975b), J. Gen. Appl. Microbiol. **21**, 225.
ANTOSZ, F. J., W. J. HAAK, M. G. WORCHA, (1987), US-Pat. 4062880.
APPLEZWEIG, N., (1978), in: Crop Resources, (D. S. SEIGLER, ed.), Academic Press, New York, 149.
ARIMA, K., G. TAMURA, M. BAE, and M. NAGASAWA, (1965), DAS 1543269.

ATRAT, P., V. DEPPMEYER, E. MÜLLER, C. HÖRHOLD, H. NAUMANN, P. HÖSEL, L. BLEI, B. GOTTSCHALDT, K.-H. BÖHME, B. SZEKALLA, B. KOCH, and W. BEITHAN, (1987), DDR-WP 248145 A1.

BALDWIN, J. E., G. A. HÖFLE, and G. W. LEVER, jr., (1974), J. Am. Chem. Soc. 96, 7125—7127.

BERG, A., K. CARLSTRÖM, J.-A. GUSTAFSSON, and M. INGELMAN-SUNDBERG, (1975), Biochem. biophys. Res. Commun. 66, 1414—1423.

BERG, A., K. CARLSTRÖM, J.-A. GUSTAFSSON, and M. INGELMAN-SUNDBERG, (1976), J. biol. Chem. 251, 2831—2838.

BERG, A., M. INGELMAN-SUNDBERG, and J.-A. GUSTAFSSON, (1979), J. biol. Chem. 254, 5264—5271.

BERG, A. and J. J. RAFTER, (1981), Biochem. J. 196, 781—786.

BERG, A., (1982), Biochem. Biophys. Res. Commun. 105, 303—311.

BERNSTEIN, S., R. J. LENHARD, W. S. ALLEN, M. HELLER, R. LITTELL, S. M. STOLAR, L. J. FELDMAN, and R. H. BLANK, (1959), J. Amer. Chem. Soc. 81, 1689—1690.

BÖHME, K.-H., C. HÖRHOLD, and W. FORBERG, (1987), DDR-WP 248596 A1.

BONDI, S., (1908), Wiener Klin. Wochenschr. 21, 271.

BONDZYNSKI, V., ST. and V. HUMNICKI, (1896), Hoppe-Seyler's Z. physiol. Chem. 22, 396—410.

BRESKVAR, K. and T. HUDNIK-PLEVNIK, (1977a), Croatica Chemica Acta 49, 207—212.

BRESKVAR, K. and T. HUDNIK-PLEVNIK, (1977b), Biochem. Biophys. Res. Commun. 74, 1192—1198.

BRESKVAR, K. and T. HUDNIK-PLEVNIK, (1981), J. Steroid Biochem. 14, 395—399.

BRESKVAR, K., (1983), J. Steroid Biochem. 18, 51—53.

BRESKVAR, K., B. ČREŠNAR, and T. HUDNIK-PLEVNIK, (1987), J. Steroid Biochem. 26, 499—503.

BROWNE, J. W., W. A. DENNY, E. R. H. JONES, G. D. MEAKINS, J. MORISWA, A. PENDLEBURY, and J. PRAGNELL, (1973), J. Chem. Soc. Perkin Trans 1, 1493—1500.

CARLSTRÖM, K., (1973), Acta Chem. Scand. 27, 1622—1627.

CEJKA, A., (1976), Eur. J. Appl. Microbiol. 3, 145—149.

CHANG, F. N. and C. J. SIH, (1964), Biochemistry 3, 1551—1557.

CHUN, Y. Y., M. IIDA, and H. IIZUKA, (1981), J. Gen. Appl. Microbiol. 27, 505—510.

CONNER, A. H., M. NAGAOKA, J. W. ROWE, and D. PERLMAN, (1976), Appl. Env. Microbiol. 32, 310—311.

CRABB, T. A., P. J. DAWSON, and R. O. WILLIAMS, (1982), J. Chem. Soc. Perkin Trans. 1, 571—574.

ČREŠNAR, B., K. BRESKVAR, and T. HUDNIK-PLEVNIK, (1985), Biochem. Biophys. Res. Commun. 133, 1057—1063.

DAVIDSON, S. J. and P. TALALAY, (1966), J. biol. Chem. 241, 909—915.

DE FLINES, J. and F. VAN DER WAARD, (1966), Dutch Patent 6605514.

DE FLINES, J., (1969), in: "Fermentation Advances", (D. PERLMAN, ed.), Academic Press, New York, London, 385—390.

DULANEY, E. L., W. J. McALEER, M. KOSLOWSKI, E. O. STAPPLEY, and J. JAGLOM, (1955), Appl. Microbiol. 3, 336—345.

DUS, K., R. GOEWERT, C. C. WEVER, and C. A. APPLEBY, (1976), Biochem. biophys. Res. Commun. 69, 437—445.

DUTTA, D., D. K. GOSH, A. K. MISHRA, and T. B. SAMANTA, (1983), Biochem. Biophys. Res. Commun. 115, 692—699.

ELIN, E. A. and L. M. KOGAN, (1966), Dokl. Akad. Nauk. SSSR 167, 1175—1180.

ELIN, E. A., L. M. KOGAN, O. S. TARASOV, and I. V. TORGOV, (1970), Khim. Prirod. Soed., 47—52.

EL-REFAI, A. M., A. R. SALLAM, and I. A. EL-KADY, (1970), Z. Allg. Microbiol. **10**, 183—187.

ERHART, G. and H. RUSCHIG, (1972), Arzneimittel (2nd ed.), Verlag Chemie, Vol. 3, 209—443.

FA, Y., (1988), 4th Symposium on Biochemical Aspects of Steroid Research, Holzhau (DDR), December 5.—10., Abstracts, p. 25.

FIESER, L. F. and M. FIESER, (1959), Steroids, Reinhold Publishing Corporation, New York, Chapman & Hall, Ltd. London.

FRIED, J., R. W. THOM, D. PERLMAN, J. E. HERZ, and A. BORMANN, (1955), Recent Prog. Horm. Res. **11**, 149—181.

FRIED, J. and J. A. EDWARDS (eds.), (1972), Organic Reactions in Steroid Chemistry, Van Nostrand Reinhold Co., New York, 127—139.

FULCO, A. J. and R. T. RUETTINGER, (1987), Life Sci. **40**, 1769—1777.

GHOSH, D. and T. B. SAMANTA, (1981), J. Steroid Biochem. **14**, 1063—1067.

GOODMAN, J. J. and L. L. SMITH, (1961), Appl. Microbiol. **9**, 372—378.

GROSSE, H.-H., L. BLEI, V. DEPPMEYER, M. MENNER, H. BOCKER, C. HÖRHOLD, and B. GOTTSCHALDT, (1987), DDR-WP 248144 A1.

HANSON, F. R., K. M. MANN, E. D. NIELSON, H. V. ANDERSON, M. P. BRUNNER, J. N. KARNEMAAT, D. R. COLINGWORTH, and W. J. HAINES, (1953), J. Amer. Chem. Soc. **75**, 5369—5370.

HAYANO, M., H. J. RINGOLD, V. STEFANOVIC, M. GUT, and R. I. DORFMAN, (1961), Biochem. Biophys. Res. Commun. **4**, 454—459.

HECHTER, O., R. P. JACOBSEN, R. JEANLOZ, H. LEVY, C. W. MARSHALL, G. PINCUS, and V. SCHENKER, (1949), J. Amer. Chem. Soc. **71**, 3261—3262.

HENCH, P. S., E. C. KENDALL, C. H. SLOCUMB, and H. F. POLLEY, (1949), Proc. Staff Meetings Mayo Clin. **24**, 181—197.

HILL, F. F., J. SCHINDLER, R. WAGNER, and W. VOELTER, (1982), Eur. J. Appl. Microbiol. Biotechnol. **15**, 25—32.

HOLLAND, H. L. and B. J. AURET, (1975), Can. J. Chem. **53**, 2041—2044.

HOLLAND, H. L., (1982), Chem. Soc. Rev. **11**, 371—395.

HOLLAND, H. L. and E. M. THOMAS, (1982), Can. J. Chem. **60**, 160—165.

HÖRHOLD, C., R. KOMEL, and H. GROH, (1979), DDR-WP 137361.

HÖRHOLD, C., G. ROSE, and G. KAUFMANN, (1981), Z. f. Allg. Mikrobiol. **21**, 289 to 293.

HÖRHOLD, C., K. UNDISZ, H. GROH, R. SAHM, W. SCHADE, and R. KOMEL, (1986), J. Basic Microbiol. **26**, 335—339.

HORVATH, J. and A. KRAMLI, (1947), Nature **160**, 639.

IIDA, M., (1986), Japan Pat. 68-198400 (A).

IIDA, M., (1987), Japan Pat. 61-219396 (A).

ITAGAKI, E., (1986), J. Biochem. **99**, 815—825.

IUPAC, (1972), Pure Appl. Chem. **31**, 285—322.

JAYANTHI, C. R., P. MADYASTHA, and K. M. MADYASTHA, (1982), Biochem. Biophys. Res. Commun. **106**, 1262—1268.

JONES, E. R. H., (1973), Pure Appl. Chem. **33**, 39—52.

KÄPPELI, O., (1986), Microbiol. Rev. **50**, 244—258.

KIESLICH, K., (1969), Synthesis, 120—134; 147—157.

KIESLICH, K., (1980a), in: "Microbial Enzymes and Bioconversion Economic Microbiology", (A. H. ROSE, ed.), Academic Press, London, Vol. 5, 369—465.

KIESLICH, K., (1980b), Biotechnology Letters **2**, 211—217.

KIM, C. H., (1937), Enzymologia **4**, 119—121.

KIM, C. H., (1939), Enzymologia **6**, 105—107.

Kon. Nederland. Gist-en Spiritusfab., Delft, (1966), DAS 1 568932.

Kon. Nederland. Gist-en Spiritusfab., Delft, (1968), DOS 1 768215.

KONDO, E. and E. MASUO, (1961), J. Gen. Appl. Microbiol. **7**, 113—117.

KRAMLI, A. and J. HORVATH, (1948), Nature **162**, 619.

KRAMLI, A. and J. HORVATH, (1949), Nature **163**, 219.

KUBOTA, M., (1987), Japan Pats. 62-226997 (A), 62-273996 (A), 62-273995 (A).

LANGECKER, H., E. SCHEIFFELE, R. GEIGER, K. PREZEWOWSKY, U. STACHE, and K. SCHMITT, (1977), in: Ullmanns Encyklopädie der technischen Chemie, 4. Aufl., Verlag Chemie, Weinheim, Vol. 13, 1—71.

LIN, Y. Y. and L. L. SMITH, (1970), Biochim. Biophys. Acta **218**, 515—525.

MADYASTHA, K. M., C. R. JAYANTHI, P. MADYASTHA, and D. SUMATHI, (1984), Can. J. Biochem. Cell. Biol. **62**, 100—107.

MAHATO, S. B. and A. MUKHERJEE, (1984), Phytochemistry **23**, 2132—2154.

MAHATO, S. B. and S. BANERJEE, (1985), Phytochemistry **24**, 1403—1421.

MAMOLI, L. and A. VERCELLONE, (1937), Ber. Dtsch. Chem. Ges. **70**, 470—471.

MAMOLI, L. and A. VERCELLONE, (1937), Ber. Dtsch. Chem. Ges. **70**, 2079—2082.

MAMOLI, A. and A. VERCELLONE, (1937), Z. physiol. Chem. **245**, 93—95.

MAMOLI, L., (1938), Ber. Dtsch. Chem. Ges. **71 B**, 2701—2703.

MAMOLI, L. and A. VERCELLONE, (1938), Ber. Dtsch. Chem. Ges. **71 B**, 1686—1687.

MARSHECK, W. J., (1971), in: Progress in Industrial Microbiol. Vol. 10, (D. J. D. HOCKENHULL, ed.), Churchill Livingstone, Edinburgh and London, 49—103.

MARTIN, C. K. A., (1984), in: Biotechnology (H.-J. REHM and G. REED, eds.), Vol. 6a, Verlag Chemie, Weinheim, 79—95.

McGREGOR, W. C., B. TAHENKIN, E. JENKIN, and R. EPPS, (1972), Biotechn. Bioeng. **14**, 831—836.

McKENZIE, B. F., V. R. MATTOX, L. L. ENGEL, and E. C. KENDALL, (1948), J. Biol. Chem. **173**, 271—281.

MEYSTRE, CH. and K. MIESCHER, (1949), Helv. Chim. Acta **32**, 1758—1764.

MONEM, A., E. LOTFY, A. R. SALLAM, and I. EL-KADY, (1969), J. Gen. Appl. Microbiol. **15**, 301—309.

MÜLLER-FROHNE, M., G.-R. JÄNIG, D. PFEIL, H. RIEMER, and K. RUCKPAUL, (1990), Biochem. Biophys. Res. Commun., (submitted).

MURRAY, H. C., (1976), Microbiology of steroids in: Industrial microbiology (B. M. MILLER and W. LITZKI, eds.), McGraw Hill Book Co, New York, 79—105.

NAGASAWA, M., N. WALAMABE, and H. HASHIBA, (1970), Agr. Biol. Chem. **34**, 838 to 847.

NARHI, L. O. and A. J. FULCO, (1986), J. biol. Chem. **261**, 7160—7170.

NEEF, G., U. EDER, A. SEEGER, and R. WIECHERT, (1980), Chem. Ber. **113**, 1184 to 1188.

NEULAND, P., K. PONSOLD, G. SCHUBERT, and M. WUNDERWALD, (1979a), DDR-Patent 142054.

NEULAND, P., K. PONSOLD, G. SCHUBERT, and M. WUNDERWALD, (1979b), DDR-Patent 144548.

NGUYEN-DANG, T., M. MAYER, and M.-M. JANOT, (1971), C.R. Acad. Sci. Ser. D. **272**, 2032.

NOBILE, A., W. CHARNEY, P. L. PERLMAN, H. L. HERZOG, C. C. PAYNE, M. E. TULLY, M. A. JEVNIK, and E. G. HERSHBERG, (1955), J. Am. Chem. Soc. **77**, 4184.

NOMINE, G., (1980), Bull. Soc. Chim. France **II**, 18—23.

ONKEN, D., (1971), Steroide, Akademie-Verlag, Berlin, Pergamon Press, Oxford, Vieweg & Sohn, Braunschweig.

ORRENIUS, S., J. WILSON, C. VON BARR, and J. SCHENKMAN, (1972), in: Biological Hydroxylation Mechanisms (G. S. OYD and R. M. S. SMELLIE, eds.), Biochemical Society Symposium **34**, 55—77.

OTTO, C., W. HERBST, R. MICHLING, and U. FRANKE, (1987), DDR-WP 250435.

PETERSON, D. H. and H. C. MURRAY, (1952), J. Amer. Chem. Soc. **74**, 1871—1872.

RAHIM, M. A. and C. J. SIH, (1966), J. biol. Chem. **241**, 3615—3623.

RUTTEN, A. A. J. J. L., M. E. FALKE, J. F. CATSBURG, R. TOPP, B. J. BLAAUBOER, I. V. HOLSTEIJN, L. DOORN, and F. X. R. V. LEEUWEN, (1987), Arch. Toxicol. **61**, 27—33.

SALLAM, L. A. R., A. M. EL-REFAI, and I. A. EL-KADY, (1971), Z. Allgem. Mikrobiol. **11**, 325—330.

SARETT, L. H., (1946), J. Biol. Chem. **162**, 601—631.

SCHENKMAN, J. B., H. REMMER, and R. W. ESTABROOK, (1967), Molec. Pharmacol. **3**, 113—123.

SCHUBERT, K., K.-H. BÖHME, and C. HÖRHOLD, (1976), DDR-Patent 132271.

SCHWALB, H., L. O. NARHI, and A. J. FULCO, (1985), Biochim. Biophys. Acta **338**, 302—311.

SEBEK, O. K. and D. PERLMAN, (1979), in: "Microbial Technology" (H. J. PEPPLER and D. PERLMAN, eds.), 2nd ed., Vol. 1, Acad. Press, New York, 483—496.

SEDLACZEK, L., J. DLUGONSKI, and A. JAWORSKI, (1984), Appl. Microbiol. Biotechnol. **20**, 166—169.

SEIDEL, L., C. HÖRHOLD, M. BIRKE, and D. NOACK, (1987), DDR-WP 249284 A1.

SHIBAHARA, M., Y. A. MOODY, and L. L. SMITH, (1970), Biochim. Biophys. Acta **202**, 172—179.

SHULL, G. M. and D. A. KITA, (1955), J. Amer. Chem. Soc. **77**, 763—764.

SIH, C. J., (1962), Biochim. Biophys. Acta **62**, 541—547.

SIH, C. J. and J. LAVAL, (1962), Biochim. Biophys. Acta **64**, 409—415.

SIH, C. J., H. H. TAI, Y. Y. TSONG, S. S. LEE, and R. G. COOMBE, (1968), Biochemistry **7**, 808—812.

SMITH, L. L., (1984), in: "Biotechnology", (H.-J. REHM and G. REED, eds.), Vol. 6a, Verlag Chemie, Weinheim, 31—78.

SÖHNGEN, N. L., (1913), Zentrbl. Bakteriol. Parasitkd. Abtlg. II **37**, 595—609.

STEVENSON, P. M., R. T. RUETTINGER, and A. J. FULCO, (1983), Biochem. biophys. Res. Commun. **112**, 927—935.

STRIJEWSKI, A., (1982), Eur. J. Biochem. **128**, 125—135.

SZENTIRMAI, A., (1988), 4th Symposium on Biochemical Aspects of Steroid Research, Holzhau, (DDR) December 5.—10., Abstracts, p. 27.

TAN, L. and P. FALARDEAU, (1970), J. Steroid Biochem. **1**, 221—227.

TAN, L. and L. L. SMITH, (1968), Biochem. Biophys. Acta **164**, 389—395.

THOMA, R. W., J. FRIED, S. BONNO, and P. GRABOWICH, (1957), J. Am. Chem. Soc. **79**, 4818.

TURFITT, G. E., (1943), Biochem. J. **37**, 115—117.

TURFITT, G. E., (1944), J. Bacteriol. **47**, 487—493.

VAN RHEENEN, V. and K. P. SHEPHARD, (1979), J. Org. Chem. **44**, 1582—1584.

WEAVER, E. A., M. E. KENNY, and M. E. WALL, (1960), Appl. Microbiol. **8**, 345—354.

WIECHERT, R., (1979), Vorlesungsreihe Schering, Heft 5, 1—16.

WILSON, J. E. and C. S. VESTLING, (1965), Arch. Biochem. Biophys. **110**, 401—410.

Wix, G., K. G. Büki, E. Tömörkeny, and G. Ambrus, (1966), Richter Gedeon Vegyeszeti Gyar RT. DAS 1 593327.
Wix, G., K. G. Büki, E. Tömörkeny, and G. Ambrus, (1968), Steroids 11, 401—413.
Wolters, B., (1985), Dtsch. Apoth. Ztg. 125, 643—647.
Yasukochi, Y. and B. S. S. Masters, (1976), J. biol. Chem. 251, 5337—5344.
Yu, C.-A. and I. C. Gunsalus, (1974), J. biol. Chem. 249, 107—110.
Zeillinger, R. and J. Spona, (1986), FEBS-Microbiol. Lett. 37, 231—237.
Zuidweg, M. H. J., W. F. v. d. Waard, and J. de Flines, (1962), Biochim. Biophys. Acta 58, 131—133.
Zuidweg, M. H. J., (1968), Biochim. Biophys. Acta 152, 144—158.

List of Authors

A. A. Akhrem
Institute of Bioorganic Chemistry
Academy of Sciences of BSSR
Zhodinskaya 5

220600 Minsk
USSR

S. Kominami
Faculty of Integrated Arts and Sciences
Department of Environmental Sciences
Hiroshima University

Hiroshima 730
Japan

J. D. Lambeth
Department of Biochemistry
Emory University Medical School

Atlanta, GA 30322
U.S.A.

R. Megges
Central Institute of Molecular Biology
Academy of Sciences of the GDR
Robert-Rössle-Straße 10

O-1115 Berlin

M. Müller-Fröhne
Central Institute of Molecular Biology
Academy of Sciences of the GDR
Robert-Rössle-Straße 10

O-1115 Berlin

Y. Nonaka
Department of Biochemistry
Osaka University Medical School
4-3-57, Nakanoshima, Kita-Ku

Osaka 530
Japan

M. Okamoto
Department of Molecular Physiological
Chemistry
Osaka University Medical School
4-3-57, Nakonoshima, Kita-Ku

Osaka 530
Japan

D. Pfeil
Central Institute of Molecular Biology
Academy of Sciences of the GDR
Robert-Rössle-Straße 10

O - 1115 Berlin

K. Ruckpaul
Central Institute of Molecular Biology
Academy of Sciences of the GDR
Robert-Rössle-Straße 10

O - 1115 Berlin

E. R. Simpson
Departments of Biochemistry and Obstetrics
and Gynecology and Cecil H. and Ida Green Center
for Reproductive Biology Sciences
University of Texas; Health Science Center

Dallas, Texas 75235
U.S.A.

S. Takemori
Faculty of Integrated Arts and Sciences
Department of Environmental Sciences
Hiroshima University

Hiroshima 730
Japan

S. A. Usanov
Institute of Bioorganic Chemistry
Academy of Sciences of BSSR
Zhodinskaya 5

220600 Minsk
USSR

R. Waterman
Departments of Biochemistry and Obstetrics
and Gynecology and Cecil H. and Ida Green Center
for Reproductive Biology Sciences
University of Texas; Health Science Center

Dallas, Texas 75 235
U.S.A.

Subject Index